Relay Feedback

Springer
*London
Berlin
Heidelberg
New York
Barcelona
Hong Kong
Milan
Paris
Singapore
Tokyo*

Qing-Guo Wang, Tong Heng Lee
and Chong Lin

Relay Feedback

Analysis, Identification and Control

With 127 Figures

Qing-Guo Wang, PhD
Tong Heng Lee, PhD
Chong Lin, PhD
Department of Computer and Electrical Engineering, National University of Singapore, 4 Engineering Drive 3, Singapore 117576, Republic of Singapore

British Library Cataloguing in Publication Data
Wang, Qing-Guo, 1957-
  Relay feedback : analysis, identification and control
  1.Relay control systems 2.Feedback control systems
  I.Title II.Lee, Tong H. III.Lin, Chong
  629.8'3
  ISBN 1852336501

Library of Congress Cataloging-in-Publication Data
Wang, Qing-Guo, 1957-
  Relay feedback : analysis, identification, and control / Qing-Guo Wang, Tong Heng Lee, andChong Lin.
    p. cm.
    Includes bibliographical references and index.
    ISBN 1-85233-650-1 (alk. paper)
    1. Relay control systems. 2. Feedback control systems. I. Lee, Tong Heng, 1958- II. Lin, Chong, 1967- III. Title.
  TJ218.5 .W36 2002
  629.8'3--dc21   2002070841

Apart from any fair dealing for the purposes of research or private study, or criticism or review, as permitted under the Copyright, Designs and Patents Act 1988, this publication may only be reproduced, stored or transmitted, in any form or by any means, with the prior permission in writing of the publishers, or in the case of reprographic reproduction in accordance with the terms of licences issued by the Copyright Licensing Agency. Enquiries concerning reproduction outside those terms should be sent to the publishers.

ISBN 1-85233-650-1 Springer-Verlag London Berlin Heidelberg
a member of BertelsmannSpringer Science+Business Media GmbH
http://www.springer.co.uk

© Springer-Verlag London Limited 2003
Printed in Great Britain

The use of registered names, trademarks etc. in this publication does not imply, even in the absence of a specific statement, that such names are exempt from the relevant laws and regulations and therefore free for general use.

The publisher makes no representation, express or implied, with regard to the accuracy of the information contained in this book and cannot accept any legal responsibility or liability for any errors or omissions that may be made.

Typesetting: Electronic text files prepared by authors
Printed and bound by Athenæum Press Ltd., Gateshead, Tyne & Wear
69/3830-543210 Printed on acid-free paper SPIN 10878188

Preface

Relay feedback has attracted considerable research attention for more than a century. The classical work of Typikin (1984) on analysis summarizes progress up to the 1960s. Early applications of relay systems ranged from stationary control of industrial processes to control of mobile objects. In the 1950s, relays were mainly used as amplifiers but such applications are now obsolete, owing to the development of electronic technology. In the 1960s, relay feedback was applied to adaptive control. One prominent example of such an application is the self-oscillating adaptive controller developed by Minneapolis Honeywell, which uses relay feedback to attain a desired amplitude margin. It was in the 1980s that Åström and Hägglund successfully applied the relay feedback method to auto-tune PID controllers for process control, and triggered a resurgence of interest in relay methods, including extensions of the method to more complex systems. Since then, new tools and powerful results have emerged.

The present monograph presents, in a single volume, a fairly comprehensive, up-to-date and detailed treatment of relay feedback theory, use of relay feedback for process identification, and use of identified models for control system design. The materials included here are based on research results of the authors and their co-workers in the domain. Both single-variable and multivariable systems are addressed. For presentation, we have made the technical development of the results as self-contained as possible. Only knowledge of linear system theory is assumed for readers. Illustrative examples of different degrees of complexity are given to facilitate understanding. Therefore, it is believed that the book can be accessed by graduate students, researchers and practising engineers.

The table of contents gives an idea of what is contained in the book. The book is organized into three parts: Part I Analysis of Relay Feedback Systems, Part II Process Identification Using Relay Feedback, and Part III Controller Design. The three parts are related but can be read independently. *Those who are only interested in relay applications can skip Part I*. A chapter-by-chapter preview of our materials is given as follows.

Part I deals with analysis of relay feedback systems and consists of four chapters. Chapter 1 considers a SISO linear system with general relay feedback and studies the existence of solutions to the system. It is shown that existence and uniqueness of solutions are always guaranteed for the system with nonzero time delay; for the delay-free case, a necessary and sufficient condition is given for the existence of solutions. Chapter 2 deals with problems of the existence of limit cycles of linear systems with relay feedback, and presents some sufficient conditions. Moreover, if a limit cycle exists, necessary and sufficient conditions are given for determining particular limit cycles. The main tool used is Brouwer's fixed point theorem and some techniques related to system theory. Chapter 3 studies the local stability problem of limit cycles for time-delay relay feedback systems with relays containing asymmetric hysteresis. It is shown that if a certain constructed matrix is Schur stable, then the local stability of the limit cycle considered is guaranteed. Chapter 4 investigates the global stability of limit cycles in relay feedback systems. A unified framework and results are presented for global convergence. The key idea is to reduce the global stability problem to the asymptotic stability of a discrete time system.

Part II studies process identification from relay feedback and consists of four chapters. Chapter 5 deals with identifying process models from steady-state responses, or limit cycles, of relay feedback systems. We show that use of the FFT in place of the describing function approximation can give more accurate frequency response estimation. More cycles of oscillations can be employed to enhance estimation robustness. Modified relays can enable estimation of more points on the process frequency response. Chapter 6 deals with identifying process frequency response by using transient responses of relay feedback systems. Chapter 7 considers the problem of converting the frequency response identified from relay feedback to a transfer function with possible dead time. Chapter 8 develops a general identification procedure applicable to various test scenarios, covering step/relay and open-loop/closed-loop types.

Part III is concerned with control system design and consists of four chapters. Chapter 9 considers control design for SISO stable processes. Internal model control design is reviewed. Its equivalent single-loop controller is derived and usually of high complexity. The model reduction technique is employed to find its approximation. Users have the option to choose between PID and high-order controllers to better suit the applications. It turns out that high-order controllers may be necessary to achieve high performance for essentially high-order processes. Chapter 10 extends the SISO methodology in Chapter 9 to the multivariable case. For internal model control, the objective closed-loop transfer functions are characterized in terms of process unavoidable non-minimum-phase elements. A multivariable controller for best achievable performance is obtained using block diagram analysis and model reduction. Chapter 11 considers control of unstable processes. The relay feedback test is used to identify an unstable process. An IMC-based single-loop controller design method is given to find the feedback controller in either PID or high-order form. Chapter 12 presents a new scheme called the Partial Internal Model Control (PIMC), which is capable of controlling both stable and unstable processes. In the PIMC, a process model is expressed as the sum of stable and anti-stable parts and only the stable part of the process model is used as the internal model. The process stable part is cancelled by the internal model and the remaining anti-stable part is stabilized and controlled using a primary controller. Chapter 13 addresses decentralized control of multivariable processes. A simple independent design method for multi-loop controllers is proposed which exploits process interactions for the improvement of loop performance.

We would like to thank C. C. Hang, Q. Bi, B. Zou, Yu Zhang, Yong Zhang, H. W. Fung and X. P. Yang for their fruitful research collaborations with us, which have led to the contributions contained in the book. We are grateful to the Centre for Intelligent Control and Department of Electrical and Computer Engineering of the National University of Singapore for providing plenty of resources for our research work.

Qing-Guo Wang
Tong Heng Lee
Chong Lin
Singapore, June 2002

Symbols

$\mathbb{R}$	field of real numbers
$\mathbb{R}^n$	n-dimensional real Euclidean space
$\mathbb{R}^{n \times m}$	space of $n \times m$ real matrices
$\mathbb{C}^{n \times m}$	space of $n \times m$ complex matrices
I	identity matrix
A^T	transpose of matrix A
A^{-1}	inverse of matrix A
$\det(A)$	determinant of matrix A
$\operatorname{rank}(A)$	rank of matrix A
$A > 0$	symmetric positive definite
$A \geq 0$	semi-definite
$A^{1/2}$	symmetric square root of $A \geq 0$, i.e., $A^{1/2} A^{1/2} = A$
$\operatorname{diag}\{A_1, \cdots, A_n\}$	diagonal matrix with A_i as its ith diagonal element
$\lambda(A)$	eigenvalues of square matrix A
$\sigma_{max}(A)$	largest singular value of matrix A
$\sigma_{min}(A)$	smallest singular value of matrix A
$\rho(A)$	spectral radius of square matrix A
$\forall$	for all
$\exists$	exist
s.t.	such that
$\in$	belong to
$\subseteq$	subset

Symbols

$\cup$	union
$\cap$	intersection
$\sum$	sum
$\lvert \cdot \rvert$	absolute value (or modulus)
$\lVert \cdot \rVert$	spectral norm
$\lVert \cdot \rVert_\infty$	induced l_∞-norm
$\to$	tend to
$f(t_-)$	$= \lim_{\epsilon \to 0+} f(t-\epsilon)$
$f(t_+)$	$= \lim_{\epsilon \to 0+} f(t+\epsilon)$
$O(t^k)$	infinitesimal of order t^k
sup	supremum
inf	infimum
$m!$	$= m(m-1) \cdots 2 \cdots 1$ for non-negative integer m
$\prod_{i=1}^{k} A_i$	$= A_k A_{k-1} \cdots A_1$

Contents

Part I. Analysis of Relay Feedback Systems

1. **Existence of Solutions** 5
   1.1 Introduction 5
   1.2 System Formulation 6
   1.3 Existence of Solutions 8
   1.4 Delay-free Case 11

2. **Existence of Limit Cycles** 21
   2.1 Introduction 21
   2.2 Sufficient Condition 22
       2.2.1 Supporting Lemmas 22
       2.2.2 Existence of Limit Cycles 29
   2.3 A Simple Existence Condition 31
   2.4 Limit Cycle Location 36

3. **Local Stability of Limit Cycles** 39
   3.1 Introduction 39
   3.2 Problem Formulation and Preliminaries 40
   3.3 Local Stability of Limit Cycles 41
   3.4 Extension 53

4. **Global Stability of Limit Cycles** 57
   4.1 Introduction 57
   4.2 Problem Formulation 58
   4.3 Supporting Lemmas 59
   4.4 Global Stability of Limit Cycles 64
   4.5 Extensions 73
   4.6 Existence of Globally Stable Limit Cycles 77
       4.6.1 Preliminaries 77
       4.6.2 Sufficient Conditions 78

Part II. Process Identification from Relay Feedback Test

5. **Relay Feedback and its Variations** 89
   5.1 Fundamentals 89
   5.2 First-order Modelling 95
   5.3 Robustness Enhancement 103
   5.4 Parasitic Relay 108
   5.5 Cascade Relay 115
   5.6 Extension to MIMO Case 126

6. **Use of Relay Transient Responses** 135
   6.1 Signal Analysis 135
   6.2 Decomposition Method 140
   6.3 Weighting Method 146
   6.4 Testing on Pilot Plants 149
   6.5 Extension to the MIMO case 160

7. **Transfer Function Modelling** 169
   7.1 From Frequency Response 170
   7.2 From Step Response 174
       7.2.1 Second-order Modelling 175
       7.2.2 nth-order Modelling 180
       7.2.3 Implementation Issues 192
       7.2.4 Simulation and Real-time Test 196
   7.3 A Hybrid Approach 202

8. **A General Identification Approach** 207
   8.1 SISO Systems 208
       8.1.1 The Method 208
       8.1.2 Simulation 211
   8.2 MIMO Systems 216
       8.2.1 The Method 217
       8.2.2 Simulation 225
   8.3 Unstable Processes 230

Part III. Controller Design

9. Single-variable Systems 239
  9.1 Design Methodology 240
  9.2 PID Controller 244
  9.3 High-order Controller 253
  9.4 Stability Analysis... 255
  9.5 Unstable Processes 261
    9.5.1 PID Controller 266
    9.5.2 High-order Controller 269

10. Multivariable Systems 273
  10.1 IMC Scheme 274
    10.1.1 Decoupling 275
    10.1.2 Analysis.. 280
    10.1.3 Design 286
    10.1.4 Simulation.. 289
  10.2 Unity Feedback System 296
    10.2.1 Design Methodology 296
    10.2.2 PID Controller 299
    10.2.3 High-order Controller 303
    10.2.4 Stability Analysis 311

11. Partial Internal Model Control 319
  11.1 Review of the IMC 320
  11.2 The Proposed PIMC Scheme............................... 322
  11.3 Internal Stability Analysis 324
  11.4 Asymptotic Tracking and Regulation...................... 326
  11.5 Primary Control Design 329
    11.5.1 PIMC Primary Controller Design.................... 329
    11.5.2 MPIMC Primary Controller Design 330
  11.6 Robustness Analysis 331
    11.6.1 Robust Stability................................... 331
    11.6.2 Practical Stability 332
  11.7 Practical Aspects 333
    11.7.1 Pre-filter Design.................................. 333
    11.7.2 Determination of G^- 335
    11.7.3 Dead Time 336

11.8 Simulation Results..336
11.9 Real-time Implementation 341
Appendix A: Controller Design for Processes with Two Unstable Poles 343
Appendix B: Formulas for Decomposition of some Typical Unstable Processes 345

12. Decentralized Control 347
12.1 The Proposed Independent Design Strategy..................348
12.2 Choice of Solutions to Controller Gain Equations 351
12.3 Rational Approximation of the Irrational Solutions 355
12.4 Controller Reduction and Performance Trade-off..............361
12.5 Stability Analysis..366
12.6 Extension to the $m \times m$ Case 369

References...375

Index...385

Part I

Analysis of Relay Feedback Systems

Introduction to Part I

Relay systems can be traced back to their classical configurations. A description of various relay extremal control systems can be found in Morosanov (1964). Applications of relay systems range from stationary control of industrial processes to control of mobile objects. In the 1950s, relays were mainly used as amplifiers but such applications are now obsolete, owing to the development of electronic technology. In the 1960s, relay feedback was applied to adaptive control (Tsypkin, 1984). One prominent example of such applications is the self-oscillating adaptive controller developed by Minneapolis Honeywell which uses relay feedback to attain a desired amplitude margin. This system was tested extensively for flight control systems, and it has been used in several missiles. It was in the eighties that Åström successfully applied the relay feedback method to auto-tune PID controllers for process control, and triggered a resurgence of interest in relay methods, including extensions of the method to more complex systems. A recent survey of relay methods is provided in Åström and Hägglund (1995).

Theoretical development is far behind the practical applications of relay feedback systems (RFS). Difficulties in the analysis occur because such systems may exhibit such phenomena as non-uniqueness of solutions, sliding motion and bifurcation, fast switching or chattering and chaos. It is known that RFSs often possess limit cycles. This property is very useful in modern control applications such as controller automatic tuning. However, to establish exact conditions for the existence of a limit cycle is not an easy task. As a result, a limit cycle is pre-assumed to exist in some practical applications and theoretical analysis. The describing function method was initially used to investigate the existence problem (Tsypkin, 1984). This can only give an approximate analysis. Exact methods have been reported recently to determine the period of limit cycles with two switches per period (Åström, 1995; Varigonda and Georgious, 2001).

If a limit cycle exists, its stability is another important property which is also assumed in engineering practice. Classical techniques such as the phase-plane approach have been employed for the stability analysis of limit cycles

(Guckenheimer and Holmes, 1983; Tsypkin, 1984). Exact methods for analyzing the limit cycle stability of relay feedback systems have attracted much attention. Local stability results have been reported in Åström (1995), Johansson *et al.* (1997) and Johansson *et al.* (1999), and are mainly based on consideration of the linear approximation of the Poincare map. The local stability property ensures that any nearby trajectory tends to the limit cycle as time goes to plus infinity. What is more meaningful in engineering practice is the global stability of limit cycles, which says that all system trajectories tend to the limit cycle as time goes to plus infinity. This problem was studied quite recently in Goncalves *et al.* (2001), and presents sufficient conditions in terms of a set of linear matrix inequalities by finding the so-called surface Lyapunov function of Poincare maps. Under specific assumptions, the result serves a class of relay feedback systems well. Other work is presented in Lin *et al.* (2001) from a different point of view, where the system setting considered therein includes a time delay and the limit cycle is not assumed to be symmetric.

Motivated by the above observations, recently we have devoted time to the analysis of linear systems with relay feedback under a general setting. These include the existence of solutions, the existence of limit cycles, and the local and global stability of limit cycles. The plant can have a time delay, the relay can contain asymmetric hysteresis, and the limit cycle can be asymmetric and have more than two switchings per period. Our results can serve a large class of relay feedback systems well. The results are for general SISO RFSs. Part I of this book is based on these latest results.

1. Existence of Solutions

This chapter considers a single-input single-output (SISO) linear system under relay feedback and studies the existence of solutions to the system.

1.1 Introduction

Relay feedback systems (RFSs) have been receiving much attention from researchers due to their extensive practical applications (Åström and Hägglund, 1995; Tsypkin, 1984). Such systems appear as a special class of differential equations with discontinuous right-hand sides. For the latter differential equations, the topic of well-posedness (i.e., existence and uniqueness of solutions) has been an important research area for several decades, and many results are summarized in Filippov (1988) from a unified standpoint. However, sometimes these existing methods are difficult to adopt to deal with relay feedback systems. So far, few results have been reported addressing the existence of solutions to RFSs. In Georgiou and Smith (2000), the well-posedness problem is studied for systems having input–output forms. In Lootsma et al. (1999), by using the theory of Linear Complementarity Problem (Cottle et al., 1992), an efficient sufficient condition is presented. The idea is to transform the system considered into a complementarity system and then solve the corresponding rational complementarity problem (Heemels et al., 1999; Van der Schaft and Schumacher, 1998). The result therein is slightly improved by Lin and Wang (2002). We notice that the feedback considered therein is connected with an ideal relay and the control input is between two values at the switching instants, and that the well-posedness result does not apply to RFSs with two-valued relay feedback. The latter system setting has recently attracted much more attention, both theoretically and practically (Åström, 1995; Goncalves et al., 2001; Varigonda and Georgious, 2001).

In this chapter, we present a detailed analysis of the existence of solutions to RFSs with two-valued relay feedback. Analyzing clearly the system solution for all time $t \geq 0$ is not only helpful for characterizing the trajectories but also meaningful for the global convergence of RFSs. The system considered here is of a general setting: the plant can have a time delay, the relay can contain asymmetric hysteresis. It is shown that under some easy-check conditions, there is a solution for all time $t \geq 0$ starting from any initial condition.

This chapter is organized as follows. Section 1.2 gives some preliminaries. Section 1.3 deals with the problem of the existence of solutions for $\tau > 0$ while Section 1.4 considers $\tau = 0$.

1.2 System Formulation

Consider a single-input single-output (SISO) plant described by

$$\dot{x}(t) = Ax(t) + bu(t - \tau)$$
$$y(t) = cx(t) \tag{1.1}$$

where $x(t) \in \mathbb{R}^n$, $y(t) \in \mathbb{R}$ and $u(t - \tau) \in \mathbb{R}$ are the state, output and control input, respectively; A, b, c are constant real matrices or vectors with appropriate dimensions; $\tau \geq 0$ stands for the time delay. The plant has relay feedback:

$$u(t) = \begin{cases} u_\beta, & \text{if } y(t) > \beta, \text{ or } y(t) \geq \alpha \text{ and } u(t_-) = u_\beta \\ u_\alpha, & \text{if } y(t) < \alpha, \text{ or } y(t) \leq \beta \text{ and } u(t_-) = u_\alpha, \end{cases} \tag{1.2}$$

where $\alpha, \beta \in \mathbb{R}$ with $\alpha < \beta$ indicating hysteresis; $u_\alpha, u_\beta \in \mathbb{R}$ and $u_\alpha \neq u_\beta$. Due to time delay $\tau \geq 0$, we have to specify the initial function $u(\tilde{t})$ for $\tilde{t} \in [-\tau, 0]$. The most natural one, which is used in practice, is:

$$u(\tilde{t}) \equiv \begin{cases} u_\beta, & \text{if } y(0) \geq \beta \\ u_\alpha, & \text{if } y(0) \leq \alpha \\ u_0 \in \mathcal{U}, & \text{if } \alpha < y(0) < \beta \end{cases} \tag{1.3}$$

where

$$\mathcal{U} := \{u_\alpha, u_\beta\}. \tag{1.4}$$

We call (1.1)–(1.3) a RFS and denote it by Σ_τ.

Define the switching planes:

$$S_\alpha := \{\xi \in \mathbb{R}^n : c\xi = \alpha\}, \tag{1.5}$$
$$S_\beta := \{\xi \in \mathbb{R}^n : c\xi = \beta\}. \tag{1.6}$$

Let

$$\mathcal{S}_{+\alpha} := \{\xi \in \mathbb{R}^n : c\xi > \alpha\}, \tag{1.7}$$
$$\mathcal{S}_{-\alpha} := \{\xi \in \mathbb{R}^n : c\xi < \alpha\}, \tag{1.8}$$
$$\mathcal{S}_{+\beta} := \{\xi \in \mathbb{R}^n : c\xi > \beta\}, \tag{1.9}$$
$$\mathcal{S}_{-\beta} := \{\xi \in \mathbb{R}^n : c\xi < \beta\}. \tag{1.10}$$

Suppose that $x(t)$ is a solution to system Σ_τ with initial condition $x(0)$ satisfying $y(0) > \alpha$ (resp. $y(0) \leq \alpha$). If the trajectory of $x(t)$ intersects $\mathcal{S}_\alpha$ (resp. $\mathcal{S}_\beta$) at x_α (resp. x_β) from $\mathcal{S}_{+\alpha}$ (resp. $\mathcal{S}_{-\beta}$), we will call the state x_α (resp. x_β) an *intersecting point*. The time instant corresponding to the intersecting point is called an *intersecting instant*. It should be stressed that in our convention, if the trajectory of $x(t)$ intersects $\mathcal{S}_\alpha$ (resp. $\mathcal{S}_\beta$) at x_α (resp. x_β) from $\mathcal{S}_{-\alpha}$ (resp. $\mathcal{S}_{+\beta}$), the state x_α (resp. x_β) is not called an intersecting point and the corresponding time instant is not the intersecting instant, since such intersecting does not cause any switch in $u(t)$. If the trajectory of $x(t)$ not only intersects but also traverses $\mathcal{S}_\alpha$ (resp. $\mathcal{S}_\beta$) at x_α (resp. x_β) from $\mathcal{S}_{+\alpha}$ (resp. $\mathcal{S}_{-\beta}$) to $\mathcal{S}_{-\alpha}$ (resp. $\mathcal{S}_{+\beta}$), we will call such an intersecting point x_α (resp. x_β) a *traversing point*. The time instant corresponding to the traversing point is called a *switching instant*. See Figures 1.1 and 1.2, where x_1, x_3 and x_4 are intersecting points (indeed x_1 and x_4 are traversing points) while x_2 and x_5 are not. It should be noted that for $\tau > 0$, the relay $u(t)$ will switch from u_β to u_α (or from u_α to u_β) at a switching instant. But at this switching instant, $u(t-\tau)$ in (1.1) remains u_β (or u_α) for a time duration of τ after which it changes to u_α (or u_β).

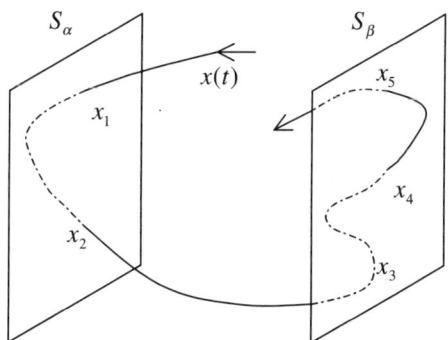

Fig. 1.1. Intersecting points x_1, x_3, x_4 and traversing points x_1, x_4

In the next section, we will study whether or not a solution $x(t)$ with initial state $x_0 \in R^n$ exists for all $t \in [0, \infty)$. We will see that a system trajectory of $x(t)$ may have two different trajectories after some intersecting instant. We

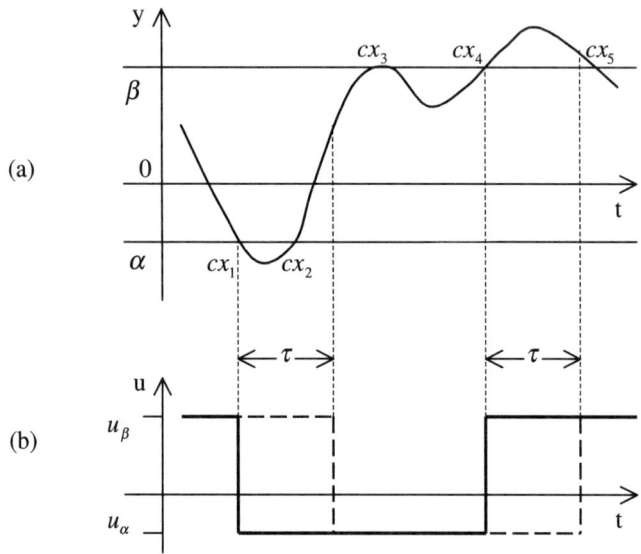

Fig. 1.2. The corresponding curves for Figure 1.1 (a) $y(t) = cx(t)$; (b) $u(t)$ (solid) and $u(t - \tau)$ (dashed)

will call these trajectories sub-trajectories of $x(t)$, and the corresponding system solutions sub-solutions.

1.3 Existence of Solutions

We see that during the time after a trajectory starts from the region $\mathcal{S}_{+\beta}$ or $\mathcal{S}_{-\beta} \cap \mathcal{S}_{+\alpha}$ or $\mathcal{S}_{-\alpha}$, but before it reaches any switching plane, the system Σ_τ is a linear system with a fixed control $u \in \mathcal{U}$, and thus the existence and uniqueness of system solution is guaranteed for some nonzero time interval. So, to determine the existence of a solution for all $t \geq 0$, it is sufficient to study the trajectory characteristic at the moment it intersects the switching plane $\mathcal{S}_\beta$ or $\mathcal{S}_\alpha$.

From (1.1)–(1.3), $u(-\tau)$ is uniquely specified by initial condition x_0. If the unique trajectory of $x(t)$ starting from x_0 does not intersect $\mathcal{S}_\beta$ or $\mathcal{S}_\alpha$ for all $t \geq 0$, then the relay remains $u(-\tau)$ forever and $x(t)$ obviously exists uniquely for all $t \geq 0$. If the trajectory of $x(t)$ intersects $\mathcal{S}_\beta$ or $\mathcal{S}_\alpha$, for $\tau > 0$, the solution $x(t)$ can exist just after the intersecting instant while for $\tau = 0$ it may not. The following lemma is useful for characterizing the trajectory of $x(t)$ at the intersecting points and analyzing the existence of solutions just after the intersecting instant.

1.3 Existence of Solutions

Lemma 1.3.1. *Consider system Σ_τ. If the trajectory of $x(t)$ starting from x_0 intersects $\mathcal{S}_\beta$ (or $\mathcal{S}_\alpha$) at intersecting point x_1 at the instant $t_1 > 0$, then there is no sliding motion along $\mathcal{S}_\beta$ (or $\mathcal{S}_\alpha$) around the intersecting instant t_1.*

Proof. We prove by contradiction. We take the case of $cx_0 \leq \alpha$ for example. (For the other cases, the proof is similar.)

Since $cx_0 \leq \alpha$, then from (1.1)–(1.3), the control remains u_α until the trajectory of $x(t)$ intersects $\mathcal{S}_\beta$ at time $t_1 > 0$. Suppose there is sliding motion along $\mathcal{S}_\beta$ just after the instant t_1. Then, for the cases $\tau > 0$ and $\tau = 0$, there is a sufficiently small $\delta > 0$, such that the control remains u_α during the small time interval $[t_1 - \delta, t_1 + \delta]$, and

$$y(t) = cx(t) \neq \beta, \quad t \in [t_1 - \delta, t_1), \tag{1.11}$$
$$y(t) = cx(t) = \beta, \quad t \in [t_1, t_1 + \delta]. \tag{1.12}$$

Since $\delta > 0$ is small, $x(t)$ can be expanded as

$$x(t) = x_1 + \sum_{i=0}^{k} \frac{1}{(i+1)!} A^i (Ax_1 + bu)(t - t_1)^{i+1} + O(t - t_1)^{k+2}, \quad \forall k = 0, 1, \ldots$$

From (1.12), it is easy to see that

$$cA^i(Ax_1 + bu) = 0, \quad \forall i = 0, 1, \ldots$$

which results in $y(t) = cx(t) \equiv \beta$ for all $t \in [t_1, t_1 + \delta]$. This contradicts (1.11). Hence, there is no sliding motion just after the instant t_1. This completes the proof.

Remark 1.3.1. We will see from Example 1.4.1 later that the trajectory of $x(t)$ may not be available just after the intersecting instant t_1 (i.e., $x(t)$ exists on $t \in [0, t_1]$ only). If the trajectory of $x(t)$ can evolve just after the intersecting instant t_1, by virtue of Lemma 1.3.1, the trajectory of $x(t)$ has at most two sub-trajectories: one traverses $\mathcal{S}_\beta$ (or $\mathcal{S}_\alpha$) at the intersecting point x_1; and one returns to $\mathcal{S}_{-\beta}$ (or $\mathcal{S}_{+\alpha}$).

It should be pointed out that Lemma 1.3.1 excludes the existence of sliding modes at the intersecting points only, but a sliding motion may occur at other points. See the following simple example.

Example 1.3.1. Consider system Σ_τ with $\tau = 0.5$ and

$$A = \begin{bmatrix} -1 & 0 \\ 0 & -1 \end{bmatrix}, \quad b = \begin{bmatrix} -1 \\ 0 \end{bmatrix}, \quad c = [1 \ 0],$$
$$u_\beta = -1, \quad u_\alpha = 1, \quad \beta = 1, \quad \alpha = -0.5.$$

10   1. Existence of Solutions

Letting $x_0 = [1\ 1]^T$ and $u(t-\tau) \equiv u_\beta$, $\forall t \in [0, \tau]$, we have $x(t) = [1\ e^{-t}]^T$ and $y(t) \equiv 1$ for all $t > 0$ and the control remains the same $u = u_\beta$ forever. So, the trajectory of $x(t)$ starting from x_0 has a sliding motion. But this trivial case does not conflict with Lemma 1.3.1, because x_0 is not an intersecting point.

For the case $\tau > 0$, let us look into the system trajectory characteristics in more detail. Due to $\tau > 0$, by Lemma 1.3.1, the trajectory of $x(t)$ starting from x_0 exists uniquely until it intersects $\mathcal{S}_\beta$ or $\mathcal{S}_\alpha$ at time t_1 at the intersecting point x_1. If $t_1 = \infty$, the solution $x(t)$ exists obviously for all $t \geq 0$; if t_1 is finite, then at x_1, the trajectory of $x(t)$ or (case 1) is tangent to $\mathcal{S}_\beta$ or $\mathcal{S}_\alpha$ and then returns to $\mathcal{S}_{-\beta}$ or $\mathcal{S}_{+\alpha}$ with the control input unchanged, or (case 2) traverses $\mathcal{S}_\beta$ or $\mathcal{S}_\alpha$ with the control input changed to another value after time duration τ. If the trajectory of $x(t)$ follows case 1, it will uniquely exist until intersecting $\mathcal{S}_\beta$ or $\mathcal{S}_\alpha$ again, at which intersecting point either case 1 happens again or it follows case 2. We concentrate on case 2 now. In this case, the phenomenon of early switching[1] may occur, which happens if $t_2 - t_1 \leq \tau$. Regarding this, we have the following statement, which says that, at most, a finite number of switchings occur during a time interval of length τ.

Lemma 1.3.2. *Let $x(t)$ be a solution to system Σ_τ with $\tau > 0$, which has its trajectory in $[0, \delta]$ with $\delta \in (0, \infty)$. Let $T \geq 0$ be any time instant satisfying $T + \tau \in [0, \delta]$. Then, during the time interval $[T, T+\tau]$, there are at most a finite number of early switchings.*

Proof. We prove by contradiction. Suppose there are an infinite number of early switchings during the time interval $[T, T+\tau]$ with the control remaining $u \in \mathcal{U}$. Then, there are consecutive traversing points $x_{\alpha i} \in \mathcal{S}_\alpha$, $x_{\beta i} \in \mathcal{S}_\beta$ and scalars $\delta_{\alpha i} > 0$, $\delta_{\beta i} > 0$, such that the trajectory of $x(t)$ spends time duration $\delta_{\alpha i}$ from $x_{\alpha i}$ to $x_{\beta i}$ and then spends $\delta_{\beta i}$ from $x_{\beta i}$ to $x_{\alpha(i+1)}$ ($i = 1, 2, \ldots$). We have

$$x_{\beta i} = e^{A\delta_{\alpha i}} x_{\alpha i} + \int_0^{\delta_{\alpha i}} e^{A(\delta_{\alpha i} - s)} bu ds,$$

giving

$$x_{\beta i} - x_{\alpha i} = (e^{A\delta_{\alpha i}} - I)x_{\alpha i} + \int_0^{\delta_{\alpha i}} e^{A(\delta_{\alpha i} - s)} bu ds.$$

Since $\sum_{i=1}^\infty (\delta_{\alpha i} + \delta_{\beta i}) \leq \tau$, it holds that $\delta_{\alpha i} \to 0$ and $\delta_{\beta i} \to 0$ as $i \to \infty$. Noting that $x_{\alpha i}$ is bounded due to the boundedness of $x(t)$ for $t \in [T, T+\tau]$, the above yields

---
[1] An early switching means that the time duration between two successive switching instants is no more than τ.

$$\lim_{i \to \infty} c(x_{\beta i} - x_{\alpha i}) = 0$$

which contradicts

$$cx_{\beta i} = \beta > \alpha = cx_{\alpha i}, \quad \forall i = 1, 2, \ldots.$$

This proves the lemma.

From the above easy analysis, the result for $\tau > 0$ can be obtained.

Theorem 1.3.1. *Consider system Σ_τ with $\tau > 0$. Then, for any initial condition $x(0) \in R^n$, there exists a unique solution $x(t)$ for all $t \in [0, \infty)$.*

Proof. Recall that for any initial condition $x(0) \in R^n$, $u(-\tau)$ is uniquely specified. So the solution $x(t)$ uniquely exists for $t \in [0, \tau]$. After that (i.e., $t > \tau$), the control input $u(t - \tau)$ is still uniquely specified due to $\tau > 0$. By virtue of Lemma 1.3.2, during any finite time duration, there are at most a finite number of switchings. These implies that $x(t)$ exists uniquely for all $t \geq 0$.

In Example 1.3.1, although the trivial case as stated there occurs, it is easy to verify that for any initial condition, there exists a unique solution for all $t \geq 0$.

1.4 Delay-free Case

For system Σ_τ with $\tau = 0$, it is not easy to study the system solutions. For a given initial condition x_0, although there exists a unique solution $x(t)$ for some time interval $[0, \delta]$ ($\delta > 0$), the trajectory of $x(t)$ may not be available just after it intersects the switching plane, or may have some sub-trajectories at an intersecting point. Let us examine the following examples.

Example 1.4.1. Consider system Σ_τ with $\tau = 0$ and

$$A = \begin{bmatrix} -1 & 0 \\ 0 & -1 \end{bmatrix}, \quad b = \begin{bmatrix} 1 \\ 1 \end{bmatrix}, \quad c = [1 \; 0],$$

$$u_\beta = -1, \quad u_\alpha = 1, \quad \beta = 0.5, \quad \alpha = -0.5.$$

We check that for any initial condition x_0, the trajectory of $x(t)$ starting from x_0 will intersect $\mathcal{S}_\beta$ or $\mathcal{S}_\alpha$ for the first time at some time t_1. However, it is easy to check that the solution $x(t)$ exists only for $t \in [0, t_1]$. After the instant t_1, no extended solution satisfying (1.1)–(1.3) is available.

12   1. Existence of Solutions

Example 1.4.2. Consider system Σ_τ with $\tau = 0$ and

$$A = \begin{bmatrix} -5 & 1 & 0 \\ -8 & 0 & 1 \\ -4 & 0 & 0 \end{bmatrix}, \quad b = \begin{bmatrix} 1 \\ 0 \\ 4 \end{bmatrix}, \quad c = [1 \ 0 \ 0],$$

$$u_\beta = 3, \quad u_\alpha = 2, \quad \beta = 1, \quad \alpha = -0.5.$$

Let the initial condition be $x_0 = [-2.2767 \ 3.5097 \ -2.4348]^T$. Then $u(0) = u_\alpha = 2$. We check that the trajectory of $x(t)$ starting from x_0 intersects $\mathcal{S}_\beta$ at the intersecting point $x_1 = [1 \ 3 \ 2]^T$ at the instant $t_1 = 0.6$. After instant t_1, there are two sub-trajectories: one traverses $\mathcal{S}_\beta$ with the relay switching to $u_\beta = 3$ (see the dashed line in Figure 1.3); the other one returns to $\mathcal{S}_{-\beta}$ with the relay remaining unchanged, and traverses $\mathcal{S}_\beta$ at the traversing point $x_2 = [1 \ 3.5934 \ 10.9780]^T$ at the instant $t_2 = 2.423$ (see the solid line in Figure 1.3).

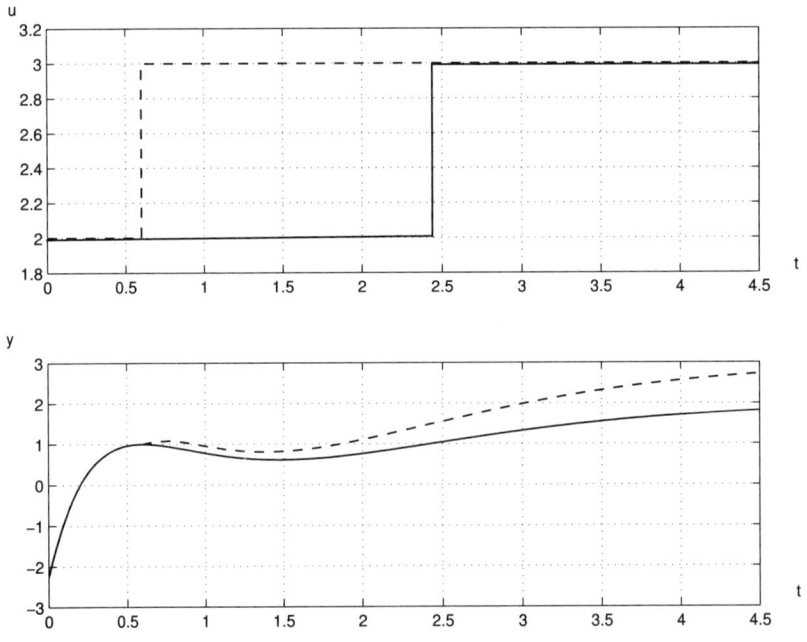

Fig. 1.3. Occurrence of two sub-trajectories for Example 1.4.2

It is seen from Example 1.4.1 that, for delay-free RFSs (i.e., $\tau = 0$), a trajectory may not evolve after it intersects $\mathcal{S}_\beta$ or $\mathcal{S}_\alpha$. In the following, we will seek conditions which ensure that the trajectory evolves just after the intersecting instants.

Recall that given any initial condition $x_0 \in R^n$, $u(0) \in \mathcal{U}$ is well-defined and there exists a unique solution $x(t)$ to (1.1)–(1.3) at least for $t \in [0, \delta_1]$ ($\delta_1 > 0$) during which it holds that $u(t) \equiv u(0)$. Obviously, if no intersecting point is available for all $t > \delta_1$, then the solution $x(t)$ is unique for all $t \geq 0$ with $u(t) \equiv u(0)$. Otherwise, δ_1 can be extended to t_1, where t_1 is the first intersecting time taken for the trajectory of $x(t)$ to go from x_0 to the first intersecting point $x_1 \in \mathcal{S}_\beta$ (or $x_1 \in \mathcal{S}_\alpha$). Note that after the instant t_1, $x(t)$ may not be available. Alternatively, if $x(t)$ can extend its trajectory just after t_1, then by Lemma 1.3.1 there is at most two sub-trajectories (Remark 1.3.1). We also denote any possible corresponding sub-solution by $x(t)$. Then $x(t)$ exists at least for $t \in [0, t_1 + \delta_2]$ for some $\delta_2 > 0$. Again, δ_2 can be extended to infinity (if there are no more intersecting points) or to $t_2 \geq \delta_2$, where t_2 is the time $x(t)$ taken going from x_1 to the next intersecting point $x_2 \in \mathcal{S}_\alpha$ (or $x_2 \in \mathcal{S}_\beta$). So, whether or not a solution $x(t)$ (perhaps a sub-solution) starting from x_0 can extend its trajectory is related to the information at the intersecting points. The following theorem gives a necessary and sufficient condition for the existence of solutions just after any intersecting instant. Let

$$\mathcal{N} = \{0, 1, \ldots, n-1\}. \tag{1.13}$$

Theorem 1.4.1. *Consider system Σ_τ with $\tau = 0$.*

(i) Suppose that a solution $x(t)$ has an intersecting point $x(t_s) \triangleq x_\beta \in \mathcal{S}_\beta$ in its trajectory. Then, the trajectory of $x(t)$ can evolve further after instant t_s if and only if either there exists $n_{12} \in \mathcal{N}$ such that

$$\begin{cases} cA^{i+1}x_\beta + cA^i bu_\beta = 0, & i = 0, 1, \ldots, n_{12} - 1, \\ cA^{n_{12}+1}x_\beta + cA^{n_{12}} bu_\beta > 0, \end{cases} \tag{1.14}$$

holds, or there exists an odd $n_{11} \in \mathcal{N}$ such that

$$\begin{cases} cA^{i+1}x_\beta + cA^i bu_\alpha = 0, & i = 0, 1, \ldots, n_{11} - 1, \\ cA^{n_{11}+1}x_\beta + cA^{n_{11}} bu_\alpha < 0 \end{cases} \tag{1.15}$$

holds, or both (1.14) and (1.15) hold.

(ii) Suppose that a solution $x(t)$ has an intersecting point $x(t_s) \triangleq x_\alpha \in \mathcal{S}_\alpha$ in its trajectory. Then, the trajectory of $x(t)$ can evolve further after instant t_s if and only if either there exists $n_{11} \in \mathcal{N}$ such that

$$\begin{cases} cA^{i+1}x_\alpha + cA^i bu_\alpha = 0, & i = 0, 1, \ldots, n_{11} - 1, \\ cA^{n_{11}+1}x_\alpha + cA^{n_{11}} bu_\alpha < 0, \end{cases} \tag{1.16}$$

holds, or there exists an odd $n_{12} \in \mathcal{N}$ such that

$$\begin{cases} cA^{i+1}x_\alpha + cA^i bu_\beta = 0, \quad i = 0, 1, \ldots, n_{12} - 1, \\ cA^{n_{12}+1}x_\alpha + cA^{n_{12}} bu_\beta > 0 \end{cases} \quad (1.17)$$

holds, or both (1.16) and (1.17) hold.

Proof. We only prove (i). For (ii), the proof is similar.

Necessity: If the solution $x(t)$ extends its existence to $t_s + \delta$ for a certain small $\delta > 0$, by virtue of Lemma 1.3.1 (or Remark 1.3.1), the trajectory of $x(t)$ has at most two sub-trajectories: (ia) one traverses $\mathcal{S}_\beta$ at the intersecting point x_β, and (ib) one returns to $\mathcal{S}_{-\beta}$ just after the instant t_s. Choosing δ sufficiently small, we have the following expansion of $x(t)$ around t_s: for (ia),

$$x(t) = \begin{cases} x_\beta + \sum_{i=0}^{n_{11}} \frac{1}{(i+1)!} A^i (Ax_\beta + bu_\alpha)(t - t_s)^{i+1} + O(t - t_s)^{n_{11}+2}, \\ \qquad \forall t \in [t_s - \delta, \, t_s), \\ x_\beta + \sum_{i=0}^{n_{12}} \frac{1}{(i+1)!} A^i (Ax_\beta + bu_\beta)(t - t_s)^{i+1} + O(t - t_s)^{n_{12}+2}, \\ \qquad \forall t \in (t_s, \, t_s + \delta], \end{cases}$$

and for (ib),

$$x(t) = x_\beta + \sum_{i=0}^{n_{11}} \frac{1}{(i+1)!} A^i (Ax_\beta + bu_\alpha)(t - t_s)^{i+1} + O(t - t_s)^{n_{11}+2},$$

$$\forall t \in [t_s - \delta, \, t_s + \delta],$$

where n_{11} and n_{12} are non-negative integers such that

$$\begin{cases} cA^{i+1}x_\beta + cA^i bu_\alpha = 0, \quad i = 0, 1, \ldots, n_{11} - 1, \\ cA^{n_{11}+1}x_\beta + cA^{n_{11}} bu_\alpha \neq 0, \\ cA^{j+1}x_\beta + cA^j bu_\beta = 0, \quad j = 0, 1, \ldots, n_{12} - 1, \\ cA^{n_{12}+1}x_\beta + cA^{n_{12}} bu_\beta \neq 0. \end{cases}$$

From the Cayley–Hamilton Theorem, for any integer $k \geq n$, A^k can be expressed as a linear combination of $I, A, \ldots, A^{n-1}$. Hence, n_{11} and n_{12} must be in the set $\mathcal{N}$. Otherwise, $cA^{i+1}x_\beta + cA^i bu = 0$ ($u \in \mathcal{U}$) for all $i \in \mathcal{N}$ will lead to $cA^{i+1}x_\beta + cA^i bu = 0$ for all non-negative integer i, and thus $cx(t) \equiv 0$ for all $t \in [t_s - \delta, \, t_s + \delta]$. This contradicts the claim that $x(t)$ has no sliding motion for $t \in [t_s - \delta, \, t_s + \delta]$. Now, since for (ia), $\begin{cases} cx_\beta = \beta, \\ cx(t) < \beta, \quad \forall t \in [t_s - \delta, \, t_s), \\ cx(t) > \beta, \quad \forall t \in (t_s, \, t_s + \delta] \end{cases}$

for (ib), $\begin{cases} cx_\beta = \beta, \\ cx(t) < \beta, \quad \forall t \in [t_s - \delta, \, t_s + \delta], \, t \neq t_s, \end{cases}$

we get

for (ia), $\begin{cases} c\dfrac{1}{(n_{11}+1)!}A^{n_{11}}(Ax_\beta + bu_\alpha)(t-t_s)^{n_{11}+1} + cO(t-t_s)^{n_{11}+2} < 0, \\ \qquad\qquad\qquad\qquad\qquad\qquad\qquad \forall t \in [t_s - \delta, \, t_s), \\ c\dfrac{1}{(n_{12}+1)!}A^{n_{12}}(Ax_\beta + bu_\beta)(t-t_s)^{n_{12}+1} + cO(t-t_s)^{n_{12}+2} > 0, \\ \qquad\qquad\qquad\qquad\qquad\qquad\qquad \forall t \in (t_s, \, t_s + \delta], \end{cases}$

for (ib), $c\dfrac{1}{(n_{11}+1)!}A^{n_{11}}(Ax_\beta + bu_\alpha)(t-t_s)^{n_{11}+1} + cO(t-t_s)^{n_{11}+2} < 0,$
$\qquad\qquad\qquad\qquad\qquad\qquad\qquad \forall t \in [t_s - \delta, \, t_s + \delta], \, t \neq t_s.$

By considering the process $t \to t_s$, we establish that, for (ia),

$\exists \, n_{11}, n_{12} \in \mathcal{N}, \text{ s.t.}, \begin{cases} cA^{i+1}x_\beta + cA^i bu_\beta = 0, \quad i = 0, 1, \ldots, n_{12} - 1, \\ cA^{n_{12}+1}x_\beta + cA^{n_{12}}bu_\beta > 0, \\ cA^{j+1}x_\beta + cA^j bu_\alpha = 0, \quad j = 0, 1, \ldots, n_{11} - 1, \\ \begin{cases} cA^{n_{11}+1}x_\beta + cA^{n_{11}}bu_\alpha > 0, \text{ if } n_{11} \text{ is even,} \\ cA^{n_{11}+1}x_\beta + cA^{n_{11}}bu_\alpha < 0, \text{ if } n_{11} \text{ is odd,} \end{cases} \end{cases}$

and for (ib),

$\exists \text{ an odd } n_{11} \in \mathcal{N}, \text{ s.t.}, \begin{cases} cA^{i+1}x_\beta + cA^i bu_\alpha = 0, \quad i = 0, 1, \ldots, n_{11} - 1, \\ cA^{n_{11}+1}x_\beta + cA^{n_{11}}bu_\alpha < 0. \end{cases}$

Combining the above with the information on $x(t)$ around the time t_β yields that, for (ia),

$\exists \, n_{12} \in \mathcal{N}, \text{ s.t.}, \begin{cases} cA^{i+1}x_\beta + cA^i bu_\beta = 0, \quad i = 0, 1, \ldots, n_{12} - 1, \\ cA^{n_{12}+1}x_\beta + cA^{n_{12}}bu_\beta > 0, \end{cases}$

and for (ib),

$\exists \text{ an odd } n_{11} \in \mathcal{N}, \text{ s.t.}, \begin{cases} cA^{i+1}x_\beta + cA^i bu_\alpha = 0, \quad i = 0, 1, \ldots, n_{11} - 1, \\ cA^{n_{11}+1}x_\beta + cA^{n_{11}}bu_\alpha < 0. \end{cases}$

This proves the necessity.

Sufficiency. Follows from reversing the proof of necessity. This completes the proof of the theorem.

Remark 1.4.1. (1.14) (resp. (1.16)) implies that there is at least one extended $x(t)$ trajectory just after the instant t_s, which traverses S_β (resp. S_α) at the intersecting point x_β (resp. x_β); (1.15) (resp. (1.17)) implies that there is at least one extended $x(t)$ trajectory just after the time t_s, which returns to $S_{-\beta}$ (resp. $S_{+\alpha}$). If and only if both (1.14) and (1.15) (resp. (1.16) and (1.17)) hold, there are two sub-trajectories of $x(t)$ just after the instant t_s.

In Example 1.4.1, although the trajectory of $x(t)$ starting from any initial condition x_0 will intersect S_β or S_α, it can be verified that at the intersecting point, neither (1.14) nor (1.15) (or, neither (1.16) nor (1.17)) is satisfied. Thus, after the first intersecting, the trajectory cannot evolve. In Example 1.4.2, at the intersecting point x_1, we check that both (1.14) and (1.15) hold and thus there are two sub-trajectories of $x(t)$ just after the instant t_1; at the intersecting point x_2, we check that (1.14) holds but (1.15) does not hold and thus there is only one trajectory of $x(t)$ just after the instant t_2, which traverses S_α.

The following is a direct corollary of Theorem 1.4.1.

Corollary 1.4.1. *Under the condition of Theorem 1.4.1, if the trajectory of $x(t)$ can evolve just after t_s and $\dot{y}(t_{s-}) \neq 0$, then the solution $x(t)$ is unique for $t \in [t_s, t_s + \delta]$ for some $\delta > 0$, and its trajectory traverses S_β (or S_α) at time $t_s > 0$.*

Proof. Follows directly from Theorem 1.4.1.

It is seen from Corollary 1.4.1 that the possible sub-trajectory only occurs at instant t, which corresponds to an intersecting point and makes $\dot{y}(t_-) = 0$.

From Theorem 1.4.1, for any initial condition $x_0 \in R^n$ (without loss of generality, let $u(0) = u_\beta$ here), let the trajectory of $x(t)$ firstly intersects S_α at the intersecting point x_{11} at the intersecting instant t_{11}. The trajectory of $x(t)$ can evolve just after the instant t_{11} if and only if Theorem 1.4.1 (ii) holds, replacing x_α by x_{11}. Evolving at x_{11}, the trajectory of $x(t)$ may (if (1.16) holds) intersect S_β at intersecting point x_{21}, or (if (1.17) holds) intersect S_α again at x_{12}. Using Theorem 1.4.1 again, the trajectory of $x(t)$ (perhaps a sub-trajectory of $x(t)$ right now) exists just after reaching x_{12} if and only if Theorem 1.4.1 (ii) holds, replacing x_α by x_{12}. If (1.17) still holds, repeat the process until such sub-trajectory of $x(t)$ traverses (if so) S_α at x_{1k_1} ($k_1 \geq 2$) with the relay switching to u_α. After leaving the point x_{1k_1}, it may intersect S_β at intersecting point x_{22}. Now, use Theorem 1.4.1 with respect to each intersecting point in S_β, and continue the process. Note that the sub-trajectories of $x(t)$ starting from any x_0 are at most countably many. For such an initial point x_0, let

$$\mathcal{X}_\alpha := \{x_{11}, \ldots, x_{1k_1}, x_{31}, \ldots, x_{3k_3}, \ldots\}, \tag{1.18}$$
$$\mathcal{X}_\beta := \{x_{21}, \ldots, x_{2k_2}, x_{41}, \ldots, x_{4k_4}, \ldots\}, \tag{1.19}$$

which are the sets of all intersecting points on S_α and S_β, respectively. Then, we have the following result.

Corollary 1.4.2. *Consider system Σ_τ with $\tau = 0$. Then any trajectory of $x(t)$ starting from x_0 exists for all $t > 0$ if and only if both of the following hold:*
(i) For any $x_\beta \in \mathcal{X}_\beta$, Theorem 1.4.1 (i) holds.
(ii) For any $x_\alpha \in \mathcal{X}_\alpha$, Theorem 1.4.1 (ii) holds.

Proof. Necessity: Obvious from Theorem 1.4.1.

Sufficiency: Using Theorem 1.4.1 repeatedly shows that the trajectory of $x(t)$ (including its sub-trajectories) can evolve further after each intersection of $\mathcal{X}_\alpha$ or $\mathcal{X}_\beta$. Let the trajectory of $x(t)$ traverse $\mathcal{X}_\alpha$ and $\mathcal{X}_\beta$ at $x_1, x_3, \ldots \in \mathcal{X}_\alpha$ and $x_2, x_4, \ldots \in \mathcal{X}_\beta$, respectively, and the time for it to go from x_i to x_{i+1} be t_i. If the number of x_i is finite, then $x(t)$ exists for all $t \geq 0$, completing the proof. If the number of x_i is infinite, to prove that $x(t)$ exists for all $t \geq 0$, it sufficient to show that

$$\lim_{k \to \infty} \sum_{i=1}^{k} t_i \longrightarrow \infty. \tag{1.20}$$

This is done in the following. For simplicity, we will use the following notation:

$$\int_{t_{11}}^{t_{12}} + \int_{t_{21}}^{t_{22}} + \ldots + \int_{t_{k1}}^{t_{k2}} f(s)ds$$
$$:= \int_{t_{11}}^{t_{12}} f(s)ds + \int_{t_{21}}^{t_{22}} f(s)ds + \ldots + \int_{t_{k1}}^{t_{k2}} f(s)ds.$$

Suppose (1.20) is not true. We show this causes a contradiction. Let

$$t_f := \sum_{i=1}^{\infty} t_i.$$

Since $t_i > 0$, $\forall i = 1, 2, \ldots$, it holds that $t_i \to 0$ as $i \to \infty$. For $k = 1, 2, \ldots$, we have

$$x_{2k} = e^{At_{2k-1}} x_{2k-1} + \int_0^{t_{2k-1}} e^{A(t_{2k-1}-s)} bu_\alpha ds,$$

$$x_{2k+1} = e^{At_{2k}} x_{2k} + \int_0^{t_{2k}} e^{A(t_{2k}-s)} bu_\beta ds.$$

So,

$$x_{2k} - x_{2k-1} = (e^{At_{2k-1}} - I)x_{2k-1} + \int_0^{t_{2k-1}} e^{A(t_{2k-1}-s)} bu_\alpha ds,$$

$$x_{2k+1} - x_{2k} = (e^{At_{2k}} - I)x_{2k} + \int_0^{t_{2k}} e^{A(t_{2k}-s)} bu_\beta ds.$$

Up to here, since it is not clear whether or not x_{2k} and/or x_{2k-1} is bounded, we cannot use a similar deduction to the proof of Lemma 1.3.2 to show a contradiction. However, using the fact

$$cx_{2k-1} = \alpha, \quad cx_{2k} = \beta, \quad \forall k = 1, 2, \ldots$$

we obtain for all $k = 1, 2, \ldots$ that

$$0 < \beta - \alpha = \begin{cases} c(e^{At_{2k-1}} - I)x_{2k-1} + \int_0^{t_{2k-1}} ce^{A(t_{2k-1}-s)} bu_\alpha ds, \\ c(e^{At_{2k}} - I)x_{2k} + \int_0^{t_{2k}} ce^{A(t_{2k}-s)} bu_\beta ds. \end{cases}$$

This shows that it must hold that

$$\|x_i\| \to \infty \text{ as } i \to \infty.$$

On the other hand, using $\int_{\underline{t}}^{\overline{t}} f(t-s)ds = \int_{t-\overline{t}}^{t-\underline{t}} f(s)ds$, we have

$$x_2 = e^{At_1}x_1 + \int_0^{t_1} e^{As} bu_\alpha ds,$$

$$x_3 = e^{A(t_2+t_1)}x_1 + \int_{t_2}^{t_2+t_1} e^{As} bu_\alpha ds + \int_0^{t_2} e^{As} bu_\beta ds,$$

$$x_4 = e^{A(t_3+t_2+t_1)}x_1 + \left(\int_{t_3+t_2}^{t_3+t_2+t_1} + \int_0^{t_3} e^{As} bu_\alpha ds\right) + \int_{t_3}^{t_3+t_2} e^{As} bu_\beta ds,$$

$$x_5 = e^{A(t_4+\cdots+t_1)}x_1 + \left(\int_{t_4+t_3+t_2}^{t_4+\cdots+t_1} + \int_{t_4}^{t_4+t_3} e^{As} bu_\alpha ds\right)$$

$$+ \left(\int_{t_4+t_3}^{t_4+t_3+t_2} + \int_0^{t_4} e^{As} bu_\beta ds\right),$$

$\ldots\ldots\ldots$

and generally, we obtain for $k = 1, 2, \ldots$

$$x_{2k} = e^{A\sum_{i=1}^{2k-1} t_i} x_1 + \left(\int_{\sum_{i=2}^{2k-1} t_i}^{\sum_{i=1}^{2k-1} t_i} + \int_{\sum_{i=4}^{2k-1} t_i}^{\sum_{i=3}^{2k-1} t_i} + \cdots + \int_0^{t_{2k-1}} e^{As} bu_\alpha ds\right)$$

$$+ \left(\int_{\sum_{i=3}^{2k-1} t_i}^{\sum_{i=2}^{2k-1} t_i} + \int_{\sum_{i=5}^{2k-1} t_i}^{\sum_{i=4}^{2k-1} t_i} + \cdots + \int_{t_{2k-1}}^{\sum_{i=2k-1}^{2k-1}-2 t_i} e^{As} bu_\beta ds\right),$$

$$x_{2k+1} = e^{A\sum_{i=1}^{2k} t_i} x_1 + \left(\int_{\sum_{i=2}^{2k} t_i}^{\sum_{i=1}^{2k} t_i} + \int_{\sum_{i=4}^{2k} t_i}^{\sum_{i=3}^{2k} t_i} + \cdots + \int_{t_{2k}}^{\sum_{i=2k-1}^{2k} t_i} e^{As} bu_\alpha ds\right)$$

$$+ \left(\int_{\sum_{i=3}^{2k} t_i}^{\sum_{i=2}^{2k} t_i} + \int_{\sum_{i=5}^{2k} t_i}^{\sum_{i=4}^{2k} t_i} + \cdots + \int_0^{t_{2k}} e^{As} bu_\beta ds\right).$$

Since
$$e^{A\sum_{i=1}^{j} t_i} \to e^{At_f} \text{ as } j \to \infty,$$
then $\|e^{A\sum_{i=1}^{j} t_i}\|$ is bounded by some scalar $M > 0$ for all $j = 1, 2, \ldots$. Thus,

$$\|x_{2k}\| \leq \|e^{A\sum_{i=1}^{2k-1} t_i} x_1\| + \int_0^{\sum_{i=1}^{2k-1} t_i} \|e^{As} bu_\alpha\| ds + \int_0^{\sum_{i=1}^{2k-1} t_i} \|e^{As} bu_\beta\| ds$$

$$\leq M\|x_1\| + \int_0^{t_f} (\|e^{As} bu_\alpha\| + \|e^{As} bu_\beta\|) ds,$$

and similarly

$$\|x_{2k+1}\| \leq M\|x_1\| + \int_0^{t_f} (\|e^{As} bu_\alpha\| + \|e^{As} bu_\beta\|) ds.$$

So, $\|x_k\|$ is bounded for all $k = 1, 2, \ldots$. This shows a contradiction. Hence (1.20) is true.

It should be pointed out that in the above proof, it is not clear *a priori* if $\lim_{k\to\infty} \sum_{i=1}^{k} t_i$ is sufficiently large and A is Hurwitz. If this is true (or, it is known *a priori* that $x(t)$ exists for all $t > 0$), then a minimum time duration $t_{min} > 0$ between any two consecutive switching instants can be computed, and thus (1.20) is guaranteed.

Remark 1.4.2. From the proof of Corollary 1.4.1, we see that there are no fast switches[2] for systems of the form (1.1)–(1.3). For the case $\alpha = \beta = 0$, fast switches may occur. See Johansson *et al.* (1999), which provides a necessary and sufficient condition for the existence of fast switches.

The following is another corollary to Theorem 1.4.1, with a strong constraint. Denote

$$\Xi_\alpha := \{\xi \in R^n : cA\xi + cbu_\beta \leq 0\}, \tag{1.21}$$
$$\Xi_\beta := \{\xi \in R^n : cA\xi + cbu_\alpha \leq 0\}. \tag{1.22}$$

Corollary 1.4.3. *Consider system Σ_τ with $\tau = 0$. Then each trajectory of $x(t)$ starting from $x_0 \in R^n$ exists for all $t > 0$ if both of the following hold:*
  (i) For any $x_\beta \in \Xi_\beta$, Theorem 1.4.1 (i) holds.
  (ii) For any $x_\alpha \in \Xi_\alpha$, Theorem 1.4.1 (ii) holds.

Proof. For any $x_0 \in R^n$, it is obvious that

$$\mathcal{X}_\alpha \subseteq \Xi_\alpha, \quad \mathcal{X}_\beta \subseteq \Xi_\beta. \tag{1.23}$$

Hence the result follows immediately from Corollary 1.4.2.

---

[2] Fast switches mean that the relay switches an infinite number of times during a finite time interval.

2. Existence of Limit Cycles

The preceding chapter studies the existence of solutions to system Σ_τ. In this chapter, we always assume this property with a default, i.e., a solution always exists for all $t \geq 0$ for any initial condition $x_0 \in \mathbb{R}^n$. This chapter will consider the systems of the same form as in Chapter 1 and study the problem on the existence of limit cycles.

2.1 Introduction

One of the important aspects of relay feedback systems, as well as many other non-linear systems, is that limit cycles may occur in the trajectories. This property is very useful in modern control applications such as automatic tuning of controllers and frequency response estimation and identification (Åström and Hägglund, 1995; Atherton, 1993; Wang et al., 1999b; Wang et al., 1999c). Such practical applications motivate the intensive investigation of limit cycle behaviours. The study consists in establishing their existence, determining their frequency and form, investigating their stability and so on. However, most analysis work has been based on the assumption that a limit cycle does exist, due to the difficulty of determining if this is really the case. Many classical results based on the describing function method are surveyed in Cook (1986) and Tsypkin (1984). The describing function method gives an approximate analysis of limit cycle behaviours for SISO systems. An exact method was reported recently in Åström (1995). The formula therein with a closed form is indeed a necessary condition for the existence of limit cycles with two switches per period. This type of periodic orbit is revisited and investigated further in Varigonda and Georgious (2001) for RFSs without time delay. For MIMO systems also, necessary conditions can be found in the literature. Loh (1994) studied limit cycle periods and characteristics by extending Tsypkin's method. The Z-transform technique is exploited in Palmor et al. (1992) and Palmor et al. (1995b) to determine limit cycle periods, which extends the idea of adopting the Z-transform for SISO systems (Åström and Hägglund, 1984a).

22   2. Existence of Limit Cycles

The existence problem is by no means the key important step in the investigation of limit cycle behaviours for a relay feedback system. In this chapter, we consider a relay feedback system of the same form as in the preceding chapter and study this problem through a state-space method. Section 2.2 presents a sufficient condition mainly based on Brouwer's fixed-point theorem. In Section 2.3, the existence problem is studied through the set of traversing points of a single system trajectory. Section 2.4 gives computation formulas for possible limit cycle location, which is naturally a necessary condition for the existence of a certain type of limit cycle.

2.2 Sufficient Condition

In this section, we consider system Σ_τ with additional constraints and seek sufficient conditions for the existence of limit cycles.

2.2.1 Supporting Lemmas

A set K is said to be *convex*, if for any $s_1, s_2 \in K$, we have $\lambda s_1 + (1-\lambda)s_2 \in K$ for all $\lambda \in [0,1]$. A subset of R^n is *compact* if and only if it is closed and bounded. For our development, we need the following fixed-point theorem (Kuttler, 1998).

Lemma 2.2.1. *(Brouwer's fixed-point theorem). Let K be a non-empty compact convex subset of R^n and let $g : K \to K$ be continuous. Then, g has a fixed point.*

We now establish some useful lemmas, from which we can derive sufficient conditions for the existence of limit cycles for system Σ_τ in the next subsection. Let $G(s) = c(sI - A)^{-1}b$. Then, we have $G(0) = -cA^{-1}b$ if A is invertible.

For system Σ_τ, if A is Hurwitz and $G(0)u_\alpha > \beta > \alpha > G(0)u_\beta$, then the trajectory of $\xi(t)$ starting from any initial condition $\xi(0)$ will intersect either S_α or S_β in a finite time. We show this by contradiction. Suppose that the trajectory of $\xi(t)$ intersects neither S_+ nor S_- in finite time. Then the trajectory of $\xi(t)$ is governed by

$$\xi(t) = e^{At}\xi(0) + \int_0^t e^{A(t-z)}bu_0 dz$$
$$= e^{At}\xi(0) + (e^{At} - I)A^{-1}bu_0$$

where $u_0 \in \mathcal{U}$ and $\mathcal{U}$ is given as in 1.4. Since $e^{At} \to 0$ as $t \to \infty$ (because A is Hurwitz), we see that

$c\xi(t) \to -cA^{-1}bu_0$ as $t \to \infty$.

Thus, for some $T > \tau \geq 0$,

$$c\xi(t) > \beta, \quad t > T, \quad \text{for } u_0 = u_\alpha,$$
$$c\xi(t) < \alpha, \quad t > T, \quad \text{for } u_0 = u_\beta,$$

which is a contradiction to the relay feedback law (1.2). So, a switch must occur at some instant $t < T$. This implies that the trajectory of $\xi(t)$ intersects either $\mathcal{S}_+$ or $\mathcal{S}_-$ in finite time.

To ensure the occurrence of consecutive switchings, we need the following assumption.

Assumption 2.1 A is Hurwitz and $G(0)u_\alpha > \beta > \alpha > G(0)u_\beta$.

Note that neither A being Hurwitz nor $G(0)u_\alpha > \beta > \alpha > G(0)u_\beta$ is a necessary condition for the occurrence of switchings. However, with our default that a solution with any initial condition $x_0 \in \mathbb{R}^n$ always exists for all $t \geq 0$, Assumption 2.1 ensures that each system trajectory makes the relay switch consecutively.

Here, we only study a class of systems satisfying the condition that for each system trajectory the time duration between any two consecutive switchings is longer than τ. Obviously, this is always true for $\tau = 0$. For $\tau > 0$, early switches may occur in some systems. When this happens, the complicated system trajectory characteristics makes system analysis very difficult. So, to avoid early switches we assume the following.

Assumption 2.2 For $\tau > 0$, any trajectory of system Σ_τ has time duration between any two consecutive switchings longer than τ.

See Chapter 4 later (indeed refer to Remark 4.3.2 for the details) for a class of systems which satisfies Assumption 2.2. In many processes, the system will exhibit monotonous behaviours, that is, $\dot{y}_s(t) \geq 0$ where $y_s(t)$ is the output step response. In this case, Assumption 2.2 is also satisfied.

The following lemma provides an invariant contraction region Ω, which any trajectory of system Σ_τ will eventually enter and remain in.

Lemma 2.2.2. *Consider system Σ_τ with A Hurwitz. Let $P > 0$ be the unique solution to the Lyapunov equation*

$$A^T P + PA = -I \tag{2.1}$$

and let

2. Existence of Limit Cycles

$$u_{max} = \max\{|u_\alpha|, |u_\beta|\},$$
$$\omega = \max_{\|\xi\| \leq 2u_{max}\|Pb\|} \|P^{\frac{1}{2}}\xi\|,$$
$$\Omega = \{\xi \in \mathbb{R}^n : \|P^{\frac{1}{2}}\xi\| \leq \omega\}. \tag{2.2}$$

Then, Ω is an invariant contraction region for system Σ_τ.

Proof. We prove by contradiction. Suppose for some solution $\xi(t)$, it holds that $\xi(t_1) \in \Omega$ and $\xi(t_3) \notin \Omega$ with $t_3 > t_1$. By continuity, there exists a time $t_2 \in [t_1, t_3)$ such that

$$\xi^T(t_2)P\xi(t_2) = \omega^2, \quad \xi^T(t)P\xi(t) > \omega^2, \quad \forall t \in (t_2, t_3], \tag{2.3}$$

which implies for $t \in (t_2, t_3]$ that

$$\|\xi(t)\| > 2u_{max}\|Pb\|. \tag{2.4}$$

Now, letting $V(x) = x^T P x$, it follows that

$$\frac{dV(x)}{dt} = x^T(A^T P + PA)x + 2x^T Pbu$$
$$= -\|x\|^2 + 2x^T Pbu.$$

Noticing that $|u| \leq u_{max}$, we obtain

$$\frac{dV(x)}{dt} \leq -\|x\|^2 + 2u_{max}\|Pb\|\|x\|.$$

It is seen that for $\|x\| > 2u_{max}\|Pb\|$, we have $\dot{V}(x) < 0$. So, from (2.4), for $t \in (t_2, t_3]$, it holds that $\frac{dV(\xi(t))}{dt} < 0$, giving

$$\xi^T(t)P\xi(t) = V(\xi(t))$$
$$= V(\xi(t_2)) + \int_{t_2}^{t} \dot{V}(\xi(z))dz$$
$$\leq V(\xi(t_2))$$
$$= \omega^2,$$

which contradicts (2.3). This completes the proof.

Remark 2.2.1. There are several ways to specify an invariant region such that all trajectories of system Σ_τ will eventually enter and remain in it. Hsu (1990) computed an invariant region for a class of relay feedback systems with the control containing an integrator. It should be noted that there may not exist a fixed time T such that all trajectories will be bounded by a fixed bound for $t \geq T$. Check the simple first-order example:

$$\dot{x}(t) = -x(t) + u(t)$$
$$y(t) = x(t)$$

with $u(t) = -1$ for $y(t) > 0$, and $u(t) = 1$ for $y(t) < 0$. Then $A = -1$ is Hurwitz. But, for any large $L > 0$ and $T > 0$ there exists an initial state $x_0 > 0$ such that $x(T) = e^{-T}(x_0 + 1) - 1 > L$. So, no such a fixed T is available.

In the ultimate contraction region Ω, a scalar $\omega_0 > 0$ can be found such that

$$\|\xi\| \leq \omega_0, \quad \forall \xi \in \Omega. \tag{2.5}$$

In the sequel, we let

$$\Omega_\alpha := \Omega \cap \mathcal{S}_\alpha,$$
$$\Omega_\beta := \Omega \cap \mathcal{S}_\beta. \tag{2.6}$$

Also, define respectively, for $\xi_0 \in \Omega_\alpha$ and $\xi_0 \in \Omega_\beta$, the following sets.

$$\mathcal{L}(\epsilon, \xi_0, \beta) = \{\xi \in \Omega_\beta : \ \|\xi - \xi(0)\| \leq \epsilon, \ \xi_0 \in \Omega_\beta, \ \epsilon > 0\}$$
$$= \{\xi \in \Omega_\beta : \ \xi = \xi(0) + \Delta, \ \xi_0 \in \Omega_\beta, \ \|\Delta\| \leq \epsilon, \ \epsilon > 0\}, \tag{2.7}$$
$$\mathcal{L}(\epsilon, \xi_0, \alpha) = \{\xi \in \Omega_\alpha : \ \|\xi - \xi(0)\| \leq \epsilon, \ \xi_0 \in \Omega_\alpha, \ \epsilon > 0\}$$
$$= \{\xi \in \Omega_\alpha : \ \xi = \xi(0) + \Delta, \ \xi_0 \in \Omega_\alpha, \ \|\Delta\| \leq \epsilon, \ \epsilon > 0\}. \tag{2.8}$$

The next lemma shows the relationship between two trajectories of system Σ_τ at the switching instants, which plays an important role in achieving the main result.

Lemma 2.2.3. *Suppose Assumptions 2.1 and 2.2 hold. If a trajectory of system Σ_τ evolves from $\xi(t_0) \in \Omega_\beta$ and traverses Ω_α at $\xi(t_0 + h)$, then there is a scalar $\epsilon > 0$ such that any trajectory evolving from the local region $\mathcal{L}(\epsilon, \xi(t_0), \beta) \subset \Omega_\beta$ will intersect Ω_α at some time $t_0 + h + \delta$, where δ satisfies*

$$|\delta|^{k+1} \leq M\|\Delta\| \tag{2.9}$$

for some scalars $M > 0$ and $k \in \mathcal{N}$, where the set $\mathcal{N}$ is as in 1.13.

In the proof of Lemma 2.2.3, the following result is needed.

Lemma 2.2.4. *Consider system Σ_τ with $\tau > 0$. Suppose a trajectory intersects the switching planes $\mathcal{S}_\alpha$ and $\mathcal{S}_\beta$ at $x_\alpha \in \mathcal{S}_\alpha$ and $x_\beta \in \mathcal{S}_\beta$, respectively. Then, x_α and x_β are traversing points if and only if there exist two even integers $n_\alpha, n_\beta \in \mathcal{N}$ such that the following hold:*

$$cA^{i+1}x_\alpha + cA^i b u_\beta = 0, \quad i = 0, 1, \ldots, n_\alpha - 1,$$
$$cA^{n_\alpha+1}x_\alpha + cA^{n_\alpha} b u_\beta < 0, \tag{2.10}$$
$$cA^{j+1}x_\beta + cA^j b u_\alpha = 0, \quad j = 0, 1, \ldots, n_\beta - 1,$$
$$cA^{n_\beta+1}x_\beta + cA^{n_\beta} b u_\alpha > 0. \tag{2.11}$$

26   2. Existence of Limit Cycles

Proof. We consider the proof of (2.11). For (2.10), the proof is similar.

Necessity: Suppose that the trajectory of $x(t)$ traverses $\mathcal{S}_\beta$ at $x_\beta \in \mathcal{S}_\beta$. Let the instant $t = t_\beta$ correspond to $x(t_\beta) = x_\beta$. For a sufficiently small $\delta > 0$, we have the following expansion of $x(t)$ in $t \in [t_\beta - \delta,\ t_\beta + \delta]$:

$$x(t) = x(t_\beta) + \sum_{i=0}^{n_\beta} \frac{1}{(i+1)!} A^i (Ax_\beta + bu_\alpha)(t - t_\beta)^{i+1} + O(t - t_\beta)^{n_\beta + 2}$$

where $n_\beta \geq 0$ is an integer such that

$$cA^{i+1} x_\beta + cA^i bu_\alpha = 0, \quad i = 0, 1, \ldots, n_\beta - 1,$$
$$cA^{n_\beta + 1} x_\beta + cA^{n_\beta} bu_\alpha \neq 0.$$

From the Cayley-Hamilton Theorem, for any integer $k \geq n$, A^k can be expressed as a linear combination of $I, A, \ldots, A^{n-1}$. Hence, $n_\beta \in \mathcal{N}$. (Otherwise, $cA^{i+1} x_\beta + cA^i bu_\alpha = 0$ for all $i \in \mathcal{N}$ will lead to $cx(t) \equiv \beta$ for all $t \in [t_0 - \delta,\ t_0 + \delta]$. This is a contradiction of x_β being a traversing point.) Now, since

$$cx_\beta = \beta,$$
$$cx(t) > \beta, \quad t \in (t_\beta,\ t_\beta + \delta],$$
$$cx(t) < \beta, \quad t \in [t_\beta - \delta,\ t_\beta),$$

we have that for $t \in (t_\beta,\ t_\beta + \delta]$

$$\frac{1}{(n_\beta + 1)!} cA^{n_\beta}(Ax_\beta + bu_\alpha)(t - t_\beta)^{n_\beta + 1} + cO(t - t_\beta)^{n_\beta + 2} > 0,$$

and for $t \in [t_\beta - \delta,\ t_\beta)$

$$\frac{1}{(n_\beta + 1)!} cA^{n_\beta}(Ax_\beta + bu_\alpha)(t - t_\beta)^{n_\beta + 1} + cO(t - t_\beta)^{n_\beta + 2} < 0.$$

Considering a process $t \to t_\beta$, we see from the last two inequalities that the non-negative integer n_β must be even and

$$cA^{n_\beta + 1} x_\beta + cA^{n_\beta} bu_\alpha > 0.$$

This proves the necessity.

Sufficiency: Sufficiency can be seen by just reversing the proof of necessity. This completes the proof of the lemma.

Proof of Lemma 2.2.3. We first assume that $\tau > 0$. Without loss of generality, we show proof with respect to the set $\mathcal{L}(\epsilon, \xi_0, \beta)$ only. As for the case with respect to the set $\mathcal{L}(\epsilon, \xi_0, \alpha)$, the proof is similar.

Let the trajectory of $\xi(t)$ traverse Ω_α at $\xi(t_0 + h)$. By virtue of Lemma 2.2.4, there exists an integer $k \in \mathcal{N}$ such that

$$cA^{i+1}\xi(t_0+h) + cA^i bu_\beta = 0, \quad i = 0, 1, \ldots, k-1,$$
$$cA^{k+1}\xi(t_0+h) + cA^k bu_\beta < 0. \tag{2.12}$$

Define
$$f(\delta) = \delta^{-k-1} c(e^{A\delta} - I)(A^{-1}bu_\beta + \xi(t_0+h)).$$

Taking into account the fact that
$$\lim_{\delta \to 0} f(\delta) = \frac{1}{(k+1)!}(cA^{k+1}\xi(t_0+h) + cA^k bu_\beta) < 0,$$

by defining $f(0) := \lim_{\delta \to 0} f(\delta) < 0$, there is a scalar r satisfying $0 < r < h - \tau$ such that $f(\delta) < 0$ is continuous on $\delta \in [-r, r]$. Define
$$M = \max_{\delta \in [-r,r]} \frac{\|ce^{A(h+\delta)}\|}{|f(\delta)|} > 0.$$

In view of the Assumptions, the trajectory of $\eta(t)$ starting from $\xi(t_0) + \Delta \in \mathcal{L}(\epsilon, \xi_0, \beta)$ will intersect Ω_α at some time $t_0 + h + \delta$, and it holds that $h + \delta > \tau$. Thus, we have

$$\xi(t_0+h) = e^{Ah}\xi(t_0) + (e^{Ah} - e^{A(h-\tau)})A^{-1}bu_\alpha + (e^{A(h-\tau)} - I)A^{-1}bu_\beta,$$
$$c\xi(t_0+h) = \alpha,$$
$$\eta(t_0+h+\delta) = e^{A(h+\delta)}(\xi(t_0) + \Delta) + (e^{A(h+\delta)} - e^{A(h+\delta-\tau)})A^{-1}bu_\alpha$$
$$+ (e^{A(h+\delta-\tau)} - I)A^{-1}bu_\beta,$$
$$c\eta(t_0+h+\delta) = \alpha.$$

After some simple manipulations, we arrive at
$$0 = ce^{A(h+\delta)}\Delta + c(e^{A\delta} - I)(A^{-1}bu_\beta + \xi(t_0+h)). \tag{2.13}$$

Next we show that if $\|\Delta\|$ is small enough, then $|\delta|$ will be less than r. We prove this by contradiction. Suppose that $\|\Delta\|$ is very small and the trajectory of $\eta(t)$ will not intersect Ω_α for all $t \in [t_0 + h - r, t_0 + h + r]$. Then, for $t \in [h - r, h + r]$,

$$\eta(t_0+t) = e^{At}(\xi(t_0) + \Delta) + (e^{At} - e^{A(t-\tau)})A^{-1}bu_\alpha + (e^{A(t-\tau)} - I)A^{-1}bu_\beta,$$
$$c\eta(t_0+t) > \alpha.$$

Since
$$\eta(t_0+h+\delta) - \xi(t_0+h) \to 0 \text{ as } \delta \to 0 \text{ and } \|\Delta\| \to 0, \tag{2.14}$$
$$cA^k(A\xi(t_0+h) + bu_\beta) < 0, \tag{2.15}$$

there exist $\epsilon_1 > 0$ and r_0 satisfying $0 < r_0 \leq r$ such that for all $|\delta| \leq r_0$ and $\|\Delta\| \leq \epsilon_1$

$$cA^k(A\eta(t_0 + h + \delta) + bu_\beta) < 0.$$

Expanding $\eta(t)$ at time t_0+h and taking into account (2.12) and $c\xi(t_0+h) = \alpha$, yields

$$\eta(t_0 + h + \delta) = \eta(t_0 + h) + \sum_{i=1}^{k} \frac{\delta^i}{i!} A^{i-1}(A\eta(t_0 + h) + bu_\beta)$$

$$+ \frac{\delta^{k+1}}{(k+1)!} A^{k+1}(A\eta(t_0 + h + \theta_\delta \delta) + bu_\beta)$$

$$= \xi(t_0 + h) + \sum_{i=1}^{k} \frac{\delta^i}{i!} A^{i-1}(A\xi(t_0 + h) + bu_\beta) + \sum_{i=0}^{k} \frac{\delta^i}{i!} A^i e^{Ah} \Delta$$

$$+ \frac{\delta^{k+1}}{(k+1)!} A^{k+1}(A\eta(t_0 + h + \theta_\delta \delta) + bu_\beta),$$

giving

$$c\eta(t_0 + h + \delta) = \alpha + \sum_{i=0}^{k} \frac{\delta^i}{i!} cA^i e^{Ah} \Delta + \frac{\delta^{k+1}}{(k+1)!} cA^{k+1}(A\eta(t_0 + h + \theta_\delta \delta) + bu_\beta)$$

for some $0 < \theta_\delta < 1$. Letting $\delta = r_0 > 0$ (and thus $|\theta_\delta \delta| \leq r_0$), and letting $\|\Delta\| \to 0$, it follows that there exists a scalar ϵ_2 satisfying $0 < \epsilon_2 \leq \epsilon_1$ such that for all $\|\Delta\| \leq \epsilon_2$

$$c\eta(t_0 + h + r_0) < \alpha.$$

This contradicts $c\eta(t_0 + t) > \alpha$ for all $t \in [h - r, h + r]$. From the above, we also see that if $\|\Delta\| \leq \epsilon_2$, then $|\delta| \leq r_0 \leq r$.

Now, from (2.13), we have

$$0 = ce^{A(h+\delta)} \Delta + \delta^{k+1} f(\delta)$$

and thus

$$\delta^{k+1} = -\frac{ce^{A(h+\delta)}}{f(\delta)} \Delta, \quad \forall \|\Delta\| \leq \epsilon_2,$$

which gives

$$|\delta|^{k+1} \leq M\|\Delta\|, \quad \forall \|\Delta\| \leq \epsilon_2. \tag{2.16}$$

This proves the lemma for $\tau > 0$.

For the case $\tau = 0$, by virtue of Theorem 1.4.1 (as stated in Remark 1.4.1), the lemma can be proved similarly by using (1.14) and (1.16). This completes the proof. □

2.2.2 Existence of Limit Cycles

This subsection presents a sufficient condition for the existence of limit cycles. The result needs the following assumption.

Assumption 2.3 For system Σ_τ, if a trajectory intersects Ω_α or Ω_β, then it traverses the switching plane $\mathcal{S}_\alpha$ or $\mathcal{S}_\beta$.

It is known from Chapter 1 that there is no sliding motion after a trajectory intersects Ω_α or Ω_β. Assumption 2.3 indeed ensures the uniqueness of system solutions in the region Ω and the vector fields point in the 'right' direction on both sides of the switching planes. Let us check a simple class of systems for the validity of Assumption 2.3.

Example 2.2.1. Consider system Σ_τ with $\tau > 0$. Suppose that there exists an invertible matrix Q such that

$$Q^{-1}AQ = \begin{bmatrix} A_{11} & 0 \\ A_{21} & A_{22} \end{bmatrix}, \quad cQ = [1 \ 0 \ \ldots \ 0] \qquad (2.17)$$

where $A_{11} \in \mathbb{R}$. For this system, if

$$A_{11}\beta + cbu_\alpha > 0,$$
$$A_{11}\alpha + cbu_\beta < 0,$$

it is easy to check that the conditions in Lemma 2.2.4 hold for $n_\alpha = n_\beta = 0$, i.e.,

$$cAx_\beta + cbu_\alpha > 0, \quad \forall x_\beta \in \mathcal{S}_\beta,$$
$$cAx_\alpha + cbu_\beta > 0, \quad \forall x_\alpha \in \mathcal{S}_\alpha.$$

It is seen that this example satisfies a very strong constraint in the sense that the traversing requirement in Assumption 2.3 holds for all $x_\alpha \in \mathcal{S}_\alpha$ and $x_\beta \in \mathcal{S}_\beta$.

With the aid of the lemmas presented in the preceding subsection, we are now in a position to give the existence result.

Theorem 2.2.1. *Suppose Assumptions 2.1, 2.2 and 2.3 hold. Then system Σ_τ has a limit cycle.*

Proof. By Assumption 2.1, the trajectory of $x(t)$ starting from $x(0) \in \Omega$ will intersect, without loss of generality, Ω_β. Suppose the intersecting point is $x(t_1)$ (at time t_1). By Assumptions 2.2 and 2.3, the trajectory of $x(t)$ will traverse Ω_β and intersect Ω_α at some point $x(t_2)$ (at time $t_2 > t_1 + \tau$), and then will leave Ω_α and traverse Ω_β again at some point $x(t_3)$ (at time $t_3 > t_2 + \tau$). Define the mapping (i.e., the Poincare map) $\Phi : \Omega_\beta \to \Omega_\beta$ as

$$\Phi(x(t_1)) = x(t_3). \tag{2.18}$$

Then Φ maps Ω_β into itself. In order to adopt Lemma 2.2.1, we need to verify the following.

(i) Φ is continuous in Ω_β. Indeed, without loss of generality, consider the three points $x(t_i)$, $i = 1, 2, 3$. By virtue of the proof of Lemma 2.2.3, for any $\alpha_3 > 0$, there exists $\alpha_2 > 0$ such that any trajectory of $\eta(t)$ evolving from $\eta(2) = x(t_2) + \Delta \in \Omega_\alpha$, where $\|\Delta\| < \alpha_2$, will intersect and traverse Ω_β, and the traversing point $\eta(3)$ satisfies $\|\eta(3) - x(t_3)\| < \alpha_3$. For this $\alpha_2 > 0$, there exists $\alpha_1 > 0$ such that any trajectory of $\xi(t)$ evolving from $\xi(1) = x(t_1) + \Delta \in \Omega_\beta$, where $\|\Delta\| < \alpha_1$, will intersect and traverse Ω_α, and the traversing point $\xi(2)$ satisfies $\|\xi(2) - x(t_2)\| < \alpha_2$. This shows that if $\|\Delta\| < \alpha_1$, then

$$\|\Phi(x(t_1) + \Delta) - \Phi(x(t_1))\| < \alpha_3$$

which proves the continuity of the mapping Φ.

(ii) Ω_β is a non-empty compact convex set. The fact that Ω_β is non-empty is obvious. The compactness of Ω_β is easily seen from the definition of compact set. The convexity is verified as follows:

$$\forall \xi_1, \xi_2 \in \Omega_\beta \Rightarrow \begin{cases} c(\lambda \xi_1 + (1-\lambda)\xi_2) = \beta \Rightarrow \lambda \xi_1 + (1-\lambda)\xi_2 \in S_\beta, \\ \|P^{\frac{1}{2}}(\lambda \xi_1 + (1-\lambda)\xi_2)\| \leq \omega \Rightarrow \lambda \xi_1 + (1-\lambda)\xi_2 \in \Omega, \end{cases}$$

where $\lambda \in [0, 1]$.

Now, by Lemma 2.2.1, there exists $x_\beta^* \in \Omega_\beta$ such that $\Phi(x_\beta^*) = x_\beta^*$, which implies that there exists a solution $x^*(t)$ (with x_β^* as its state) and time $h_1^* < h_2^*$ such that

$$\Phi(x^*(h_1^*)) = x^*(h_2^*). \tag{2.19}$$

After $t = h_2^*$, the trajectory of $x^*(t)$ will be the same as from $t = h_1^*$ to $t = h_2^*$. This shows that the trajectory of $x^*(t)$ is a limit cycle with a period $h_2^* - h_1^*$. This completes the proof.

Remark 2.2.2. The result of Theorem 2.2.1 is for system Σ_τ with hysteresis. For systems with no delay and hysteresis (i.e., $\tau = 0$ and $\alpha = \beta = 0$), fast switches may occur, and there are possibly infinite (but countable) numbers of switching times during a finite time interval. (See Theorem 1 in Johansson et al. (1999).) In the case of this phenomenon, we say that the system trajectory has an accumulation point. (See the example in Remark 15 in Lootsma et al.

(1999), which is taken from Filippov (1988).) Note that under Assumption 2.3, with no accumulation points, the result in Theorem 2.2.1 still holds for system Σ_τ with $\tau = 0$, $\alpha = \beta = 0$, A Hurwitz and $G(0)u_\alpha > 0 > G(0)u_\beta$.

2.3 A Simple Existence Condition

We have studied in Section 2.2.2 the existence of limit cycles for relay feedback systems of the form Σ_τ. In practice, from the simulations, the trajectories of many such systems, whether A is Hurwitz or not, often tend to a certain oscillation with the relay switching consecutively. This section studies the existence problem through a single system trajectory, and establishes a simple condition to determine the existence of limit cycles. The limit cycles are not merely confined to be symmetric or to have two switches a period.

Let a trajectory of system Σ_τ traverse $\mathcal{S}_\alpha$ (resp. $\mathcal{S}_\beta$) at x_α (resp. x_β), and then traverse $\mathcal{S}_\beta$ (resp. $\mathcal{S}_\alpha$) at x_β (resp. x_α). Define the mapping T_α (resp. T_β) as

$$T_\alpha : x_\alpha \longrightarrow x_\beta \quad (\text{resp. } T_\beta : x_\beta \longrightarrow x_\alpha). \tag{2.20}$$

It is seen that such defined T_α and T_β are one-to-one continuous mappings. (Note that when a trajectory intersects the switching plane but does not traverse it, such an intersecting point is not a traversing point.) If a system trajectory keeps traversing an alternative switching plane after it traverses each switching plane, i.e., in the sequence $\mathcal{S}_\alpha, \mathcal{S}_\beta, \mathcal{S}_\alpha, \ldots$, in an infinite way, then the trajectory is said to cause regular relay switchings. Our analysis begins with a trajectory which meets the following assumption.

Assumption 2.4 The trajectory of $x(t)$ of system Σ_τ starting from x_0 causes regular relay switchings and the time difference between each two successive switchings is longer than τ.

Before the analysis we can arbitrarily choose a point $x_0 \in \mathbb{R}^n$, and check if the resulting trajectory meets the assumption. If not, we change the x_0 until a trajectory meeting Assumption 2.4 is found. If no such trajectory exists, theoretically, there does not exist a limit cycle which causes regular relay switchings and for which the time difference between each two successive switchings is longer than τ.

Suppose now that a trajectory which satisfies Assumption 2.4 is available. In more detail, without loss of generality, let $cx_0 > \beta$ (and thus $u = u_\beta$) at the starting point. Suppose the trajectory of $x(t)$ traverses $\mathcal{S}_\alpha$ for the first time at

$x_{1\alpha}$. After a time longer than τ, the trajectory traverses S_β for the first time at $x_{1\beta}$. After another time longer than τ, the trajectory traverses S_α for the second time at $x_{2\alpha}$, and then, after another time longer than τ, it traverses S_β for the second time at $x_{2\beta}$. The process continues forever. See Figure 2.1.

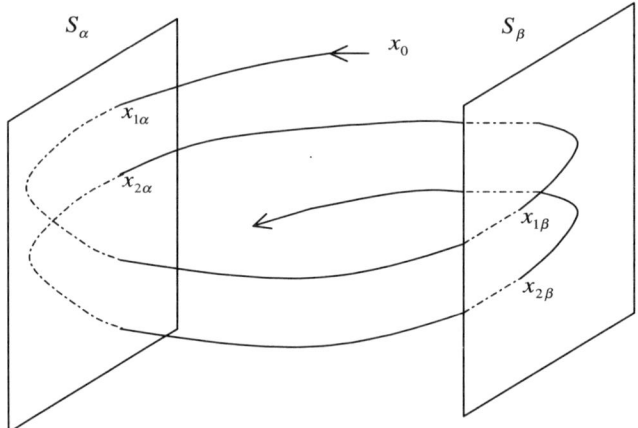

Fig. 2.1. The trajectory of $x(t)$

Let
$$\Gamma_\alpha := \{x_{1\alpha}, x_{2\alpha}, \cdots\}, \tag{2.21}$$
$$\Gamma_\beta := \{x_{1\beta}, x_{2\beta}, \cdots\}. \tag{2.22}$$

It is obvious that the two sets Γ_α and Γ_β are at most countable. It is seen that $T_\alpha(x_{1\alpha}) = x_{1\beta}$ and $T_\beta(x_{1\beta}) = x_{2\alpha}$. If $x_{2\alpha} = x_{1\alpha}$, then $x_{2\beta} = T_\alpha(x_{2\alpha}) = x_{1\beta}$. This results in $x_{i\alpha} = x_{1\alpha}$ and $x_{i\beta} = x_{1\beta}$ for all $i = 2, 3, \cdots$, and thus such a trajectory is a limit cycle with two switchings per period after the first switching instant. Similarly, if $x_{2\beta} \neq x_{1\beta}$, then the trajectory of $x(t)$ is a limit cycle with two switchings per period after the second switching instant. Now, for some natural number $N > 1$, suppose $x_{1\alpha}, x_{2\alpha}, \cdots, x_{(N-1)\alpha}$ are different and $x_{N\alpha} = x_{N_0\alpha}$ for some $1 \leq N_0 < N$. Then it is easy to verify that $x_{N_0\beta}, x_{2\beta}, \cdots, x_{(N-1)\beta}$ are different and $x_{N\beta} = x_{N_0\beta}$. This implies that the trajectory of $x(t)$ evolving from the traversing point $x_{N_0\alpha}$ is a limit cycle with $2(N - N_0)$ switchings per period.

The following is a summary of the above analysis.

Proposition 2.3.1. *Suppose Assumption 2.4 holds. If the set Γ_α is finite, i.e., there exist two natural numbers N and N_0 with $1 \leq N_0 < N$ such that*

2.3 A Simple Existence Condition

$x_{1\alpha}, x_{2\alpha}, \cdots, x_{(N-1)\alpha}$ are different and $x_{N\alpha} = x_{N_0\alpha}$, then the trajectory of $x(t)$ evolving from the traversing point $x_{N_0\alpha}$ is a limit cycle with $2(N - N_0)$ switchings per period. Moreover, all the traversing points are $x_{i\alpha}$ and $x_{i\beta}$, $i = N_0, \ldots, (N-1)$.

In practice, if the given system has at most a finite number of limit cycles, the chance of the trajectory starting from the arbitrarily chosen x_0 to be a limit cycle is nearly zero. This is to say that the sets Γ_α and Γ_β are not finite in general. Our main concern in the following will focus on the case that Γ_α and Γ_β are infinite (indeed countable).

Suppose that Γ_α is infinite. Then Γ_β is also infinite. If Γ_α has limit points, obviously, these limit points belong to $\mathcal{S}_\alpha$. It will be clear later whether or not they are elements of Γ_α. Let one such limit point be $\xi_{1\alpha} \in \mathcal{S}_\alpha$. Then there exists a subset

$$\Gamma_{k\alpha} := \{x_{k_1\alpha}, x_{k_2\alpha}, \cdots, x_{k_i\alpha}, \cdots\} \subseteq \Gamma_\alpha \tag{2.23}$$

such that

$$\lim_{i \to \infty} x_{k_i\alpha} = \xi_{1\alpha}. \tag{2.24}$$

Lemma 2.3.1. $\lim_{i \to \infty} x_{k_i\beta}$ exists, and if denoted by $\xi_{1\beta}$, then $T_\alpha(\xi_{1\alpha}) = \xi_{1\beta}$.

Proof. Since T_α is continuous, then

$$\lim_{i \to \infty} x_{k_i\beta} = \lim_{i \to \infty} T_\alpha(x_{k_i\alpha}) = T_\alpha(\lim_{i \to \infty} x_{k_i\alpha}) = T_\alpha(\xi_{1\alpha}) = \xi_{1\beta}.$$

This proves the lemma.

Lemma 2.3.1 says that if a system trajectory traverses $\mathcal{S}_\alpha$ at $\xi_{1\alpha}$, then this trajectory must intersect and traverse $\mathcal{S}_\beta$ at $\xi_{1\beta}$. Moreover, the time taken by this trajectory in going from $\xi_{1\alpha}$ to $\xi_{1\beta}$ is longer than τ.

Remark 2.3.1. The boundedness of $x(t)$ guarantees the existence of limit points of Γ_α, as a bounded infinite set has a convergent sequence.

We now give the result for the case that Γ_α has a unique limit point.

Theorem 2.3.1. *Suppose that Assumption 2.4 holds and Γ_α is infinite. If Γ_α has a unique limit point, say $\xi_{1\alpha}$, or, if Γ_β has a unique limit point, say $\xi_{1\beta}$, then there is a limit cycle with two switchings per period, which traverses $\mathcal{S}_\alpha$ (resp. $\mathcal{S}_\beta$) at $\xi_{1\alpha}$ (resp. $\xi_{1\beta}$).*

Proof. We first assume that Γ_α has a unique limit point $\xi_{1\alpha}$. By Lemma 2.3.1, $T_\alpha(\xi_{1\alpha})$ (set to be $\xi_{1\beta}$) is a limit point of Γ_β. Following the proof of Lemma 2.3.1, $T_\beta(\xi_{1\beta})$ is defined, which is a limit point of Γ_α. So, we have $T_\beta(\xi_{1\beta}) = \xi_{1\alpha}$ due to the uniqueness of limit points of Γ_α. This shows that the trajectory which traverses $\mathcal{S}_\alpha$ (resp. $\mathcal{S}_\beta$) at $\xi_{1\alpha}$ (resp. $\xi_{1\beta}$) is a limit cycle with two switchings per period.

Now, suppose $\xi_{2\beta}$ is another limit point of Γ_β. Then, we get $T_\beta(\xi_{2\beta}) = \xi_{1\alpha}$ since $\xi_{1\alpha}$ is the unique limit point of Γ_α. Thus, $\xi_{2\beta}$ is the traversing point of the limit cycle, implying that $\xi_{2\beta} = \xi_{1\beta}$.

Similarly, if we assume that Γ_β has a unique limit point, the rest of the proof follows a similar line. This completes the proof.

In Theorem 2.3.1, it is seen that $\xi_{1\alpha} \notin \Gamma_\alpha$ and $\xi_{1\beta} \notin \Gamma_\beta$. Otherwise, the set Γ_α would be finite which is a contradiction. The result in Theorem 2.3.1 can be extended to the case that Γ_α has multiple limit points.

Theorem 2.3.2. *Suppose that Assumption 2.4 holds and Γ_α is infinite. Let f be a natural number. If Γ_α has f limit points, say $\xi_{1\alpha}, \cdots, \xi_{f\alpha}$, or if Γ_β has f limit points, say $\xi_{1\beta}, \cdots, \xi_{f\beta}$, then there is at least one limit cycle with $2f_1$ switchings per period where f_1 is a natural number satisfying $1 \leq f_1 \leq f$.*

Proof. We first assume that Γ_α has f limit points, say $\xi_{1\alpha}, \cdots, \xi_{f\alpha}$. From Lemma 2.3.1, for some $\xi_{m_1\alpha} \in \{\xi_{1\alpha}, \cdots, \xi_{f\alpha}\}$, we have $T_\alpha(\xi_{m_1\alpha})$ is a limit point of Γ_β. Then there is a trajectory, say x_1^*, which traverses $\mathcal{S}_\alpha$ at $\xi_{m_1\alpha}$ and spends a time longer than τ to reach the traversing point $T_\alpha(\xi_{m_1\alpha}) \in \mathcal{S}_\beta$. After that, each time x_1^* traverses $\mathcal{S}_\alpha$ or $\mathcal{S}_\beta$, the traversing point is a limit point of Γ_α or Γ_β. Since Γ_α has a finite number of limit points, Proposition 2.3.1 concludes that x_1^* is a limit cycle with $2f_1$ switchings per period where f_1 is some natural number satisfying $1 \leq f_1 \leq f$. Moreover, letting $\xi_{m_i\alpha} \in \{\xi_{1\alpha}, \cdots, \xi_{f\alpha}\}$, $i = 1, 2, \cdots, f_1$, be the different f_1 traversing points of x_1^* on $\mathcal{S}_\alpha$, we see that $T_\alpha(\xi_{m_i\alpha})$, $i = 1, 2, \cdots, f_1$, are the different f_1 traversing points of x_1^* on $\mathcal{S}_\beta$.

Let $\xi_{p_1\alpha} \in \{\xi_{1\alpha}, \cdots, \xi_{f\alpha}\}$ and $\xi_{p_1\alpha} \neq \xi_{m_i\alpha}$, $i = 1, 2, \cdots, f_1$. A similar deduction yields that, with $\xi_{p_1\alpha}$ as a traversing point, another limit cycle, say x_2^*, exists with $2f_2$ switchings per period where f_2 is some natural number satisfying $1 \leq f_2 \leq f - f_1$. It is obvious that the $2f_2$ traversing points of x_2^* are different from those of x_1^*.

Repeating the above deduction, we see that there are a finite number of limit cycles which contain the $2f$ limit points

$$\xi_{i\alpha} \in \mathcal{S}_\alpha, \quad \text{and} \quad T_\alpha(\xi_{i\alpha}) \in \mathcal{S}_\beta, \quad i = 1, 2, \cdots, f$$

as the traversing points.

Now suppose that Γ_β has another limit point, say $\xi_{(f+1)\beta}$. Then, it is easy to show that $\xi_{(f+1)\beta}$ belongs to one of the above limit cycles, and hence $\xi_{(f+1)\beta} = T_\alpha(\xi_{i\alpha})$ for some $i = 1, 2, \cdots, f$. This proves that Γ_β has only f limit points.

Assume next that Γ_β has f limit points. A similar proof shows that Γ_α also has f limit points, and there exists at least one limit cycle. This completes the proof.

It is seen from the proof of Theorem 2.3.2 that all the $2f$ limit points are the traversing points of certain limit cycles. So, we have $\xi_{i\alpha} \notin \Gamma_\alpha$ and $\xi_{i\beta} \notin \Gamma_\beta$ for all $i = 1, 2, \cdots, f$.

So far, we have given existence results about limit cycles for system Σ_τ with $\tau > 0$. Systems of the form Σ_τ with no delays (i.e., $\tau = 0$) can be analyzed similarly. However, we need a slight modification. This is because, unlike the case $\tau > 0$, the trajectories of system Σ_τ with $\tau = 0$ may not be smooth anymore at the intersecting points when they intersect $\mathcal{S}_\alpha$ (resp. $\mathcal{S}_\beta$). In the following, we will use Σ_0 to denote the system Σ_τ with $\tau = 0$. Arbitrarily choose a point $x_0 \in R^n$. Assumption 2.4 is reduced as below.

Assumption 2.5 The trajectory of $x(t)$ of system Σ_0 starting from x_0 causes regular relay switchings.

Similar to (2.21) and (2.22), let Γ_α and Γ_β be the two sets of traversing points on $\mathcal{S}_\alpha$ and $\mathcal{S}_\beta$, respectively, which are at most countable. For Γ_α and Γ_β finite, a simple result analogous to Proposition 2.3.1 is as follows.

Proposition 2.3.2. *Consider system Σ_0 and assume Assumption 2.5 holds. If the set Γ_α is finite, i.e., there exist two natural numbers N and N_0 with $1 \leq N_0 < N$ such that $x_{1\alpha}, x_{2\alpha}, \cdots, x_{(N-1)\alpha}$ are different and $x_{N\alpha} = x_{N_0\alpha}$, then the trajectory evolving from the traversing point $x_{N_0\alpha}$ is a limit cycle with $2(N - N_0)$ switchings per period. Moreover, all the traversing points are $x_{i\alpha}$ and $x_{i\beta}$, $i = N_0, \ldots, (N-1)$.*

For Γ_α and Γ_β infinite, we need the following assumption. Denote $\dot{y}(t_-)$ the left derivative of $y(t)$ at time t, i.e., $\dot{y}(t_-) = \lim_{\epsilon > 0, \epsilon \to 0} \dot{y}(t - \epsilon)$.

Assumption 2.6 There is a small scalar $\epsilon_0 > 0$ such that the trajectory of $x(t)$ starting from x_0 satisfies $|\dot{y}(t_-)| \geq \epsilon_0$ at all traversing points.

We have the following result which is analogous to Theorem 2.3.2.

Proposition 2.3.3. *Consider Σ_0 and assume Assumptions 2.5 and 2.6 hold. Let Γ_α be infinite (so Γ_β is infinite). If Γ_α has a finite number of limit points, or if Γ_β has a finite number of limit points (say f), then there is at least one limit cycle with $2f_1$ switchings per period where f_1 is a natural number satisfying $1 \leq f_1 \leq f$. Moreover, the f_1 traversing points of the limit cycle on S_α belong to the set of limit points of Γ_α.*

Proof. The proof is similar to that of Theorem 2.3.2, and thus is omitted.

2.4 Limit Cycle Location

If there is a limit cycle for system Σ_τ, say x^*, then its trajectory is governed by a periodic solution to the system. This section gives the formulas for specifying the locations of the limit cycle in a closed form. Assume that the trajectory of the limit cycle x^* makes the relay switch two times per period. (For more than two switches per period, more formulas are needed.) Let the traversing points be $x_\alpha^* \in S_\alpha$ and $x_\beta^* \in S_\beta$, respectively, and the period be $(\tau + h_\alpha) + (\tau + h_\beta)$ with $h_\alpha > 0$ and $h_\beta > 0$, where $\tau + h_\alpha$ (resp. $\tau + h_\beta$) is the time taken for the trajectory of x^* to move from x_α^* to x_β^* (resp. from x_β^* to x_α^*). Then, the parameters should satisfy the following conditions:

$$x_\beta^* = e^{A(\tau+h_\alpha)} x_\alpha^* + \int_0^\tau e^{A(\tau+h_\alpha-s)} bu_\beta ds + \int_0^{h_\alpha} e^{A(h_\alpha-s)} bu_\alpha ds,$$

$$x_\alpha^* = e^{A(\tau+h_\beta)} x_\beta^* + \int_0^\tau e^{A(\tau+h_\beta-s)} bu_\alpha ds + \int_0^{h_\beta} e^{A(h_\beta-s)} bu_\beta ds,$$

$$cx_\beta^* = \beta,$$

$$cx_\alpha^* = \alpha. \qquad (2.25)$$

Obviously, the following should also be satisfied:

$$c(e^{At} x_\alpha^* + \int_0^t e^{A(t-s)} bu_\beta ds) \leq \beta, \quad \forall t \in [0, \tau],$$

$$c(e^{At} x_\alpha^* + \int_0^\tau e^{A(t-s)} bu_\beta ds + \int_0^{t-\tau} e^{A(t-s)} bu_\alpha ds) \leq \beta, \quad \forall t \in [\tau, \tau + h_\alpha],$$

$$c(e^{At} x_\beta^* + \int_0^t e^{A(t-s)} bu_\alpha ds) \geq \alpha, \quad \forall t \in [0, \tau],$$

$$c(e^{At} x_\beta^* + \int_0^\tau e^{A(t-s)} bu_\alpha ds + \int_0^{t-\tau} e^{A(t-s)} bu_\beta ds) \geq \alpha, \quad \forall t \in [\tau, \tau + h_\beta].$$

Noting the fact that $I - e^{At}$ is invertible (for $t \neq 0$) if and only if $\lambda(A) \neq j2k\pi t^{-1}$ for any integer k, we see that if A has no roots in the imaginary axis, (2.25) gives

$$\alpha = c(I - e^{A(\tau+h_\alpha+\tau+h_\beta)})^{-1} \left(\int_{h_\beta}^{h_\alpha+\tau+h_\beta} e^{As} bu_\alpha ds \right.$$

$$\left. + \int_{h_\alpha+\tau+h_\beta}^{\tau+h_\alpha+\tau+h_\beta} e^{As} bu_\beta ds + \int_0^{h_\beta} e^{As} bu_\beta ds \right),$$

$$\beta = c(I - e^{A(\tau+h_\alpha+\tau+h_\beta)})^{-1} \left(\int_{h_\alpha}^{h_\alpha+\tau+h_\beta} e^{As} bu_\beta ds \right.$$

$$\left. + \int_{h_\alpha+\tau+h_\beta}^{\tau+h_\alpha+\tau+h_\beta} e^{As} bu_\alpha ds + \int_0^{h_\alpha} e^{As} bu_\alpha ds \right), \quad (2.26)$$

where we use the fact that $\int_{\underline{t}}^{\overline{t}} f(t-s)ds = \int_{t-\overline{t}}^{t-\underline{t}} f(s)ds$. Once h_α and h_β have been computed from (2.26) using a numerical method, x_α^* and x_β^* are obtained as

$$x_\alpha^* = (I - e^{A(\tau+h_\alpha+\tau+h_\beta)})^{-1} \left(\int_{h_\beta}^{h_\alpha+\tau+h_\beta} e^{As} bu_\alpha ds \right.$$

$$\left. + \int_{h_\alpha+\tau+h_\beta}^{\tau+h_\alpha+\tau+h_\beta} e^{As} bu_\beta ds + \int_0^{h_\beta} e^{As} bu_\beta ds \right),$$

$$x_\beta^* = (I - e^{A(\tau+h_\alpha+\tau+h_\beta)})^{-1} \left(\int_{h_\alpha}^{h_\alpha+\tau+h_\beta} e^{As} bu_\beta ds \right.$$

$$\left. + \int_{h_\alpha+\tau+h_\beta}^{\tau+h_\alpha+\tau+h_\beta} e^{As} bu_\alpha ds + \int_0^{h_\alpha} e^{As} bu_\alpha ds \right). \quad (2.27)$$

If $\alpha + \beta = 0$, $u_\alpha + u_\beta = 0$ and the limit cycle x^* is symmetric, the expression in (2.26) is simpler as only one equation with one variable (i.e., h_α or h_β where $h_\alpha = h_\beta$) is left. As a result, the traversing point $x_\alpha^* = -x_\beta^*$ can be computed accordingly through (2.27). This is also discussed in Åström (1995) and Varigonda and Georgious (2001).

3. Local Stability of Limit Cycles

This chapter is concerned with the local stability of limit cycles for linear systems with relay feedback, for the cases where the linear system includes a time delay in its dynamics and the relay can possess asymmetric hysteresis. The limit cycle considered can be asymmetric, have more than two switchings per period, and zero output derivatives at the switching instants. It will be shown that if a certain constructed matrix is Schur stable, then the local stability of the considered limit cycle is guaranteed.

3.1 Introduction

We have studied in the preceding chapters problems concerning the existence of solutions and the existence of limit cycles for relay feedback systems. Another important analysis topic is the stability of limit cycles of relay feedback systems. This includes *local* stability and *global* stability of limit cycles. Local stability ensures that all nearby trajectories converge to the limit cycle as time tends to plus infinity, while global stability means that all trajectories converge to the limit cycle as time tends to plus infinity. The phase-plane approach is a classical technique employed for the stability analysis of limit cycles (Guckenheimer and Holmes, 1983; Tsypkin, 1984). Exact methods have also been reported in the literature (Åström, 1995; Goncalves *et al.*, 2001; Johansson *et al.*, 1997; Johansson *et al.*, 1999; Lin *et al.*, 2000). Åström (1995) gives elegant criteria for the local stability of symmetric and asymmetric limit cycles by considering the linear approximation of the Poincare map. Johansson *et al.* (1997) and Johansson *et al.* (1999) focus on the fast switches of relay feedback systems and present local stability results for limit cycles with sliding motion. A method to compute a local stability bound is presented in Goncalves *et al.* (1998). It gives sufficient conditions for the locally stable region in terms of the existence of solutions to linear matrix inequalities. Another discussion of local stability is given in Lin *et al.* (2000). For the global stability of limit cycles, a recent paper (Goncalves *et al.*, 2001) obtains sufficient conditions in terms of a set of linear matrix inequalities by finding the so-called surface Lyapunov function of Poincare maps. As seen in the stability analysis, a limit cycle is always assumed to exist *a priori*.

In this chapter we consider the local stability of limit cycles for a time-delay relay feedback system with the relay containing asymmetric hysteresis. The relay is not required merely to switch twice per period and the assumed limit cycle is not confined to be symmetric. Additionally, it is not required that the trajectory of the limit cycle be non-tangent with the switching planes at the switching instants. From an engineering point of view, time-delay systems are of considerable interest (most industrial processes have time delay). Theoretically, a non-zero time delay ensures that a system trajectory evolves uniquely at the intersecting points. Also, non-zero time delay makes it possible to relax the non-tangent condition of the trajectory of the limit cycle at the traversing points. This relies on continuity at the intersecting points, and intuitively it is the 'overshoot' effect. Otherwise, if the considered system is delay-free, the non-tangent condition of the trajectory of the limit cycle at the traversing points has to be assumed, like the case considered in Åström (1995), Johansson *et al.* (1997) and Lin *et al.* (2000). Such a condition avoids the occurrence of multiple trajectories after a trajectory intersects the switching planes, and makes the local stability analysis simpler. We will includes this case for delay-free systems in Remark 3.3.3.

This chapter is organized as follows. In Section 3.2, the considered system and problem are formulated. Section 3.3 presents sufficient conditions for the local stability of a limit cycle with two switches per period. Section 3.4 extends the result to the local stability of limit cycles with more than two switches per period.

3.2 Problem Formulation and Preliminaries

Consider a single-input single-output plant described by

$$\dot{x}(t) = Ax(t) + bu(t - \tau)$$
$$y(t) = cx(t) \tag{3.1}$$

where $x(t) \in \mathbb{R}^n$, $y(t) \in \mathbb{R}$ and $u(t-\tau) \in \mathbb{R}$ are the state, output and control input, respectively; A, b, c are constant real matrices or vectors with appropriate dimensions; $\tau > 0$ denotes the time delay. The plant has relay feedback:

$$u(t) = \begin{cases} u_\beta, & \text{if } y(t) > \beta, \text{ or } y(t) \geq \alpha \text{ and } u(t_-) = u_\beta, \\ u_\alpha, & \text{if } y(t) < \alpha, \text{ or } y(t) \leq \beta \text{ and } u(t_-) = u_\alpha, \end{cases} \quad (3.2)$$

where $\alpha, \beta \in \mathbb{R}$ with $\alpha \leq \beta$ denotes the hysteresis; $u_\alpha, u_\beta \in \mathbb{R}$ and $u_\alpha \neq u_\beta$. Due to time delay $\tau > 0$, we have to specify the initial function $u(\tilde{t})$ for $\tilde{t} \in [-\tau, 0]$. For the convenience of our analysis, in this chapter we let

$$u(\tilde{t}) \equiv \begin{cases} u_\beta, & \text{if } y(0) > \alpha \\ u_\alpha, & \text{if } y(0) \leq \alpha. \end{cases} \quad (3.3)$$

We still call (3.1)–(3.3) a relay feedback system and denote it by Σ.

As in the preceding chapters, define the switching planes:

$$S_\alpha := \{\xi \in \mathbb{R}^n : c\xi = \alpha\}, \quad (3.4)$$
$$S_\beta := \{\xi \in \mathbb{R}^n : c\xi = \beta\}. \quad (3.5)$$

Let

$$S_{+\alpha} := \{\xi \in \mathbb{R}^n : c\xi > \alpha\}, \quad (3.6)$$
$$S_{-\alpha} := \{\xi \in \mathbb{R}^n : c\xi < \alpha\}, \quad (3.7)$$

and let $S_{+\beta}$ and $S_{-\beta}$ be defined similarly. Let $\xi_\alpha \in S_\alpha$, $\xi_\beta \in S_\beta$, and let

$$\begin{aligned} S_{(\epsilon,\xi_\alpha)} &:= \{\xi \in S_\alpha : \|\xi - \xi_\alpha\| \leq \epsilon\} \\ &= \{\xi \in S_\alpha : \xi = \xi_\alpha + \Delta, \|\Delta\| \leq \epsilon\}, \end{aligned} \quad (3.8)$$

$$\begin{aligned} S_{(\epsilon,\xi_\beta)} &=: \{\xi \in S_\beta : \|\xi - \xi_\beta\| \leq \epsilon\} \\ &= \{\xi \in S_\beta : \xi = \xi_\beta + \Delta, \|\Delta\| \leq \epsilon\}. \end{aligned} \quad (3.9)$$

If $\xi \in S_{(\epsilon,\xi_\alpha)}$ (or $\xi \in S_{(\epsilon,\xi_\beta)}$), we say that ξ is within ϵ distance of ξ_α (or ξ_β).

3.3 Local Stability of Limit Cycles

In the local stability analysis for limit cycles of system Σ, we assume that there exists a limit cycle x^* of the following form.

Form 3.1. The limit cycle x^*, which traverses S_α and S_β at the traversing points $x_\alpha^* \in S_\alpha$ and $x_\beta^* \in S_\beta$, respectively, makes the relay switch twice per period. The limit cycle period is $(\tau + h_\alpha) + (\tau + h_\beta)$ with $h_\alpha > 0$ and $h_\beta > 0$, where $\tau + h_\alpha$ (resp. $\tau + h_\beta$) is the time taken for the trajectory of x^* to move from x_α^* to x_β^* (resp. from x_β^* to x_α^*).

For illustration, see Figures 3.1 and 3.2 where $x^*(t)$ denotes the system solution corresponding to the limit cycle x^*. For determining the locations of the limit cycle in Form 3.1, computation formulas have been given in Section 2.4 of Chapter 2. Numerical methods have to be exploited to specify the parameters of traversing points and periods of the limit cycle.

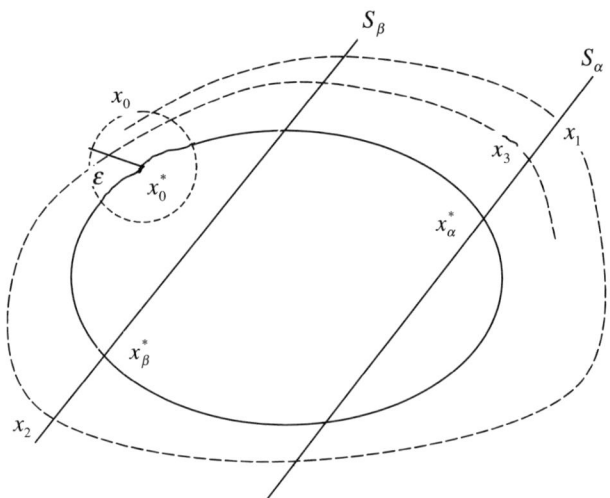

Fig. 3.1. The trajectories of $x^*(t)$ and $x(t)$ starting from $x_0 \in \mathcal{R}_\epsilon$

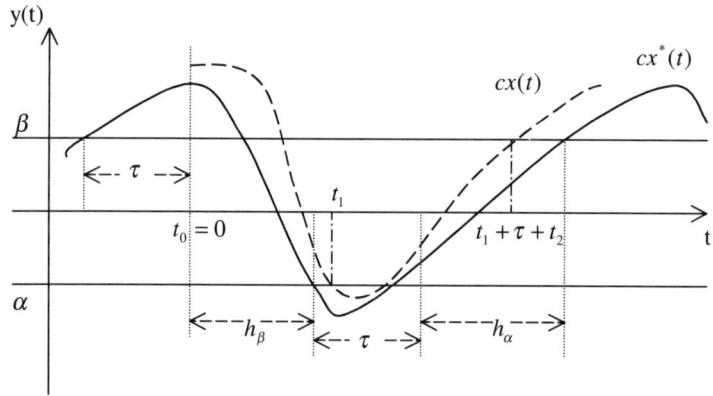

Fig. 3.2. The trajectories of $cx^*(t)$ (solid) and $cx(t)$ (dashed) corresponding to $x^*(t)$ and $x(t)$, respectively

Without loss of generality, we set $t_0 = 0$ corresponding to the time instant when the trajectory of x^* makes the relay switch from u_α to u_β; see Figures 3.1 and 3.2. Define

$$\mathcal{R}_\epsilon =: \{\xi \in R^n : \|\xi - x_0^*\| \leq \epsilon\}$$
$$= \{\xi \in R^n : \xi = x_0^* + \Delta,\ \Delta \in R^n,\ \|\Delta\| \leq \epsilon\}. \tag{3.10}$$

Since $cx_0^* > \alpha$, let a scalar ϵ_1 satisfy

$$0 < \epsilon_1 < \frac{cx_0^* - \alpha}{\|c\|}. \tag{3.11}$$

Then, any trajectory of system Σ starting from $\mathcal{R}_{\epsilon_1}$ will satisfy $cx_0 = c(x_0^* + \Delta) > \alpha$, implying (from (3.3)) that the relay will be $u(-\tau) = u_\beta$ at the starting moment.

To achieve our stability result, we need to establish some lemmas first. As in Chapter 1, let

$$\mathcal{N} = \{0, 1, \ldots, n - 1\}. \tag{3.12}$$

The first lemma specifies two positive integers $n_\alpha, n_\beta \in \mathcal{N}$, which will be used in the development.

Lemma 3.3.1. *For the limit cycle in Form 3.1, there exist two even integers $n_\alpha, n_\beta \in \mathcal{N}$ such that the following hold*

$$cA^{i+1}x_\alpha^* + cA^i bu_\beta = 0,\quad i = 0, 1, \ldots, n_\alpha - 1,$$
$$cA^{n_\alpha+1}x_\alpha^* + cA^{n_\alpha} bu_\beta < 0, \tag{3.13}$$
$$cA^{j+1}x_\beta^* + cA^j bu_\alpha = 0,\quad j = 0, 1, \ldots, n_\beta - 1,$$
$$cA^{n_\beta+1}x_\beta^* + cA^{n_\beta} bu_\alpha > 0. \tag{3.14}$$

Proof. This is obvious from Lemma 2.2.4.

It is seen that if $n_\alpha = n_\beta = 0$, then the trajectory of the limit cycle is non-tangent with the switching planes, $\mathcal{S}_\alpha$ and $\mathcal{S}_\beta$, at the traversing points. The conditions in Lemma 3.1 also ensure that the vector fields point in the 'right' direction on both side of the switching planes, e.g., the nearby trajectories traverse the switching planes. Now, we analyze the trajectory of $x(t)$ starting from a point near to x_0^*. The following lemma is with respect to the first switching instant.

Lemma 3.3.2. *For any given $\delta > 0$, there exists $\epsilon_\delta > 0$ such that the trajectory of $x(t)$ starting from $\mathcal{R}_{\epsilon_\delta}$ will intersect $\mathcal{S}_\alpha$ at instant $t_{\delta 1}$ and will leave $\mathcal{S}_\alpha \cup \mathcal{S}_{+\alpha}$ for $\mathcal{S}_{-\alpha}$ after instant t_δ, where $t_{\delta 1}$ and t_δ satisfy $h_\beta - \delta < t_{\delta 1} \leq t_\delta < h_\beta + \delta$.*

Proof. Intuitively, this is true by continuity. The detailed proof is given below.

For $t \in [0, \tau + h_\beta]$, the trajectory of x^* is governed by

$$x^*(t) = e^{At}x_0^* + \int_0^t e^{A(t-s)}bu_\beta ds. \tag{3.15}$$

For some δ_1 satisfying $0 < \delta_1 < \min\{h_\beta, \tau, \delta\}$, it holds that

$cx^*(t) > \alpha,\quad \text{for } t \in [0, h_\beta),$

$cx^*(t) < \alpha,\quad \text{for } t \in (h_\beta, h_\beta + \delta_1],$

$cx^*(h_\beta) = \alpha.$

The trajectory of $x(t)$ starting from $x_0 = x_0^* + \Delta \in \mathcal{R}_{\epsilon_1}$ is governed by

$$x(t) = e^{At}(x_0^* + \Delta) + \int_0^t e^{A(t-s)}bu_\beta ds \tag{3.16}$$

for small $t > 0$. By continuity, there is a scalar ϵ_{δ_1} satisfying $0 < \epsilon_{\delta_1} \leq \epsilon_1$ such that (3.16) holds for all $\|\Delta\| \leq \epsilon_{\delta_1}$ and meanwhile

$cx(t) > \alpha,\quad \forall t \in [0, h_\beta - \delta_1].$

This implies that for all $\|\Delta\| \leq \epsilon_{\delta_1}$, the trajectory of $x(t)$ will not intersect $\mathcal{S}_\alpha$ for $t \in [0, h_\beta - \delta_1]$. Since $cx^*(h_\beta + \delta_1) < \alpha$, there exists ϵ_δ satisfying $0 < \epsilon_\delta \leq \epsilon_{\delta_1}$ such that

$$cx^*(h_\beta + \delta_1) + ce^{A(h_\beta + \delta_1)}\Delta < \alpha,\quad \forall \|\Delta\| \leq \epsilon_\delta. \tag{3.17}$$

Next we show that for all $\|\Delta\| \leq \epsilon_\delta$, the trajectory of $x(t)$ must intersect $\mathcal{S}_\alpha$ at some instant t_{δ_1} and will leave $\mathcal{S}_\alpha \cup \mathcal{S}_{+\alpha}$ for $\mathcal{S}_{-\alpha}$ after instant t_δ, where t_{δ_1} and t_δ satisfy $h_\beta - \delta < t_{\delta_1} \leq t_\delta < h_\beta + \delta$. We prove this by contradiction. If this is not the case, then for all $t \in [h_\beta - \delta_1, h_\beta + \delta_1]$, the relay remains u_β and the trajectory of $x(t)$ will be governed by (3.16) satisfying

$$cx(t) \geq \alpha. \tag{3.18}$$

However, taking into account (3.17), we have from (3.15) and (3.16) that

$$cx(h_\beta + \delta_1) = cx^*(h_\beta + \delta_1) + ce^{A(h_\beta + \delta_1)}\Delta < \alpha, \tag{3.19}$$

which contradicts (3.18). This completes the proof.

Remark 3.3.1. In Lemma 3.3.2, we see that t_{δ_1} is an intersecting instant. It is easy to show that there is no sliding motion in the time interval $[t_{\delta_1}, t_\delta]$, and thus t_δ is a traversing instant. However, the type of motion characteristic between the intersecting instant and the traversing instant does not affect our investigation on the stability of the limit cycle. Our concern in the development is the traversing instant which plays an important role in the local stability analysis.

For $t \in \mathbb{R}$, define

$$F_\alpha(t) := (e^{At} - I)x_\alpha^* + \int_0^t e^{As} bu_\beta ds,$$
$$f_\alpha(t) := cF_\alpha(t),$$
$$F_\beta(t) := (e^{At} - I)x_\beta^* + \int_0^t e^{As} bu_\alpha ds,$$
$$f_\beta(t) := cF_\beta(t). \tag{3.20}$$

Using (3.13) and (3.14), it is easy to verify that

$$\lim_{t \to 0} \frac{f_\alpha(t)}{t^{n_\alpha+1}} = \frac{1}{(n_\alpha+1)!} \left(cA^{n_\alpha+1} x_\alpha^* + cA^{n_\alpha} bu_\beta \right) < 0,$$
$$\lim_{t \to 0} \frac{f_\beta(t)}{t^{n_\beta+1}} = \frac{1}{(n_\beta+1)!} \left(cA^{n_\beta+1} x_\beta^* + cA^{n_\beta} bu_\alpha \right) > 0. \tag{3.21}$$

By defining

$$\left. \frac{f_\alpha(t)}{t^{n_\alpha+1}} \right|_{t=0} := \lim_{t \to 0} \frac{f_\alpha(t)}{t^{n_\alpha+1}},$$
$$\left. \frac{f_\beta(t)}{t^{n_\beta+1}} \right|_{t=0} := \lim_{t \to 0} \frac{f_\beta(t)}{t^{n_\alpha+1}},$$

there exist two scalars $r_\alpha^* > 0$ and $r_\beta^* > 0$ such that $\frac{f_\alpha(t)}{t^{n_\alpha+1}} < 0$ and $\frac{f_\beta(t)}{t^{n_\alpha+1}} > 0$ are continuous on $t \in [-r_\alpha^*, r_\alpha^*]$ and $t \in [-r_\beta^*, r_\beta^*]$, respectively. Let

$$r_{min} = \min\{h_\alpha, h_\beta, r_\alpha^*, r_\beta^*\}. \tag{3.22}$$

Lemma 3.3.2 characterizes the trajectory of $x(t)$ at the first traversing instant. To study the local stability of x^*, we need to verify whether or not successive switchings can occur. The next lemma is useful, and characterizes a fixed scalar $\epsilon_{\delta_0} > 0$ such that any trajectory evolving from the traversing points within ϵ_{δ_0} distance of x_α^* (or x_β^*) will intersect and traverse $\mathcal{S}_\beta$ (or $\mathcal{S}_\alpha$).

Lemma 3.3.3. *For any given $\delta_0 \in (0, r_{min}]$, there exists a fixed scalar $\epsilon_{\delta_0} > 0$ such that the trajectory of $x(t)$ evolving from any traversing point in $\mathcal{S}_{(\epsilon_{\delta_0}, x_\alpha^*)}$ (or $\mathcal{S}_{(\epsilon_{\delta_0}, x_\beta^*)}$) (here, set the traversing instant to be zero) will traverse $\mathcal{S}_\beta$ (or $\mathcal{S}_\alpha$), and the traversing instant $\tau + t_{trav}$ satisfies $h_\alpha - \delta_0 < t_{trav} < h_\alpha + \delta_0$ (or $h_\beta - \delta_0 < t_{trav} < h_\beta + \delta_0$).*

Proof. Firstly, consider the trajectory of $x(t)$ evolving from traversing points in $\mathcal{S}_\alpha$. The trajectory of $x^*(t)$ is governed by

$$x^*(t) = e^{At} x_\alpha^* + \int_0^t e^{A(t-s)} bu_\beta ds, \quad \forall t \in [0, \tau],$$
$$x^*(t) = e^{At} x_\alpha^* + \int_0^\tau e^{A(t-s)} bu_\beta ds + \int_0^{t-\tau} e^{A(t-\tau-s)} bu_\alpha ds, \quad \forall t \in [\tau, \tau + h_\alpha].$$

The trajectory of $x(t)$ evolving from $x_\alpha = x_\alpha^* + \Delta \in S_\alpha$ with small $\|\Delta\|$ is governed by

$$x(t) = e^{At}(x_\alpha^* + \Delta) + \int_0^t e^{A(t-s)} b u_\beta ds, \quad \forall t \in [0, \tau],$$

$$x(t) = e^{At}(x_\alpha^* + \Delta) + \int_0^\tau e^{A(t-s)} b u_\beta ds + \int_0^{t-\tau} e^{A(t-\tau-s)} b u_\alpha ds,$$
$$\forall t \in [\tau, \tau + \delta_2],$$

$$cx(t) < \beta, \quad \forall t \in [0, \tau + \delta_2],$$

where $\delta_2 > 0$ is a sufficiently small scalar. Similar to the proof of Lemma 3.3.2, by comparing the two trajectories of $x(t)$ and $x^*(t)$, it can be shown that there exists $\epsilon_{\delta_0 1}$ such that the trajectory of $x(t)$ evolving from $x_\alpha = x_\alpha^* + \Delta \in S_{(\epsilon_{\delta_0 1}, x_\alpha^*)}$ will traverse S_β. Moreover, the traversing instant $\tau + t_{trav}$ satisfies $h_\alpha - \delta_1 < t_{trav} < h_\alpha + \delta_1$, and thus $h_\alpha - \delta_0 < t_{trav} < h_\alpha + \delta_0$.

Next, consider the trajectory of $x(t)$ evolving from traversing points in S_β. Similarly, for the given $\delta_0 > 0$, there exists $\epsilon_{\delta_0 2}$ such that the trajectory of $x(t)$ evolving from $S_{(\epsilon_{\delta_0 2}, x_\beta^*)}$ will traverse S_α, and the traversing instant $\tau + t_{trav}$ satisfies $h_\beta - \delta_0 < t_{trav} < h_\beta + \delta_0$. Now, let $\epsilon_{\delta_0} = \min\{\epsilon_{\delta_0 1}, \epsilon_{\delta_0 2}\}$. Then, the result follows immediately.

The following lemma characterizes the first traversing point of $x(t)$ starting from a region close to x_0^*.

Lemma 3.3.4. *There exists ϵ_2 satisfying $0 < \epsilon_2 \leq \epsilon_1$ such that any trajectory of system Σ starting from $x_0 = x_0^* + \Delta \in \mathcal{R}_{\epsilon_2}$ will intersect and traverse S_α, and the traversing instant t_1 and the traversing point $x(t_1)$ satisfy*

$$x(t_1) - x_\alpha^* = \left(I - \frac{F_\alpha(t_1 - h_\beta)c}{f_\alpha(t_1 - h_\beta)}\right) e^{At_1} \Delta. \quad (3.23)$$

Proof. From Lemma 3.3.2, there exists ϵ_2 satisfying $0 < \epsilon_2 \leq \epsilon_1$ such that the trajectory of $x(t)$ starting from $\mathcal{R}_{\epsilon_2}$ will traverse S_α at some instant t_1, where t_1 satisfies $|t_1 - h_\beta| < r_{min}$. Since

$$x_\alpha^* = e^{Ah_\beta} x_0^* + \int_0^{h_\beta} e^{A(h_\beta - s)} b u_\beta ds,$$

$$cx_\alpha^* = \alpha,$$

$$x(t_1) = e^{At_1}(x_0^* + \Delta) + \int_0^{t_1} e^{A(t_1 - s)} b u_\beta ds,$$

$$cx(t_1) = \alpha, \quad (3.24)$$

after some simple manipulations, we have

$$ce^{At_1}\Delta + c(e^{A(t_1-h_\beta)} - I)x_\alpha^* + c\int_0^{t_1-h_\beta} e^{As}bu_\beta ds = 0,$$

or,

$$ce^{At_1}\Delta + f_\alpha(t_1 - h_\beta) = 0.$$

Noting that $(t_1 - h_\beta)^{n_\alpha+1} f_\alpha^{-1}(t_1 - h_\beta)$ is well-defined for $|t_1 - h_\beta| \leq r_{min}$, the above equation yields that

$$(t_1 - h_\beta)^{n_\alpha+1} = -\frac{(t_1 - h_\beta)^{n_\alpha+1} ce^{At_1}}{f_\alpha(t_1 - h_\beta)} \Delta. \tag{3.25}$$

Using (3.24), we get

$$\begin{aligned} x(t_1) - x_\alpha^* &= e^{At_1}\Delta + (e^{A(t_1-h_\beta)} - I)x_\alpha^* + \int_0^{t_1-h_\beta} e^{As}bu_\beta ds \\ &= e^{At_1}\Delta + \frac{F_\alpha(t_1 - h_\beta)}{(t_1 - h_\beta)^{n_\alpha+1}}(t_1 - h_\beta)^{n_\alpha+1} \\ &= \left(I - \frac{F_\alpha(t_1 - h_\beta)c}{f_\alpha(t_1 - h_\beta)}\right) e^{At_1}\Delta. \end{aligned}$$

This completes the proof.

Let $\Delta_1 = x(t_1) - x_\alpha^*$. By virtue of Lemma 3.3.2, $|t_1 - h_\beta|$ can be made arbitrarily small by choosing $\|\Delta\|$ sufficiently small, and thus $\|\Delta_1\|$ can be arbitrarily small. (See Remark 3.3.2 later for an alternative depiction.) The next lemma concerns the second traversing point of $x(t)$ close to x_β^*.

Lemma 3.3.5. *There exists ϵ_3 satisfying $0 < \epsilon_3 \leq \epsilon_2$ such that any trajectory of system Σ starting from $x_0 = x_0^* + \Delta \in \mathcal{R}_{\epsilon_3}$ will intersect and traverse S_β after the first traversing instant t_1, and the second traversing point $x(t_1 + \tau + t_2)$ satisfies*

$$x(t_1 + \tau + t_2) - x_\beta^* = \left(I - \frac{F_\beta(t_2 - h_\alpha)c}{f_\beta(t_2 - h_\alpha)}\right) e^{A(\tau+t_2)}\Delta_1, \tag{3.26}$$

where $\tau + t_2$ with $t_2 > 0$ is the time taken for the trajectory to move from $x(t_1)$ to $x(t_1 + \tau + t_2)$.

Proof. Similar to the proof of Lemma 3.3.2, it is easy to show that if Δ_1 is small, then the trajectory evolving from $x(t_1) = \Delta_1 + x_\alpha^* \in S_\alpha$ will traverse S_β, and the time $\tau + t_2$ can be made approaching $\tau + h_\alpha$. Thus, there exists ϵ_3 satisfying $0 < \epsilon_3 \leq \epsilon_2$ such that any trajectory of system Σ starting from $\mathcal{R}_{\epsilon_3}$ will make the time duration $\tau + t_2$ satisfy $|t_2 - h_\alpha| < r_{min}$. Since

$$x_\beta^* = e^{A(\tau+h_\alpha)}x_\alpha^* + \int_0^\tau e^{A(\tau+h_\alpha-s)}bu_\beta ds + \int_0^{h_\alpha} e^{A(h_\alpha-s)}bu_\alpha ds,$$

$$cx_\beta^* = \beta,$$

$$x(t_1+\tau+t_2) = e^{A(\tau+t_2)}x(t_1) + \int_0^\tau e^{A(\tau+t_2-s)}bu_\beta ds + \int_0^{t_2} e^{A(t_2-s)}bu_\alpha ds,$$

$$cx(t_1+\tau+t_2) = \beta, \qquad (3.27)$$

after some manipulations, we have

$$ce^{A(\tau+t_2)}\Delta_1 + c(e^{A(t_2-h_\alpha)} - I)x_\beta^* + c\int_0^{t_2-h_\alpha} e^{As}bu_\alpha ds = 0,$$

or

$$ce^{A(\tau+t_2)}\Delta_1 + f_\beta(t_2-h_\alpha) = 0.$$

Noting that $(t_2-h_\alpha)^{n_\beta+1}f_\beta^{-1}(t_2-h_\alpha)$ is well-defined for $|t_2-h_\alpha| \leq r_{min}$, we arrive at

$$(t_2-h_\alpha)^{n_\beta+1} = -\frac{(t_2-h_\alpha)^{n_\beta+1}ce^{A(\tau+t_2)}}{f_\beta(t_2-h_\alpha)}\Delta_1. \qquad (3.28)$$

Using (3.27), we obtain

$$x(t_1+\tau+t_2) - x_\beta^* = e^{A(\tau+t_2)}\Delta_1 + (e^{A(t_2-h_\alpha)} - I)x_\beta^* + \int_0^{t_2-h_\alpha} e^{As}bu_\alpha ds$$

$$= e^{A(\tau+t_2)}\Delta_1 + \frac{F_\beta(t_2-h_\alpha)}{(t_2-h_\alpha)^{n_\beta+1}}(t_2-h_\alpha)^{n_\beta+1}$$

$$= \left(I - \frac{F_\beta(t_2-h_\alpha)c}{f_\beta(t_2-h_\alpha)}\right)e^{A(\tau+t_2)}\Delta_1.$$

This completes the proof.

Remark 3.3.2. Let

$$M_1 = \max_{t \in [-r_{min}, r_{min}]} \frac{t^{n_\alpha+1}\|ce^{Ah_\beta}e^{At}\|}{-f_\alpha(t)} > 0,$$

$$M_2 = \max_{t \in [-r_{min}, r_{min}]} \frac{t^{n_\beta+1}\|ce^{A(\tau+h_\alpha)}e^{At}\|}{f_\beta(t)} > 0. \qquad (3.29)$$

From (3.25) and (3.28), we see that $|t_1-h_\beta|^{n_\alpha+1} \leq M_1\|\Delta\|$ and $|t_2-h_\alpha|^{n_\beta+1} \leq M_2\|\Delta_1\|$ hold under $\|\Delta\| \leq \epsilon_3$. Hence, $|t_1-h_\beta| \to 0$ as $\|\Delta\| \to 0$, and this leads to $\|\Delta_1\| \to 0$ and $|t_2-h_\alpha| \to 0$.

Similarly, by choosing x_0 close enough to x_0^*, the trajectory starting from x_0 can make the relay switch for the third time, fourth time, and so on. However, we could not specify a local region $\mathcal{R}_\epsilon$ in this way such that any trajectory starting from it will make the relay switch consecutively forever. To achieve this, we need the following lemma which is related to a Schur stability problem for discrete-time systems.

Lemma 3.3.6. *Given a positive integer p, suppose that $A_i, \Theta_{ij} \in R^{n \times n}$ ($i = 1, 2, \ldots, p$; $j = 1, 2, \ldots$) and $\rho(A_1 A_2 \cdots A_p) < 1$. Then, there exists $\theta_0 > 0$ such that for all Θ_{ij} satisfying $\|\Theta_{ij} - A_i\| \leq \theta_0$, it holds $\|\prod_{j=1}^{k}(\Theta_{1j}\Theta_{2j}\cdots\Theta_{pj})\| \to 0$ as $k \to \infty$.*

Proof. Due to $\rho(A_1 A_2 \cdots A_p) < 1$, it is easy to see that there is a scalar $\theta > 0$ such that for all $\Theta_i \in R^{n \times n}$ satisfying $\|\Theta_i\| \leq \theta$, it holds that $\|\prod_{j=1}^{k}(A_1 A_2 \cdots A_p + \Theta_j)\| \to 0$ as $k \to \infty$. For this $\theta > 0$, there exists $\theta_0 > 0$ such that if $\|\Theta_{ij} - A_i\| \leq \theta_0$, then the matrix $\Theta_{1j}\Theta_{2j}\cdots\Theta_{pj}$ can be expressed as

$$\Theta_{1j}\Theta_{2j}\cdots\Theta_{pj} = A_1 A_2 \cdots A_p + \Omega_j$$

where Ω_j satisfies $\|\Omega_j\| \leq \theta$. This proves the lemma.

With the above lemmas proven, we are now in a position to present the main result for the stability of the limit cycle $x^*(t)$.

Theorem 3.3.1. *The limit cycle $x^*(t)$ in Form 3.1 is locally stable if*

$$\rho(W_1 W_2) < 1, \tag{3.30}$$

where

$$W_1 = \left(I - \frac{A^{n_\alpha}(Ax_\alpha^* + bu_\beta)c}{cA^{n_\alpha}(Ax_\alpha^* + bu_\beta)}\right) e^{A(\tau + h_\beta)},$$

$$W_2 = \left(I - \frac{A^{n_\beta}(Ax_\beta^* + bu_\alpha)c}{cA^{n_\beta}(Ax_\beta^* + bu_\alpha)}\right) e^{A(\tau + h_\alpha)}. \tag{3.31}$$

Here, n_α and n_β are even integers as given in Lemma 3.3.1.

Proof. Suppose $\rho(W_1 W_2) < 1$. By virtue of Lemma 3.3.6, there exists a scalar $\theta_0 > 0$ such that for all $\Theta_{ij} \in R^{n \times n}$ ($i = 1, 2$; $j = 1, 2, \ldots$) satisfying $\|\Theta_{ij} - W_i\| \leq \theta_0$, it holds that $\|\prod_{j=1}^{k}(\Theta_{1j}\Theta_{2j})\| \to 0$ as $k \to \infty$. In other words, there is a positive integer N_0 such that for all $\Theta_{ij} \in R^{n \times n}$ satisfying $\|\Theta_{ij} - W_i\| \leq \theta_0$,

$$\|\prod_{j=1}^{N_0+k}(\Theta_{1j}\Theta_{2j})\| < 1, \quad \forall k = 0, 1, 2, \ldots \tag{3.32}$$

For $j = 1, 2, \ldots$, let

$$W(\delta_{1j}, x_\alpha^*) = \left(I - \frac{F_\alpha(\delta_{1j})c}{f_\alpha(\delta_{1j})}\right) e^{A(\tau + h_\beta + \delta_{1j})},$$

$$W(\delta_{2j}, x_\beta^*) = \left(I - \frac{F_\beta(\delta_{2j})c}{f_\beta(\delta_{2j})}\right) e^{A(\tau + h_\alpha + \delta_{2j})}, \tag{3.33}$$

where $\delta_{1j}, \delta_{2j} \in \mathbb{R}$. It is seen that $W(\delta_{1j}, x_\alpha^*) \to W_1$ and $W(\delta_{2j}, x_\beta^*) \to W_2$ as $\delta_{ij} \to 0$ for $i = 1, 2$ and $j = 1, 2, \ldots$. Thus, for the above $\theta_0 > 0$, there exists $\delta_0 > 0$ such that the following hold.

$$\|W(\delta_{1j}, x_\alpha^*) - W_1\| \leq \theta_0, \quad \forall |\delta_{1j}| \leq \delta_0, \quad j = 1, 2, \ldots,$$
$$\|W(\delta_{2j}, x_\beta^*) - W_2\| \leq \theta_0, \quad \forall |\delta_{2j}| \leq \delta_0, \quad j = 1, 2, \ldots. \tag{3.34}$$

Let $\delta_{min} = \min\{\delta_0, r_{min}\}$ where r_{min} is as in 3.22. For this $\delta_{min} > 0$, by Lemma 3.3.3, there exists a fixed scalar $\epsilon_{min} > 0$ such that the trajectory of $x(t)$ evolving from any traversing point in $\mathcal{S}_{(\epsilon_{min}, x_\alpha^*)}$ (or $\mathcal{S}_{(\epsilon_{min}, x_\beta^*)}$) will traverse $\mathcal{S}_\beta$ (or $\mathcal{S}_\alpha$) by spending time $\tau + t_{trav}$, where t_{trav} satisfies $h_\alpha - \delta_{min} < t_{trav} < h_\alpha + \delta_{min}$ (or $h_\beta - \delta_{min} < t_{trav} < h_\beta + \delta_{min}$).

Now, let

$$w = \max\{\|W_1\|, \|W_2\|\},$$
$$\bar{\epsilon}_{min} = \min\{\epsilon_{min}, \frac{\epsilon_{min}}{(w+\theta_0)^{2N_0}}\}. \tag{3.35}$$

By virtue of Lemmas 3.3.2 and 3.3.4, there exists a scalar ϵ satisfying $0 < \epsilon \leq \epsilon_1$, such that any trajectory starting from $x_0 = x_0^* + \Delta \in \mathcal{R}_\epsilon$ will traverse $\mathcal{S}_\alpha$ with the traversing instant t_1 and the traversing point x_1 satisfying (3.23). Moreover, it holds that $\|x_1 - x_\alpha^*\| \leq \bar{\epsilon}_{min} \leq \epsilon_{min}$. We show next that for this $\epsilon > 0$, $\mathcal{R}_\epsilon$ is a local stable region. This is two-folded, i.e., any trajectory starting from $\mathcal{R}_\epsilon$ will make the relay switch consecutively and converge to the limit cycle x^* as t tends to plus infinity. Figures 3.1 and 3.2 illustrate this.

Since $\|x_1 - x_\alpha^*\| \leq \bar{\epsilon}_{min} \leq \epsilon_{min}$, the second traversing will occur at $\mathcal{S}_\beta$. By virtue of Lemma 3.3.5 and the above analysis, the second traversing point x_2 and the time $\tau + t_2$ for the trajectory of $x(t)$ to move from x_1 to x_2 satisfy (see (3.26))

$$|t_2 - h_\alpha| \leq \delta_{min}, \tag{3.36}$$

$$x_2 - x_\beta^* = \left(I - \frac{F_\beta(t_2 - h_\alpha)c}{f_\beta(t_2 - h_\alpha)}\right) e^{A(\tau + t_2)}(x_1 - x_\alpha^*)$$
$$= W(t_2 - h_\alpha, x_\beta^*)(x_1 - x_\alpha^*). \tag{3.37}$$

From (3.36), we see that (3.34) holds, yielding $\|W(t_2 - h_\alpha, x_\beta^*)\| \leq w + \theta_0$. Thus, (3.37) gives

$$\|x_2 - x_\beta^*\| \leq (w + \theta_0)\bar{\epsilon}_{min} \leq \epsilon_{min}, \tag{3.38}$$

which implies that the third traversing will occur at $\mathcal{S}_\alpha$. Let the third traversing point be x_3 and the time for $x(t)$ to move from x_2 to x_3 be $\tau + t_3$. Then, by virtue of Lemma 3.3.3 and from a similar deduction to that of Lemma 3.3.5, we have

$$|t_3 - h_\beta| \leq \delta_{min},$$
$$x_3 - x_\alpha^* = \left(I - \frac{F_\alpha(t_3 - h_\beta)c}{f_\alpha(t_3 - h_\beta)}\right) e^{A(\tau+t_3)}(x_2 - x_\beta^*)$$
$$= W(t_3 - h_\beta, x_\alpha^*)(x_2 - x_\beta^*).$$

Also, we have
$$\|x_3 - x_\alpha^*\| \leq (w + \theta_0)^2 \bar{\epsilon}_{min} \leq \epsilon_{min},$$

which implies that the fourth traversing will occur at $\mathcal{S}_\beta$. The process continues. Let the $(2N_0 + 1)$th traversing point at $\mathcal{S}_\alpha$ be x_{2N_0+1} and the time for the trajectory of $x(t)$ to move from x_{2N_0} to x_{2N_0+1} be $\tau + t_{2N_0+1}$. Then,

$$|t_{2N_0+1} - h_\beta| \leq \delta_{min},$$
$$x_{2N_0+1} - x_\alpha^* = \left(I - \frac{F_\alpha(t_{2N_0+1} - h_\beta)c}{f_\alpha(t_{2N_0+1} - h_\beta)}\right) e^{A(\tau+t_{2N_0+1})}(x_{2N_0} - x_\beta^*)$$
$$= W(t_{2N_0+1} - h_\beta, x_\alpha^*)(x_{2N_0} - x_\beta^*).$$

Additionally, we have
$$\|x_{2N_0+1} - x_\alpha^*\| \leq (w + \theta_0)^{2N_0} \bar{\epsilon}_{min} \leq \epsilon_{min},$$

which implies that the $(2N_0 + 2)$th traversing will occur at $\mathcal{S}_\beta$. Let the $(2N_0 + 2)$th traversing point at $\mathcal{S}_\beta$ be x_{2N_0+2} and the time for the trajectory of $x(t)$ to move from x_{2N_0+1} to x_{2N_0+2} be $\tau + t_{2N_0+2}$. Then,

$$|t_{2N_0+2} - h_\alpha| \leq \delta_{min},$$
$$x_{2N_0+2} - x_\beta^* = W(t_{2N_0+2} - h_\alpha, x_\beta^*)(x_{2N_0+1} - x_\alpha^*)$$
$$= W(t_{2N_0+2} - h_\alpha, x_\beta^*)$$
$$\times \left(\prod_{j=1}^{N_0} W(t_{2j+1} - h_\beta, x_\alpha^*) W(t_{2j} - h_\alpha, x_\beta^*)\right)(x_1 - x_\alpha^*).$$

Taking into account (3.32), it is easy to see that
$$\left\|\prod_{j=1}^{N_0} W(t_{2j+1} - h_\beta, x_\alpha^*) W(t_{2j} - h_\alpha, x_\beta^*)\right\| < 1,$$

which leads to
$$\|x_{2N_0+2} - x_\beta^*\| \leq \|W(t_{2N_0+2} - h_\alpha, x_\beta^*)\| \|x_1 - x_\alpha^*\| \leq (w + \theta_0)\bar{\epsilon}_{min} \leq \epsilon_{min}.$$

This implies that the $(2N_0 + 3)$th traversing will occur at $\mathcal{S}_\alpha$. Continuing the process and noting (3.32) we conclude that for any $k \geq 1$, the $(2N_0 + 2k)$th and the $(2N_0 + 2k + 1)$th traversing will occur. Let the $(2N_0 + 2k)$th and the

$(2N_0 + 2k + 1)$th traversing points be $x_{2N_0+2k} \in \mathcal{S}_\beta$ and $x_{2N_0+2k+1} \in \mathcal{S}_\alpha$, respectively, and the time for the trajectory of $x(t)$ to move from x_{2N_0+2k-1} to x_{2N_0+2k} and from x_{2N_0+2k} to x_{2N_0+2k+1} be $\tau + t_{2N_0+2k}$ and $\tau + t_{2N_0+2k+1}$, respectively. Then, we have

$$|t_{2N_0+2k} - h_\alpha| \leq \delta_{min},$$
$$x_{2N_0+2k} - x_\beta^* = W(t_{2N_0+2k} - h_\alpha, x_\beta^*)(x_{2N_0+2k-1} - x_\alpha^*),$$
$$\|x_{2N_0+2k} - x_\beta^*\| \leq \epsilon_{min},$$
$$|t_{2N_0+2k+1} - h_\beta| \leq \delta_{min},$$
$$x_{2N_0+2k+1} - x_\alpha^* = W(t_{2N_0+2k+1} - h_\beta, x_\alpha^*)(x_{2N_0+2k} - x_\beta^*),$$
$$\|x_{2N_0+2k+1} - x_\alpha^*\| \leq \epsilon_{min}.$$

This shows that the relay will switch consecutively.

To end the proof, it is sufficient to show that

$$\|x_{2k+1} - x_\alpha^*\| \to 0 \quad \text{as } k \to 0. \tag{3.39}$$

To see this, note that

$$x_{2k+1} - x_\alpha^* = \left(\prod_{j=1}^{k} W(t_{2j+1} - h_\beta, x_\alpha^*)W(t_{2j} - h_\alpha, x_\beta^*)\right)(x_1 - x_\alpha^*),$$
$$\|W(t_{2j} - h_\alpha, x_\beta^*) - W_2\| \leq \theta_0,$$
$$\|W(t_{2j+1} - h_\beta, x_\alpha^*) - W_1\| \leq \theta_0.$$

Again, using Lemma 3.3.6, from the statement at the very beginning of the proof, (3.39) is fulfilled. This completes the proof of the theorem.

Theorem 3.3.1 presents a criterion to check the local stability of the limit cycle x^*. The idea is, under the given conditions, to find a scalar $\epsilon > 0$ such that any trajectory of system Σ starting from the region $\mathcal{R}_\epsilon$ of the form (3.10) will tend to the limit cycle x^* and make the relay switch consecutively. The scalar ϵ is a local stability bound around the initial condition x_0^*. Since $\rho(W_2 W_1) = \rho(W_1 W_2)$, we see that $\rho(W_2 W_1) < 1$ is also a sufficient condition for the local stability of the limit cycle in Form 3.1. This can also be verified by letting $t_0 = 0$ correspond to a time instant when the relay switches from u_β to u_α.

Remark 3.3.3. Note that the results in this chapter are for relay feedback systems with time delay $\tau > 0$. If $\tau = 0$, the method used in this chapter is not applicable due to possible occurrence of multiple trajectories at the switching instants. We should make it clear that for the case $\tau = 0$, if both n_α and n_β are computed to be zero, then the result in Theorem 3.3.1 still holds true. Indeed,

$n_\alpha = n_\beta = 0$ corresponds to the case that the trajectory of the limit cycle is non-tangent with the switching planes at the switching instants, like the case considered in Åström (1995), Johansson et al. (1997) and Lin et al. (2000).

3.4 Extension

In this section, we consider the local stability of a limit cycle with $2q$ ($q \geq 1$) switches per period. The limit cycle considered is as follows.

Form 3.2. The limit cycle x^* of system Σ, which traverses $\mathcal{S}_\alpha$ and $\mathcal{S}_\beta$ at the traversing points $x^*_{\alpha i} \in \mathcal{S}_\alpha$ and $x^*_{\beta i} \in \mathcal{S}_\beta$ ($i = 1, 2, \ldots, q$), respectively, makes the relay switch $2q$ times per period. The period is $\sum_{i=1}^{q}(\tau + h_{\alpha i} + \tau + h_{\beta i})$ with $h_{\alpha i} > 0$ and $h_{\beta i} > 0$ ($i = 1, 2, \ldots, q$), where $\tau + h_{\alpha i}$ (resp. $\tau + h_{\beta i}$) is the time taken for the trajectory of x^* to move from $x^*_{\alpha i}$ to $x^*_{\beta i}$ (resp. from $x^*_{\beta i}$ to $x^*_{\alpha(i+1)}$).

Note that $x^*_{\alpha(q+1)} = x^*_{\alpha 1}$ in Form 3.2. In the following, we give an extension result to Theorem 3.3.1. The techniques used remain the same, except that more analysis steps and more complicated formulas are involved.

Similarly to Lemma 3.3.1, there exist $2q$ even integers $n_{\alpha l}, n_{\beta l} \in \mathcal{N}$, $l = 1, 2, \ldots, q$, such that

$$cA^{i+1}x^*_{\alpha l} + cA^i bu_\beta = 0, \quad i = 0, 1, \ldots, n_{\alpha l} - 1,$$
$$cA^{n_{\alpha l}+1}x^*_{\alpha l} + cA^{n_{\alpha l}} bu_\beta < 0,$$
$$cA^{i+1}x^*_{\beta l} + cA^i bu_\alpha = 0, \quad i = 0, 1, \ldots, n_{\beta l} - 1,$$
$$cA^{n_{\beta l}+1}x^*_{\beta l} + cA^{n_{\beta l}} bu_\alpha > 0, \quad (3.40)$$

hold for all $l = 1, 2, \ldots, q$. The extended local stability result in this section is as follows.

Theorem 3.4.1. *The limit cycle x^* in Form 3.2 is locally stable if for some $k \in \{1, 2, \ldots, 2q\}$, it holds that*

$$\rho(W_k W_{k-1} \cdots W_1 W_{2q} W_{2q-1} \cdots W_{k+1}) < 1, \quad (3.41)$$

where, for $l = 1, 2, \ldots, q$,

$$W_{2l-1} = \left(I - \frac{A^{n_{\alpha l}}(Ax^*_{\alpha l} + bu_\beta)c}{cA^{n_{\alpha l}}(Ax^*_{\alpha l} + bu_\beta)}\right) e^{A(\tau + h_{\beta(l-1)})},$$

$$W_{2l} = \left(I - \frac{A^{n_{\beta l}}(Ax^*_{\beta l} + bu_\alpha)c}{cA^{n_{\beta l}}(Ax^*_{\beta l} + bu_\alpha)}\right) e^{A(\tau + h_{\alpha l})}. \quad (3.42)$$

Here, $h_{\beta 0} = h_{\beta q}$.

Proof. By virtue of Lemmas 3.3.2–3.3.5, based on Lemma 3.3.6, the proof follows a similar line to that of Theorem 3.3.1, and thus is omitted here.

Theorem 3.4.1 presents q conditions for the local stability of the limit cycle in Form 3.2. If any one of the q conditions is true, a stability bound $\bar{\epsilon} > 0$ can be found such that any trajectory of system Σ starting from the region $\mathcal{R}_{\bar{\epsilon}}$ of the form (3.10) will tend to the limit cycle and make the relay switch consecutively.

Finally, we give a numerical example to show the effectiveness and the use of the sufficient conditions presented.

Example 3.4.1. Consider system Σ with

$$A = \begin{bmatrix} 1 & 0 & 0 \\ -1 & -2 & 1 \\ 1 & 0 & -1 \end{bmatrix}, \quad b = \begin{bmatrix} 1 \\ 1 \\ 1 \end{bmatrix}, \quad c = [1 \ 0 \ 0],$$

$\tau = 0.1$, $\alpha = -0.1$, $\beta = 0.2$, $u_\alpha = 2$, $u_\beta = -1$.

We check that A is not Hurwitz with $\lambda(A) = \{1, -2, -1\}$, but the system has a limit cycle with four switches per period which meets Form 3.2. The limit cycle is shown in Figure 3.3. The period and the four traversing points are computed to be

$$h_{\alpha 1} = 0.25, \quad h_{\alpha 2} = 0.25, \quad h_{\beta 1} = 0.65, \quad h_{\beta 2} = 1.05,$$

$$x_{\alpha 1} = \begin{bmatrix} -0.1 \\ -0.5 \\ -0.43 \end{bmatrix}, \quad x_{\alpha 2} = \begin{bmatrix} -0.1 \\ -0.39 \\ -0.3 \end{bmatrix}, \quad x_{\beta 1} = \begin{bmatrix} 0.2 \\ 0.23 \\ 0.3 \end{bmatrix}, \quad x_{\beta 2} = \begin{bmatrix} 0.2 \\ 0.3 \\ 0.44 \end{bmatrix}.$$

Now, we use Theorem 3.4.1 to check whether or not this limit cycle is locally stable.

It is easy to obtain from (3.40) that

$$n_{\alpha 1} = n_{\alpha 2} = n_{\beta 1} = n_{\beta 2} = 0.$$

We further compute from (3.42) that

$$W_1 = \begin{bmatrix} 0 & 0 & 0 \\ -1.5653 & 0.1003 & 0.2164 \\ -0.5028 & 0 & 0.3166 \end{bmatrix}, \quad W_2 = \begin{bmatrix} 0 & 0 & 0 \\ -1.3156 & 0.4966 & 0.2081 \\ -0.8684 & 0 & 0.7047 \end{bmatrix},$$

$$W_3 = \begin{bmatrix} 0 & 0 & 0 \\ -1.2486 & 0.2231 & 0.2492 \\ -0.7173 & 0 & 0.4724 \end{bmatrix}, \quad W_4 = \begin{bmatrix} 0 & 0 & 0 \\ -1.3156 & 0.4966 & 0.2081 \\ -0.7781 & 0 & 0.7047 \end{bmatrix}.$$

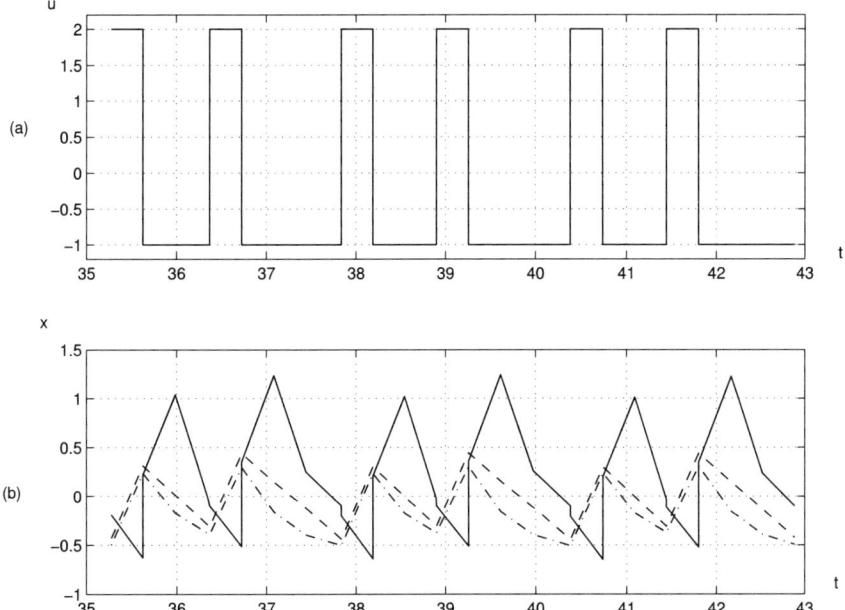

Fig. 3.3. (a) The control $u(t)$; (b) the trajectories of $x^*(t) = [x_1^*(t) \quad x_2^*(t) \quad x_3^*(t)]^T$ ($x_1^*(t)$ solid, $x_2^*(t)$ dash-dot, $x_3^*(t)$ dashed)

So,

$$\lambda(W_4 W_3 W_2 W_1) = \lambda \left(\begin{bmatrix} 0 & 0 & 0 \\ -0.1764 & 0.0055 & 0.0688 \\ -0.1180 & 0 & 0.0743 \end{bmatrix} \right) = \{0,\ 0.0055,\ 0.0743\},$$

which gives $\rho(W_4 W_3 W_2 W_1) < 1$. Hence, we conclude from Theorem 3.4.1 that the limit cycle is locally stable.

4. Global Stability of Limit Cycles

In this chapter, we study the global stability of limit cycles for relay feedback systems with the same form as in Chapters 1 and 2. We first assume there exists a certain type of limit cycles and investigate its global stability. The key idea is to reduce the global stability problem to the asymptotic stability of a discrete time system. We then, in Section 4.6, study the global convergence to a limit cycle without assuming its existence, by using the contraction mapping principle.

4.1 Introduction

In engineering practice, to verify the global stability of a limit cycle for a relay feedback system is more meaningful than the local stability. This is because global stability ensures that starting from any set point the system will be driven to converge to the periodic motions. However, theoretically, the analysis of the global stability of limit cycles is hard work. Megretski (1996) addressed the global stability of oscillations by considering systems in input–output forms, which suit processes having an impulse response sufficiently close, in a certain sense, to a second-order non-minimum phase process. The sufficient condition presented therein is based on inspection of the open-loop step response and its variance. Goncalves *et al.* (2001) investigated the global stability of symmetric limit cycles with two switchings per period for delay-free systems. The sufficient conditions are given in terms of a set of linear matrix inequalities by finding the so-called surface Lyapunov function of Poincare maps. Under some assumptions, the result serves a class of relay feedback systems well.

Experience indicates that a wide class of systems has a unique globally stable limit cycle. It is known that the uniqueness of a limit cycle is necessary for its global stability. However, how to check the uniqueness of a limit cycle is a difficulty. The available method is to compute the period of a possible limit cycle through numerical procedures, and then confirm that only one corresponds to a limit cycle (Åström, 1995; Goncalves et al., 2001). If so, the traversing points can be computed straightforwardly with the known period. With this information, we are able to check the global stability of the limit cycle (Goncalves et al., 2001; Lin et al., 2001). Thus, how to determine that a globally stable limit cycle exists without any information of a limit cycle deserves a study.

In this chapter, we concentrate on the analysis of the global stability of limit cycles for SISO RFSs. Section 4.2 gives some preliminaries. Section 4.3 establishes several useful lemmas for attaining the main results. Section 4.4 studies the global stability of a limit cycle with two switchings per period. Section 4.5 presents an extension to the results in Section 4.4 for the case of more than two switchings per period. Section 4.6 studies the problem of the global convergence to a limit cycle without the assumption that a type of limit cycle exists.

4.2 Problem Formulation

The system considered remains the same as in Chapters 1 and 2. The plant is described by

$$\dot{x}(t) = Ax(t) + bu(t - \tau)$$
$$y(t) = cx(t) \tag{4.1}$$

where $x(t) \in \mathbb{R}^n$, $y(t) \in \mathbb{R}$ and $u(t - \tau) \in \mathbb{R}$ are the state, output and control input, respectively; A, b, c are constant real matrices or vectors with appropriate dimensions; $\tau \geq 0$ denotes the time delay. The plant has relay feedback:

$$u(t) = \begin{cases} u_\beta, & \text{if } y(t) > \beta, \text{ or } y(t) \geq \alpha \text{ and } u(t_-) = u_\beta, \\ u_\alpha, & \text{if } y(t) < \alpha, \text{ or } y(t) \leq \beta \text{ and } u(t_-) = u_\alpha, \end{cases} \tag{4.2}$$

where $\alpha, \beta \in \mathbb{R}$ with $\alpha < \beta$ denotes the hysteresis; $u_\alpha, u_\beta \in \mathbb{R}$ and $u_\alpha \neq u_\beta$. The initial function $u(\tilde{t})$ for $\tilde{t} \in [-\tau, 0]$ is:

$$u(\tilde{t}) \equiv \begin{cases} u_\beta, & \text{if } y(0) \geq \beta \\ u_\alpha, & \text{if } y(0) \leq \alpha \\ u_0 \in \mathcal{U}, & \text{if } \alpha < y(0) < \beta \end{cases} \tag{4.3}$$

where

$$\mathcal{U} := \{u_\alpha, u_\beta\}. \tag{4.4}$$

As in Chapters 1 and 2, we call (4.1)–(4.3) a RFS and denote it by Σ_τ. The notation $\mathcal{S}_\alpha$ and $\mathcal{S}_\beta$ denote the switching planes as in Chapter 1, and the sets $\mathcal{S}_{+\alpha}$, $\mathcal{S}_{-\alpha}$, $\mathcal{S}_{+\beta}$ and $\mathcal{S}_{-\beta}$ are the same as defined there.

We see that in the region $\mathcal{S}_{+\beta}$ or $\mathcal{S}_{-\beta} \cap \mathcal{S}_{+\alpha}$ or $\mathcal{S}_{-\alpha}$, the system Σ_τ is a linear system with a fixed control $u \in \mathcal{U}$, and thus the existence and uniqueness of the system solution is guaranteed. Since a trajectory may intersect the switching planes, the situation may be changeable. It has been shown in Chapter 1 that for system Σ_τ with $\tau > 0$, there exists a unique solution $x(t)$ for all $t \geq 0$ for any initial condition x_0. However, this is not true for $\tau = 0$. Let's examine the following examples.

Example 4.2.1. Consider system Σ_τ with $\tau = 0$ and

$$A = \begin{bmatrix} -1 & 0 \\ 0 & -1 \end{bmatrix}, \quad b = \begin{bmatrix} 1 \\ 1 \end{bmatrix}, \quad c = [1\ 0],$$

$$u_\beta = -1, \quad u_\alpha = 1, \quad \beta = 0.5, \quad \alpha = -0.5.$$

We check that the trajectory of $x(t)$ starting from any initial condition x_0 intersects $\mathcal{S}_\beta$ or $\mathcal{S}_\alpha$ for the first time at some time t_1 with a possible switch. However, it is easy to verify that the trajectory of $x(t)$ exists only for $t \in [0, t_1]$. After time t_1, the trajectory cannot evolve any further.

Example 1.4.2 in Chapter 1 shows the occurrence of sub-solutions or sub-trajectories.

For the global convergence of system Σ_τ, the existence of solutions to system Σ_τ should be guaranteed for all $t \geq 0$ for any initial condition $x_0 \in \mathbb{R}^n$. In this chapter, we always assume this property by default.

4.3 Supporting Lemmas

From Lemma 1.3.1 in Chapter 1, we see that there is no sliding motion around the intersecting point of a trajectory after it intersects $\mathcal{S}_\beta$ (or $\mathcal{S}_\alpha$). By virtue of this fact, with our default that a solution $x(t)$ always exists with any initial condition x_0, we see that just after the intersecting instant, the trajectory of $x(t)$ has at most two sub-trajectories: one traverses $\mathcal{S}_\beta$ (or $\mathcal{S}_\alpha$) at the intersecting point; one returns to $\mathcal{S}_{-\beta}$ (or $\mathcal{S}_{+\alpha}$). Note that for $\tau = 0$, the two sub-trajectories may occur simultaneously. For $\tau > 0$, there is only one trajectory, and if no early switches occur, the solution $x(t)$ exists uniquely for all $t > 0$. This characteristic will be useful for the global convergence analysis.

60   4. Global Stability of Limit Cycles

In the analysis of the global stability of a limit cycle, we always assume the existence of a certain type of limit cycle. Here and in Section 4.4, the limit cycle x^* of system Σ_τ is assumed to have two switchings per period, whose trajectory is transversal[1] to $\mathcal{S}_\alpha$ and $\mathcal{S}_\beta$ at the traversing points $x_\alpha^* \in \mathcal{S}_\alpha$ and $x_\beta^* \in \mathcal{S}_\beta$, respectively. The period of the limit cycle is $T_\alpha + T_\beta$ where $T_\alpha > 0$ (resp. $T_\beta > 0$) is the time taken for the trajectory of x^* to move from x_α^* to x_β^* (resp. from x_β^* to x_α^*). See Figures 4.1 and 4.2 where $x^*(t)$ denotes the system solution corresponding to the limit cycle x^*.

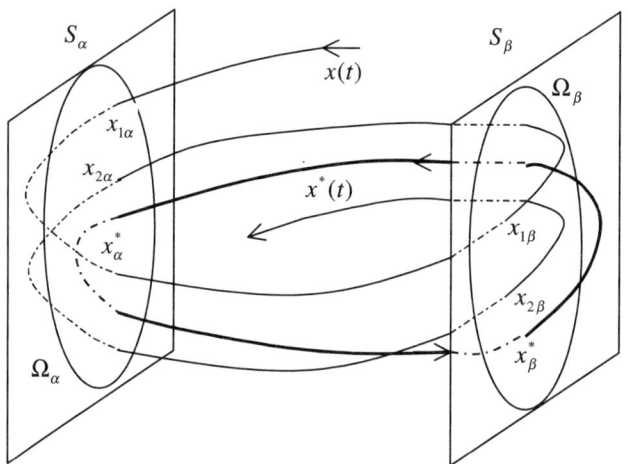

Fig. 4.1. The trajectories of $x^*(t)$ and $x(t)$

Obviously, the uniqueness of a limit cycle is a prerequisite for its global stability. However, uniqueness is not used in this chapter to deduce the sufficient conditions for the global stability of a limit cycle; on the other hand, our conditions that ensure the global stability of a limit cycle imply the uniqueness of the limit cycle.

Due to the transversal condition of the limit cycle, the following is satisfied:

$$cAx_\alpha^* + cbu_\beta < 0, \tag{4.5}$$
$$cAx_\beta^* + cbu_\alpha > 0. \tag{4.6}$$

Recall that if Assumption 2.2 is satisfied, the limit cycle period satisfies $T_\alpha > \tau$, $T_\beta > \tau$, which can be computed by (see (2.26) in Chapter 2):

---
[1] A trajectory of $x(t)$ is said to be *transversal* to $\mathcal{S}_\alpha$ and $\mathcal{S}_\beta$ at time t_T if $cx(t_{T-}) \neq 0$.

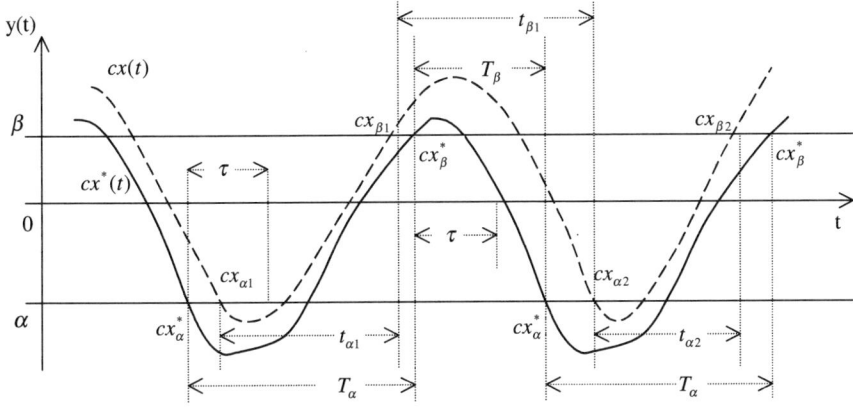

Fig. 4.2. The trajectories of $y^*(t) = cx^*(t)$ and $y(t) = cx(t)$

$$\alpha = c(I - e^{A(T_\alpha+T_\beta)})^{-1}(e^{A(T_\alpha+T_\beta-\tau)} - e^{A(T_\beta-\tau)})A^{-1}b(u_\alpha - u_\beta) - cA^{-1}bu_\beta,$$
$$\beta = c(I - e^{A(T_\alpha+T_\beta)})^{-1}(e^{A(T_\alpha+T_\beta-\tau)} - e^{A(T_\alpha-\tau)})A^{-1}b(u_\beta - u_\alpha) - cA^{-1}bu_\alpha.$$

Once T_α and T_β have been obtained, x_α^* and x_β^* are given as

$$x_\alpha^* = (I - e^{A(T_\alpha+T_\beta)})^{-1}(e^{A(T_\alpha+T_\beta-\tau)} - e^{A(T_\beta-\tau)})A^{-1}b(u_\alpha - u_\beta) - A^{-1}bu_\beta,$$
$$x_\beta^* = (I - e^{A(T_\alpha+T_\beta)})^{-1}(e^{A(T_\alpha+T_\beta-\tau)} - e^{A(T_\alpha-\tau)})A^{-1}b(u_\beta - u_\alpha) - A^{-1}bu_\alpha.$$

From Lemma 2.2.2, any trajectory of system Σ_τ with A Hurwitz will eventually enter and remain in the region Ω of the form (2.2). So, under Assumption 2.1, to study the global stability of the limit cycle x^* is equivalent to considering the system trajectories in the region Ω. Let

$$\omega_{0\alpha}^{min} = \frac{\beta - \alpha}{\|c\|(\omega_0 + \|A^{-1}bu_\alpha\|)},$$

$$\omega_{0\alpha}^{max} = \frac{G(0)u_\alpha - \beta}{\|c\|(\omega_0 + \|A^{-1}bu_\alpha\|)},$$

$$\omega_{0\beta}^{min} = \frac{\beta - \alpha}{\|c\|(\omega_0 + \|A^{-1}bu_\beta\|)},$$

$$\omega_{0\beta}^{max} = \frac{\alpha - G(0)u_\beta}{\|c\|(\omega_0 + \|A^{-1}bu_\beta\|)},$$

$$\omega_{\tau\alpha}^{max} = \frac{G(0)u_\alpha - \beta}{\|c\|(\omega_0 + 2\|A^{-1}bu_\beta\| + \|A^{-1}bu_\alpha\|)},$$

$$\omega_{\tau\beta}^{max} = \frac{\alpha - G(0)u_\beta}{\|c\|(\omega_0 + 2\|A^{-1}bu_\alpha\| + \|A^{-1}bu_\beta\|)}. \tag{4.7}$$

The following lemma gives lower and upper bounds of the time between any two consecutive switchings in the region Ω. These bounds will be used in establishing the main result for the global stability of the limit cycle x^*.

4. Global Stability of Limit Cycles

Lemma 4.3.1. *Consider system Σ_τ and assume Assumption 2.1 holds. Then, in the region Ω, the time t_α (resp. t_β) taken by a trajectory to go from the traversing point $x_\alpha \in \Omega_\alpha$ (resp. $x_\beta \in \Omega_\beta$) to the traversing point $x_\beta \in \Omega_\beta$ (resp. $x_\alpha \in \Omega_\alpha$) satisfies*

(i) *for $\tau = 0$,*

$$t_{0\alpha}^{min} \leq t_\alpha \leq t_{0\alpha}^{max}, \tag{4.8}$$
$$t_{0\beta}^{min} \leq t_\beta \leq t_{0\beta}^{max}, \tag{4.9}$$

where $t_{0\alpha}^{min}$, $t_{0\alpha}^{max}$, $t_{0\beta}^{min}$ and $t_{0\beta}^{max}$ are such that the following hold:

$$\begin{aligned}
\|e^{At} - I\| &< \omega_{0\alpha}^{min}, & \forall t \in [0, t_{0\alpha}^{min}), \\
\|e^{At}\| &< \omega_{0\alpha}^{max}, & \forall t \in (t_{0\alpha}^{max}, \infty), \\
\|e^{At} - I\| &< \omega_{0\beta}^{min}, & \forall t \in [0, t_{0\beta}^{min}), \\
\|e^{At}\| &< \omega_{0\beta}^{max}, & \forall t \in (t_{0\beta}^{max}, \infty);
\end{aligned} \tag{4.10}$$

(ii) *for $\tau > 0$, under Assumption 2.2, it holds that*

$$\tau < t_\alpha \leq t_{\tau\alpha}^{max}, \tag{4.11}$$
$$\tau < t_\beta \leq t_{\tau\beta}^{max}, \tag{4.12}$$

where $t_{\tau\alpha}^{max}$ and $t_{\tau\beta}^{max}$ are such that the following hold:

$$\begin{aligned}
\|e^{A(t-\tau)}\| &< \omega_{\tau\alpha}^{max}, & \forall t \in (t_{\tau\alpha}^{max}, \infty), \\
\|e^{A(t-\tau)}\| &< \omega_{\tau\beta}^{max}, & \forall t \in (t_{\tau\beta}^{max}, \infty).
\end{aligned} \tag{4.13}$$

Proof. (i) We first show (4.8). Since

$$x_\beta = e^{At_\alpha} x_\alpha + \int_0^{t_\alpha} e^{A(t_\alpha - s)} b u_\alpha \, ds,$$

we have

$$x_\beta + A^{-1} b u_\alpha = e^{At_\alpha} (x_\alpha + A^{-1} b u_\alpha).$$

Recall the fact that

$$c x_\alpha = \alpha, \quad c x_\beta = \beta, \quad -c A^{-1} b = G(0), \quad \|x_\alpha\| \leq \omega_0, \quad \|x_\beta\| \leq \omega_0.$$

If $t_{0\alpha}^{min} > t_\alpha$, then

$$0 < \beta - \alpha = c(e^{At_\alpha} - I)(x_\alpha + A^{-1} b u_\alpha) \leq \|c\| \omega_{0\alpha}^{min} (\omega_0 + \|A^{-1} b u_\alpha\|) < \beta - \alpha$$

which is a contradiction. If $t_\alpha > t_{0\alpha}^{max}$, then

$$0 < G(0) u_\alpha - \beta = -c e^{At_\alpha} (x_\alpha + A^{-1} b u_\alpha)$$
$$\leq \|c\| \omega_{0\alpha}^{max} (\omega_0 + \|A^{-1} b u_\alpha\|) < G(0) u_\alpha - \beta$$

which is also a contradiction. This proves (4.8). For (4.9), the proof follows a similar line.

(ii) For $\tau > 0$, under Assumption 2.2, it is obvious that $t_\alpha > \tau$ and $t_\beta > \tau$. Next we show $t_\alpha \leq t_{\tau\alpha}^{max}$ (resp. $t_\beta \leq t_{\tau\beta}^{max}$). We prove this by contradiction. Since

$$x_\beta = e^{At_\alpha}x_\alpha + (e^{At_\alpha} - e^{A(t_\alpha-\tau)})A^{-1}bu_\beta + (e^{A(t_\alpha-\tau)} - I)A^{-1}bu_\alpha,$$

(resp. $x_\alpha = e^{At_\beta}x_\beta + (e^{At_\beta} - e^{A(t_\beta-\tau)})A^{-1}bu_\alpha + (e^{A(t_\beta-\tau)} - I)A^{-1}bu_\beta,$)

and

$$cx_\alpha = \alpha, \quad cx_\beta = \beta, \quad -cA^{-1}b = G(0), \quad \|x_\alpha\| \leq \omega_0, \quad \|x_\beta\| \leq \omega_0,$$

we have

$$0 < G(0)u_\alpha - \beta = -ce^{At_\alpha}x_\alpha - c(e^{At_\alpha} - e^{A(t_\alpha-\tau)})A^{-1}bu_\beta$$
$$-ce^{A(t_\alpha-\tau)}A^{-1}bu_\alpha.$$

(resp. $0 < \alpha - G(0)u_\beta = ce^{At_\beta}x_\beta + c(e^{At_\beta} - e^{A(t_\beta-\tau)})A^{-1}bu_\alpha$
$+ce^{A(t_\beta-\tau)}A^{-1}bu_\beta.$)

If $t_\alpha > t_{\tau\alpha}^{max}$ (resp. $t_\beta > t_{\tau\beta}^{max}$), then

$$\|e^{A(t_\alpha-\tau)}\| < \omega_{\tau\alpha}^{max} \quad \text{and} \quad \|e^{At_\alpha}\| < \omega_{\tau\alpha}^{max},$$

(resp. $\|e^{A(t_\beta-\tau)}\| < \omega_{\tau\beta}^{max}$ and $\|e^{At_\beta}\| < \omega_{\tau\beta}^{max},$)

giving

$$G(0)u_\alpha - \beta < \omega_{\tau\alpha}^{max}\|c\|(\omega_0 + 2\|A^{-1}bu_\beta\| + \|A^{-1}bu_\alpha\|) \leq G(0)u_\alpha - \beta.$$

(resp. $\alpha - G(0)u_\beta < \omega_{\tau\beta}^{max}\|c\|(\omega_0 + 2\|A^{-1}bu_\alpha\| + \|A^{-1}bu_\beta\|) \leq \alpha - G(0)u_\beta.$)

This shows a contradiction. Hence we have (4.11) and (4.12). This completes the proof.

Remark 4.3.1. Lemma 4.3.1 gives lower and upper bounds for the time between any two consecutive switchings in the region Ω. In fact, we are interested in reducing the lengths of the four intervals $[t_{0\alpha}^{min}, t_{0\alpha}^{max}]$, $[t_{0\beta}^{min}, t_{0\beta}^{max}]$, $(\tau, t_{\tau\alpha}^{max}]$ and $(\tau, t_{\tau\beta}^{max}]$. To achieve this, taking the interval $[t_{0\alpha}^{min}, t_{0\alpha}^{max}]$ for example, we need only to choose the new $t_{0\alpha}^{min}$ and $t_{0\alpha}^{max}$, denoted by $\tilde{t}_{0\alpha}^{min}$ and $\tilde{t}_{0\alpha}^{max}$, as follows.

$$\tilde{t}_{0\alpha}^{min} = \sup\{t_{0\alpha}^{min}\}, \quad t_{0\alpha}^{min} \text{ satisfying (4.10)}, \tag{4.14}$$

$$\tilde{t}_{0\alpha}^{max} = \inf\{t_{0\alpha}^{max}\}, \quad t_{0\alpha}^{max} \text{ satisfying (4.10)}. \tag{4.15}$$

Similar refined arguments apply to the other three time intervals. The proof of Lemma 4.3.1 also provides a method to compute a bound $\tau_{min} > 0$ on the time delay τ such that, for system Σ_τ with $0 < \tau < \tau_{min}$, the time between any two consecutive switchings of a trajectory is longer than τ provided that the trajectory enters the region Ω. This is done in the following remark.

Remark 4.3.2. Let $t_{0\alpha}^{min}$ and $t_{0\beta}^{min}$ be as in Lemma 4.3.1 (or as the refined versions in Remark 4.3.1). It is not hard to see that for system Σ_τ with $\tau < \min\{t_{0\alpha}^{min}, t_{0\beta}^{min}\}$, no early switches will occur when a trajectory enters the region Ω. (Otherwise, it will lead to a contradiction by virtue of the proof of Lemma 4.3.1 (i).) Now, let τ_{min} be such that

$$\|e^{At} - I\| < \varepsilon_0, \quad \forall t \in [0, \tau_{min}],$$

where

$$\varepsilon_0 = \frac{\beta - \alpha}{\|c\|(\omega_0 + \omega_{\alpha\beta})},$$

$$\omega_{\alpha\beta} = \max\{2\|A^{-1}bu_\beta\| + \|A^{-1}bu_\alpha\|, \; 2\|A^{-1}bu_\alpha\| + \|A^{-1}bu_\beta\|\}.$$

Then we conclude that for system Σ_τ with $\tau < \tau_{min}$, when a trajectory enters the region Ω, the time between any two consecutive switchings is longer than τ. Indeed, we notice that $\tau_{min} < \min\{t_{0\alpha}^{min}, t_{0\beta}^{min}\}$, and So, $\tau < \tau_{min}$ ensures that no early switches occur when a trajectory enters the region Ω. Due to the possible occurrence of early switches before the trajectory enters Ω, it may spend a time h from $x_\alpha \in \Omega_\alpha$ to $x_\beta \in \Omega_\beta$ (or, from $x_\beta \in \Omega_\beta$ to $x_\alpha \in \Omega_\alpha$). Without loss of generality, suppose the relay switches at most one time during the time h. (The following verification of a contradiction is still valid if the relay switches more than once during the time h.) We show $h > \tau$. Otherwise, if $h \leq \tau < \tau_{min}$, this will lead to a contradiction. In fact, it is seen that

$$x_\beta - x_\alpha$$
$$= (e^{Ah} - I)x_\alpha + (e^{Ah} - e^{A(h-t_h)})A^{-1}bu_1 + (e^{A(h-t_h)} - I)A^{-1}bu_2$$

(or, $\quad x_\alpha - x_\beta$
$$= (e^{Ah} - I)x_\beta + (e^{Ah} - e^{A(h-t_h)})A^{-1}bu_1 + (e^{A(h-t_h)} - I)A^{-1}bu_2)$$

holds for some $t_h \in [0, h]$ and $u_i \in \mathcal{U}$, $i = 1, 2$, with $u_1 \neq u_2$. Noticing that $cx_\alpha = \alpha$, $cx_\beta = \beta$, $\|x_\alpha\| \leq \omega_0$ and $\|x_\beta\| \leq \omega_0$, the above yields

$$\beta - \alpha < \|c\|\varepsilon_0\omega_0 + 2\varepsilon_0\|c\|\|A^{-1}bu_1\| + \varepsilon_0\|c\|\|A^{-1}bu_2\| \leq \beta - \alpha,$$

which is a contradiction.

4.4 Global Stability of Limit Cycles

We turn now to the global stability of x^* under Assumption 2.1 and Assumption 2.2 (for $\tau > 0$ only). Starting from any initial condition $x_0 \in \mathbb{R}^n$, the trajectory of $x(t)$ will eventually enter the region Ω and will traverse S_α and S_β consecutively. Let the first traversing point in Ω_α and Ω_β be $x_{\alpha 1}$ and $x_{\beta 1}$, respectively. Note that $x_{\alpha 1}$ and $x_{\beta 1}$ are determined by $x_0 \in \mathbb{R}^n$. For the trajectory of $x(t)$ with first traversing points $x_{\alpha 1} \in \Omega_\alpha$ and $x_{\beta 1} \in \Omega_\beta$, let the consecutive traversing point in Ω_α and Ω_β be $x_{\alpha i}$ and $x_{\beta i}$, $i = 2, 3, \ldots$, respectively; let the time taken for $x(t)$ to go from $x_{\alpha i}$ to $x_{\beta i}$ be $t_{\alpha i}$, and that from $x_{\beta i}$ to $x_{\alpha(i+1)}$ be $t_{\beta i}$. See Figures 4.1 and 4.2 for illustrations. By Lemma 4.3.1, it is obvious that

$$\text{for } \tau = 0, \quad \begin{cases} t_{0\alpha}^{min} \leq t_{\alpha i} \leq t_{0\alpha}^{max}, \\ t_{0\beta}^{min} \leq t_{\beta i} \leq t_{0\beta}^{max}, \end{cases} \quad i = 1, 2, \ldots \tag{4.16}$$

$$\text{for } \tau > 0, \quad \begin{cases} \tau < t_{\alpha i} \leq t_{\tau\alpha}^{max}, \\ \tau < t_{\beta i} \leq t_{\tau\beta}^{max}, \end{cases} \quad i = 1, 2, \ldots \tag{4.17}$$

For simplicity, let

$$t_{\alpha min} = \begin{cases} t_{0\alpha}^{min}, & \text{for } \tau = 0, \\ \tau, & \text{for } \tau > 0, \end{cases} \quad t_{\alpha max} = \begin{cases} t_{0\alpha}^{max}, & \text{for } \tau = 0, \\ t_{\tau\alpha}^{max}, & \text{for } \tau > 0, \end{cases}$$

$$t_{\beta min} = \begin{cases} t_{0\beta}^{min}, & \text{for } \tau = 0, \\ \tau, & \text{for } \tau > 0, \end{cases} \quad t_{\beta max} = \begin{cases} t_{0\beta}^{max}, & \text{for } \tau = 0, \\ t_{\tau\beta}^{max}, & \text{for } \tau > 0. \end{cases} \tag{4.18}$$

Now, define

$$F_\beta(t) := (e^{At} - I)(x_\beta^* + A^{-1}bu_\alpha),$$
$$f_\beta(t) := cF_\beta(t),$$
$$F_\alpha(t) := (e^{At} - I)(x_\alpha^* + A^{-1}bu_\beta),$$
$$f_\alpha(t) := cF_\alpha(t), \tag{4.19}$$

and

$$t^{-1}f_\beta(t)|_{t=0} := \lim_{t \to 0} t^{-1}f_\beta(t) = cAx_\beta^* + cbu_\alpha > 0,$$
$$t^{-1}f_\alpha(t)|_{t=0} := \lim_{t \to 0} t^{-1}f_\alpha(t) = cAx_\alpha^* + cbu_\beta < 0,$$
$$t^{-1}F_\beta(t)|_{t=0} := \lim_{t \to 0} t^{-1}F_\beta(t) = Ax_\beta^* + bu_\alpha,$$
$$t^{-1}F_\alpha(t)|_{t=0} := \lim_{t \to 0} t^{-1}F_\alpha(t) = Ax_\alpha^* + bu_\beta. \tag{4.20}$$

Then, $t^{-1}f_\beta(t)$, $t^{-1}f_\alpha(t)$, $t^{-1}F_\beta(t)$ and $t^{-1}F_\alpha(t)$ are continuous on t. Since $\frac{f_\beta(t-T_\alpha)}{t-T_\alpha} \neq 0$ and $\frac{f_\alpha(t-T_\beta)}{t-T_\beta} \neq 0$ for $t = T_\alpha$ and $t = T_\beta$, respectively, then they will still be valid in some time intervals containing T_α or T_β. We assume that the intervals can be extended to $[t_{\alpha min}, t_{\alpha max}]$ and $[t_{\beta min}, t_{\beta max}]$, respectively, which is equivalent to the following.

4. Global Stability of Limit Cycles

Assumption 4.1 Let

$$f_\beta(t - T_\alpha) \neq 0, \quad \forall t \in [t_{\alpha min}, t_{\alpha max}], \ t \neq T_\alpha, \tag{4.21}$$

$$f_\alpha(t - T_\beta) \neq 0, \quad \forall t \in [t_{\beta min}, t_{\beta max}], \ t \neq T_\beta. \tag{4.22}$$

Recall that we need only confine the study to the region Ω. So, in what follows, for a trajectory of $x(t)$, without loss of generality, let $t = 0$ correspond to the first traversing point $x_{\alpha 1} \in \Omega_\alpha$. Since for $\tau \geq 0$,

$$x(t) = e^{At}x_{\alpha 1} + \int_0^t e^{A(t-s)}bu_\beta ds$$
$$= e^{At}(x_{\alpha 1} + A^{-1}bu_\beta) - A^{-1}bu_\beta, \quad \forall t \in [0, \tau], \tag{4.23}$$

$$x(t) = e^{A(t-\tau)}x(\tau) + \int_0^{t-\tau} e^{A(t-\tau-s)}bu_\alpha ds$$
$$= e^{At}(x_{\alpha 1} + A^{-1}bu_\beta) + e^{A(t-\tau)}A^{-1}b(u_\alpha - u_\beta) - A^{-1}bu_\alpha, \quad \forall t \in [\tau, t_{\alpha 1}],$$

we have

$$x_\beta^* = e^{AT_\alpha}(x_\alpha^* + A^{-1}bu_\beta) + e^{A(T_\alpha-\tau)}A^{-1}b(u_\alpha - u_\beta) - A^{-1}bu_\alpha, \tag{4.24}$$

$$x_{\beta 1} = e^{At_{\alpha 1}}(x_{\alpha 1} + A^{-1}bu_\beta) + e^{A(t_{\alpha 1}-\tau)}A^{-1}b(u_\alpha - u_\beta) - A^{-1}bu_\alpha. \tag{4.25}$$

Multiplying (4.24) by $e^{A(t_{\alpha 1}-T_\alpha)}$ and combining with (4.25), we get

$$x_{\beta 1} - x_\beta^* = e^{At_{\alpha 1}}(x_{\alpha 1} - x_\alpha^*) + (e^{A(t_{\alpha 1}-T_\alpha)} - I)(x_\beta^* + A^{-1}bu_\alpha). \tag{4.26}$$

By Assumption 4.1, noting that $cx_\beta^* = cx_{\beta 1} = \beta$, then

$$t_{\alpha 1} - T_\alpha = -ce^{At_{\alpha 1}}(x_{\alpha 1} - x_\alpha^*)\frac{t_{\alpha 1} - T_\alpha}{f_\beta(t_{\alpha 1} - T_\alpha)}. \tag{4.27}$$

Since $t^{-1}F_\beta(t)$ is well defined for $t \in [t_{\alpha min} - T_\alpha, t_{\alpha max} - T_\alpha]$, substituting (4.27) into (4.26) gives

$$x_{\beta 1} - x_\beta^* = e^{At_{\alpha 1}}(x_{\alpha 1} - x_\alpha^*) + (t_{\alpha 1} - T_\alpha)\frac{F_\beta(t_{\alpha 1} - T_\alpha)}{t_{\alpha 1} - T_\alpha}$$
$$= e^{At_{\alpha 1}}(x_{\alpha 1} - x_\alpha^*) - ce^{At_{\alpha 1}}(x_{\alpha 1} - x_\alpha^*)\frac{F_\beta(t_{\alpha 1} - T_\alpha)}{f_\beta(t_{\alpha 1} - T_\alpha)}$$
$$= \left(I - \frac{F_\beta(t_{\alpha 1} - T_\alpha)c}{f_\beta(t_{\alpha 1} - T_\alpha)}\right)e^{At_{\alpha 1}}(x_{\alpha 1} - x_\alpha^*). \tag{4.28}$$

Similarly, from

$$x_\alpha^* = e^{AT_\beta}(x_\beta^* + A^{-1}bu_\alpha) + e^{A(T_\beta-\tau)}A^{-1}b(u_\beta - u_\alpha) - A^{-1}bu_\beta,$$

$$x_{\alpha 2} = e^{At_{\beta 1}}(x_{\beta 1} + A^{-1}bu_\alpha) + e^{A(t_{\beta 1}-\tau)}A^{-1}b(u_\beta - u_\alpha) - A^{-1}bu_\beta,$$

$$cx_\alpha^* = cx_{\alpha 2} = \alpha, \tag{4.29}$$

we obtain

$$x_{\alpha 2} - x_\alpha^* \tag{4.30}$$
$$= e^{At_{\beta 1}}(x_{\beta 1} - x_\beta^*) + (e^{A(t_{\beta 1} - T_\beta)} - I)(x_\alpha^* + A^{-1}bu_\beta)$$
$$= \left(I - \frac{F_\alpha(t_{\beta 1} - T_\beta)c}{f_\alpha(t_{\beta 1} - T_\beta)}\right) e^{At_{\beta 1}}(x_{\beta 1} - x_\beta^*)$$
$$= \left(I - \frac{F_\alpha(t_{\beta 1} - T_\beta)c}{f_\alpha(t_{\beta 1} - T_\beta)}\right) e^{At_{\beta 1}} \left(I - \frac{F_\beta(t_{\alpha 1} - T_\alpha)c}{f_\beta(t_{\alpha 1} - T_\alpha)}\right) e^{At_{\alpha 1}}(x_{\alpha 1} - x_\alpha^*).$$

Continuing the process, we have
$$x_{\alpha(i+1)} - x_\alpha^*$$
$$= \left(I - \frac{F_\alpha(t_{\beta i} - T_\beta)c}{f_\alpha(t_{\beta i} - T_\beta)}\right) e^{At_{\beta i}} \left(I - \frac{F_\beta(t_{\alpha i} - T_\alpha)c}{f_\beta(t_{\alpha i} - T_\alpha)}\right) e^{At_{\alpha i}}(x_{\alpha i} - x_\alpha^*),$$
$$i = 1, 2, \ldots. \tag{4.31}$$

Thus, noting $\|x_{\alpha 1} - x_\alpha^*\|$ is bounded by $2\omega_0$, where ω_0 is as in (2.5) of Chapter 2, the global stability of the limit cycle x^* has been reduced to the asymptotic stability of a discrete-time system. That is, we have already established the following result.

Theorem 4.4.1. *Consider system Σ_τ and suppose Assumptions 2.1, 2.2 (for $\tau > 0$ only) and 4.1 hold. The limit cycle x^* is globally stable if the following linear time-varying discrete system is asymptotically stable:*
$$X_{i+1} = W(t_{\beta i}, t_{\alpha i}) X_i, \quad i = 1, 2, \ldots \tag{4.32}$$

where
$$W(t_{\beta i}, t_{\alpha i}) = W(t_{\beta i}) W(t_{\alpha i}), \tag{4.33}$$
$$W(t_{\beta i}) = \left(I - \frac{F_\alpha(t_{\beta i} - T_\beta)c}{f_\alpha(t_{\beta i} - T_\beta)}\right) e^{At_{\beta i}}, \quad t_{\beta i} \in [t_{\beta min}, t_{\beta max}],$$
$$W(t_{\alpha i}) = \left(I - \frac{F_\beta(t_{\alpha i} - T_\alpha)c}{f_\beta(t_{\alpha i} - T_\alpha)}\right) e^{At_{\alpha i}}, \quad t_{\alpha i} \in [t_{\alpha min}, t_{\alpha max}].$$

Viewing (4.32) as a perturbed system due to the uncertainty of $t_{\alpha i}$ and $t_{\beta i}$, we next solve the robust asymptotic stability problem. We use three methods to do this, and later in Remark 4.4.1 we compare the three methods. It is easy to verify from (4.20) that $W(T_\beta, T_\alpha)$ is well-defined and is given as
$$W(T_\beta, T_\alpha) = W(T_\beta) W(T_\alpha) \tag{4.34}$$

where
$$W(T_\beta) = \left(I - \frac{(Ax_\alpha^* + bu_\beta)c}{c(Ax_\alpha^* + bu_\beta)}\right) e^{AT_\beta},$$
$$W(T_\alpha) = \left(I - \frac{(Ax_\beta^* + bu_\alpha)c}{c(Ax_\beta^* + bu_\alpha)}\right) e^{AT_\alpha}. \tag{4.35}$$

A. Norm constraint method

The following result is straightforward.

Proposition 4.4.1. *If there exists a scalar $\theta < 1$ and an integer $N \geq 1$ such that*

$$\|\prod_{i=1}^{N} W(t_{\beta i}, t_{\alpha i})\| \leq \theta, \quad \forall t_{\beta i} \in [t_{\beta min}, t_{\beta max}], \ t_{\alpha i} \in [t_{\alpha min}, t_{\alpha max}], \quad (4.36)$$

then system (4.32) is asymptotically stable.

To reduce conservativeness, it would be better for N to be chosen large. However, this makes computation difficult.

B. Quadratic stability method

This method says that if there exists a positive definite matrix $P > 0$ such that

$$P - W^T(t_{\beta i}, t_{\alpha i}) P W(t_{\beta i}, t_{\alpha i}) > 0, \quad \forall i = 1, 2, \ldots, \quad (4.37)$$

then (4.32) is asymptotically stable. To achieve this, the following is a sufficient condition.

Proposition 4.4.2. *If there exists a positive definite matrix $P > 0$ such that*

$$P - W^T(T_\beta, T_\alpha) P W(T_\beta, T_\alpha) > 0 \quad (4.38)$$

and

$$\|W(t_\beta, t_\alpha) - W(T_\beta, T_\alpha)\| < \sigma_{min} \begin{bmatrix} P & W^T(T_\beta, T_\alpha) \\ W(T_\beta, T_\alpha) & P^{-1} \end{bmatrix},$$

$$\forall t_\beta \in [t_{\beta min}, t_{\beta max}], \ t_\alpha \in [t_{\alpha min}, t_{\alpha max}], \quad (4.39)$$

then system (4.32) is asymptotically stable.

Proof. If the condition holds, then

$$\begin{bmatrix} P & W^T(t_\beta, t_\alpha) \\ W(t_\beta, t_\alpha) & P^{-1} \end{bmatrix} > 0$$

for all $t_\beta \in [t_{\beta min}, t_{\beta max}]$ and $t_\alpha \in [t_{\alpha min}, t_{\alpha max}]$, which, using the Schur complement, is equivalent to the condition

$$P - W^T(t_\beta, t_\alpha) P W(t_\beta, t_\alpha) > 0$$

for all $t_\beta \in [t_{\beta min}, t_{\beta max}]$ and $t_\alpha \in [t_{\alpha min}, t_{\alpha max}]$. This implies (4.37), and thus system (4.32) is asymptotically stable.

Also, conservativeness can be reduced if we use the following result.

Proposition 4.4.3. *If there exists a positive definite matrix $P > 0$ and an integer $N \geq 1$ such that*

$$P - (W^N(T_\beta, T_\alpha))^T P W^N(T_\beta, T_\alpha) > 0 \tag{4.40}$$

and

$$\left\| \prod_{i=1}^{N} W(t_{\beta i}, t_{\alpha i}) - W^N(T_\beta, T_\alpha) \right\| < \sigma_{min} \begin{bmatrix} P & (W^N(T_\beta, T_\alpha))^T \\ W^N(T_\beta, T_\alpha) & P^{-1} \end{bmatrix},$$

$$\forall t_{\beta i} \in [t_{\beta min}, t_{\beta max}],\ t_{\alpha i} \in [t_{\alpha min}, t_{\alpha max}],\ i = 1, 2, \ldots N, \tag{4.41}$$

then system (4.32) is asymptotically stable.

Proof. The proof follows a similar line to that of Proposition 4.4.2.

C. Parameter-dependent Lyapunov function method

The idea of this method is to solve the robust stability problem by using the parameter-dependent Lyapunov function $V(x) = x^T P(x) x$. The following lemma is useful for the development.

Lemma 4.4.1. *Let $\rho(W(T_\beta, T_\alpha)) < 1$ and let ρ_0 satisfy $\rho(W(T_\beta, T_\alpha)) < \rho_0 < 1$. Then,*

(i) there exists $\delta_0 > 0$ such that $\rho(W(T_\beta, T_\alpha) + \Delta) \leq \rho_0$ for all $\|\Delta\| \leq \delta_0$;

(ii) there exists $m > 0$ such that

$$\|(W(T_\beta, T_\alpha) + \Delta)^k\| \leq m \left(\frac{1 + \rho_0}{2} \right)^k, \quad \forall \|\Delta\| \leq \delta_0,\ \forall k = 1, 2, \ldots. \tag{4.42}$$

Proof. (i) Since $\rho(W(T_\beta, T_\alpha)) < \rho_0 < 1$, we have $\rho(\rho_0^{-1} W(T_\beta, T_\alpha)) < 1$. It is known that $\min_{|z|=1} \sigma_{min}(zI - \rho_0^{-1} W(T_\beta, T_\alpha))$ is exactly the complex stability radius of $\rho_0^{-1} W(T_\beta, T_\alpha)$ (Mori, 1990). Therefore, if

$$\|\rho_0^{-1} \Delta\| \leq \min_{|z|=1} \sigma_{min}(zI - \rho_0^{-1} W(T_\beta, T_\alpha)),$$

then

$$\rho(\rho_0^{-1} W(T_\beta, T_\alpha) + \rho_0^{-1} \Delta) \leq 1,$$

which yields

$$\rho(W(T_\beta, T_\alpha) + \Delta) \leq \rho_0.$$

Hence a suitable δ_0 is given by

$$\rho_0 \min_{|z|=1} \sigma_{min}(zI - \rho_0^{-1} W(T_\beta, T_\alpha)).$$

(ii) By the Schur–Toeplitz theorem for triangularization (see Lancaster and Tismenetsky (1985), p.176), for each $\Delta \in \mathbb{R}^{n \times n}$ satisfying $\|\Delta\| \leq \delta_0$, there exists a unitary matrix $U_\Delta \in C^{n \times n}$ such that

$$Z_\Delta := U_\Delta^T (W(T_\beta, T_\alpha) + \Delta) U_\Delta := (z_{ij}), \quad i,j = 1, 2, \ldots, n$$

is an upper triangular matrix where the diagonal entries z_{ii} are the eigenvalues of $W(T_\beta, T_\alpha) + \Delta$. By virtue of (i),

$$|z_{ii}| \leq \rho_0$$
$$|z_{ij}| \leq \|Z_\Delta\| = \|W(T_\beta, T_\alpha) + \Delta\| \leq \|W(T_\beta, T_\alpha)\| + \delta_0, \quad \text{for } i < j.$$

Let

$$d = \min\left\{1, \frac{1 - \rho_0}{2(n-1)(\|W(T_\beta, T_\alpha)\| + \delta_0)}\right\} > 0,$$
$$D = \text{diag}\{1, d, \ldots, d^{n-1}\},$$
$$\bar{Z}_\Delta = D^{-1} Z_\Delta D := (\bar{z}_{ij}), \quad i,j = 1, 2, \ldots, n.$$

Then we have

$$\|\bar{Z}_\Delta\|_\infty = \left\| \begin{bmatrix} \bar{z}_{11} & d\bar{z}_{12} & \cdots & d^{n-1}\bar{z}_{1n} \\ 0 & \bar{z}_{22} & \cdots & d^{n-2}\bar{z}_{2n} \\ \vdots & \vdots & \ddots & \vdots \\ 0 & 0 & \cdots & \bar{z}_{nn} \end{bmatrix} \right\|_\infty$$

$$\leq \rho_0 + (\|W(T_\beta, T_\alpha)\| + \delta_0)(d + \ldots + d^{n-1})$$
$$\leq \rho_0 + (\|W(T_\beta, T_\alpha)\| + \delta_0)(n-1)d$$
$$\leq 0.5(1 + \rho_0)$$
$$< 1.$$

Since for $M \in C^{n \times n}$, $\|D^{-1} M D\|_\infty$ is also a matrix norm (see Lancaster and Tismenetsky (1985), p.359), there exists a scalar $m > 0$ such that $\|M\| \leq m \|D^{-1} M D\|_\infty$ for all $M \in C^{n \times n}$. Therefore,

$$\|(W(T_\beta, T_\alpha) + \Delta)^k\| = \|Z_\Delta^k\| \leq m \|\bar{Z}_\Delta^k\|_\infty \leq m \|\bar{Z}_\Delta\|_\infty^k \leq m \left(\frac{1 + \rho_0}{2}\right)^k.$$

This proves the lemma.

The proof of Lemma 4.4.1 gives exact formulas for computing δ_0 and m. It would be better if we could obtain ρ_0 and δ_0 as large as they could be, and m as small as it could be. The next result is based on the parameter-dependent Lyapunov function method.

4.4 Global Stability of Limit Cycles

Proposition 4.4.4. *System (4.32) is asymptotically stable, if $\rho(W(T_\beta, T_\alpha)) < 1$ and*

$$\|W(t_\beta, t_\alpha) - W(T_\beta, T_\alpha)\| < \delta,$$
$$\forall t_\beta \in [t_{\beta min}, t_{\beta max}], \ t_\alpha \in [t_{\alpha min}, t_{\alpha max}], \quad (4.43)$$

where

$$\delta = \min\{\delta_0, \ \delta_1\},$$
$$\delta_1 = \left(4(\|W(T_\beta, T_\alpha)\| + \delta_0)^3 p^2\right)^{-1},$$
$$p = m\left(1 - \left(\frac{1+\rho_0}{2}\right)^2\right)^{-1}, \quad (4.44)$$

with ρ_0, δ_0 and m as in Lemma 4.4.1.

Proof. If (4.43) holds, then by Lemma 4.4.1, $\rho(W(t_{\beta i}, t_{\alpha i})) \leq \rho_0 < 1$ holds for all $t_{\beta i} \in [t_{\beta min}, t_{\beta max}]$ and $t_{\alpha i} \in [t_{\alpha min}, t_{\alpha max}]$. So, for each $i = 1, 2, \ldots$, we have that

$$P_i - W^T(t_{\beta i}, t_{\alpha i}) P_i W(t_{\beta i}, t_{\alpha i}) = I$$

for some $P_i > 0$. It is known (Robbe and Sadkane, 2000) that

$$P_i = \sum_{k=0}^{\infty} (W^T(t_{\beta i}, t_{\alpha i}))^k (W(t_{\beta i}, t_{\alpha i}))^k.$$

Let the parameter-dependent Lyapunov function be $V = X_i^T P_i X_i$. We show that $V(X_{i+1}) - V(X_i) < 0$. To do this, it is sufficient to show that

$$P_i - W^T(t_{\beta i}, t_{\alpha i}) P_{i+1} W(t_{\beta i}, t_{\alpha i}) = I + W^T(t_{\beta i}, t_{\alpha i})(P_i - P_{i+1}) W(t_{\beta i}, t_{\alpha i}) > 0.$$

With some manipulation, we see that

$$(P_{i+1} - P_i) - W^T(t_{\beta(i+1)}, t_{\alpha(i+1)})(P_{i+1} - P_i) W(t_{\beta(i+1)}, t_{\alpha(i+1)}) = Q,$$

where

$$Q = W^T(t_{\beta(i+1)}, t_{\alpha(i+1)}) P_i (W(t_{\beta(i+1)}, t_{\alpha(i+1)}) - W(t_{\beta i}, t_{\alpha i}))$$
$$+ (W(t_{\beta(i+1)}, t_{\alpha(i+1)}) - W(t_{\beta i}, t_{\alpha i}))^T P_i W(t_{\beta i}, t_{\alpha i}).$$

So,

$$P_{i+1} - P_i = \sum_{k=0}^{\infty} (W^T(t_{\beta(i+1)}, t_{\alpha(i+1)}))^k Q (W(t_{\beta(i+1)}, t_{\alpha(i+1)}))^k.$$

Taking into account the fact, for all $i = 1, 2, \ldots$, that

4. Global Stability of Limit Cycles

$$\|P_i\| \leq \sum_{k=0}^{\infty} \|(W^T(t_{\beta i}, t_{\alpha i}))^k\| \|(W(t_{\beta i}, t_{\alpha i}))^k\| \leq \sum_{k=0}^{\infty} m\left(\frac{1+\rho_0}{2}\right)^{2k} = p,$$

$$\|Q\| \leq 2(\|W(T_\beta, T_\alpha)\| + \delta_0)\|P_i\|\|W(t_{\beta(i+1)}, t_{\alpha(i+1)}) - W(t_{\beta i}, t_{\alpha i})\|$$
$$\leq 4(\|W(T_\beta, T_\alpha)\| + \delta_0)p\delta,$$

we arrive at

$$\|P_{i+1} - P_i\| \leq \|Q\| \sum_{k=0}^{\infty} \|(W^T(t_{\beta(i+1)}, t_{\alpha(i+1)}))^k\| \|(W(t_{\beta(i+1)}, t_{\alpha(i+1)}))^k\|$$
$$< (\|W(T_\beta, T_\alpha)\| + \delta_0)^{-2}.$$

Hence, $I + W^T(t_{\beta i}, t_{\alpha i})(P_i - P_{i+1})W(t_{\beta i}, t_{\alpha i}) > 0$ is satisfied. This completes the proof.

Similarly to the former two methods (Propositions 4.4.1 and 4.4.3), to reduce conservativeness, we can use an integer N to check the condition $\rho(W^N(T_\beta, T_\alpha)) < 1$ and establish a corresponding result. To save the space, we do not list it here.

Remark 4.4.1. In general, method A is simple but is the most conservative one among the three. Method B is more conservative than method C, since it needs to find a single matrix $P > 0$ while method C uses the parameter-dependent Lyapunov function. However, based on our extensive simulations, it is guessed that if Assumptions 2.1 and 2.2 (for $\tau > 0$ only) are satisfied, either there are more than one limit cycle (thus not globally stable), or the limit cycle is globally stable, which can be checked by Proposition 4.4.1 or 4.4.3. This is largely owing to the observations that $W(t_{\beta i})$ and $W(t_{\alpha i})$ in (4.32) have special structures, e.g.,

- $cW(t_{\beta i}) = cW(t_{\alpha i}) \equiv 0$, $\forall t_{\beta i}, t_{\alpha i}$;
- $\text{rank}(F_\beta c/f_\beta) = \text{rank}(F_\alpha c/f_\alpha) = 1$;
- $e^{At} \to 0$ as $t \to +\infty$.

These observations make our checking methods effective and we expect that a wide class of systems have a globally stable limit cycle. The next section extends the result to the case that a limit cycle makes the relay switch more than twice per period, and gives a simple numerical example to illustrate the use of the result.

Remark 4.4.2. It is easy to see that if we let $t = 0$ correspond to the first traversing point $x_{\beta 1} \in \Omega_\beta$, similar analysis leads to a parallel result to Theorem 4.4.1. That is, under Assumptions 2.1, 2.2 (for $\tau > 0$ only) and 4.1, the limit cycle x^* is globally stable, if the following system is asymptotically stable.

$$\tilde{X}_{i+1} = W(t_{\alpha i}, t_{\beta i})\tilde{X}_i, \quad i = 1, 2, \ldots \tag{4.45}$$

where

$$W(t_{\alpha i}, t_{\beta i}) = W(t_{\alpha i})W(t_{\beta i}),$$
$$t_{\beta i} \in [t_{\beta min}, t_{\beta max}], \quad t_{\alpha i} \in [t_{\alpha min}, t_{\alpha max}], \tag{4.46}$$

and $W(t_{\alpha i})$ and $W(t_{\beta i})$ are as in Theorem 4.4.1. Accordingly, sufficient results analogous to Propositions 4.4.1–4.4.3 can be established.

So far in this section we have studied the global stability problem for system Σ_τ. The limit cycle considered may not be symmetric. Because time-delay systems are of considerable interest from an engineering point of view (most industrial processes have time delay), the method presented in this section gives a unified approach for both $\tau = 0$ and $\tau > 0$. For delay-free systems ($\tau = 0$) of the form Σ_τ with $\beta = -\alpha$ and $u_\alpha = -u_\beta = 1$, Goncalves et al. (2001) studies the global stability problem for symmetric limit cycles with two switchings per period. The method involves finding a single matrix $P > 0$ for the so-called surface Lyapunov function of Poincare maps, which is related to method B in this section. Also, some other techniques (e.g. S-procedure) are used to reduce the conservativeness.

4.5 Extensions

In the preceding section, the limit cycle is assumed to be of two switchings per period. This section studies the global stability of a limit cycle with more than two switchings per period, and shows that the corresponding results can be obtained similarly. Let the limit cycle, still denoted by x^*, make the relay switch $2q$ times per period, where q is a positive integer. The limit cycle x^* is transversal to S_α and S_β at the traversing points $x^*_{\alpha i} \in S_\alpha$ and $x^*_{\beta i} \in S_\beta$, $i = 1, 2, \ldots, q$, respectively. The period of x^* is $\sum_{i=1}^{q}(T_{\alpha i} + T_{\beta i})$ where $T_{\alpha i} > 0$ (resp. $T_{\beta i} > 0$) is the time taken for the trajectory of the limit cycle to go from $x^*_{\alpha i}$ to $x^*_{\beta i}$ (resp. from $x^*_{\beta i}$ to $x^*_{\alpha(i+1)}$).

It is seen that $T_{\alpha l} \in [t_{\alpha min}, t_{\alpha max}]$ and $T_{\beta l} \in [t_{\beta min}, t_{\beta max}]$, $l = 1, 2, \ldots, q$, where $t_{\alpha min}$, $t_{\alpha max}$, $t_{\beta min}$ and $t_{\beta max}$ are given by (4.18). As in Section 4.4, for $l = 1, 2, \ldots, q$, define

$$F_{\beta l}(t) := (e^{At} - I)(x^*_{\beta l} + A^{-1}bu_\alpha),$$
$$f_{\beta l}(t) := cF_{\beta l}(t),$$
$$F_{\alpha l}(t) := (e^{At} - I)(x^*_{\alpha l} + A^{-1}bu_\beta),$$
$$f_{\alpha l}(t) := cF_{\alpha l}(t), \tag{4.47}$$

and

$$t^{-1}f_{\beta l}(t)|_{t=0} := \lim_{t \to 0} t^{-1}f_{\beta l}(t) = cAx^*_{\beta l} + cbu_\alpha > 0,$$
$$t^{-1}f_{\alpha l}(t)|_{t=0} := \lim_{t \to 0} t^{-1}f_{\alpha l}(t) = cAx^*_{\alpha l} + cbu_\beta < 0,$$
$$t^{-1}F_{\beta l}(t)|_{t=0} := \lim_{t \to 0} t^{-1}F_{\beta l}(t) = Ax^*_{\beta l} + bu_\alpha,$$
$$t^{-1}F_{\alpha l}(t)|_{t=0} := \lim_{t \to 0} t^{-1}F_{\alpha l}(t) = Ax^*_{\alpha l} + bu_\beta. \tag{4.48}$$

Then, $t^{-1}f_{\beta l}(t)$, $t^{-1}f_{\alpha l}(t)$, $t^{-1}F_{\beta l}(t)$ and $t^{-1}F_{\alpha l}(t)$ are continuous on t. Since $\frac{f_{\beta l}(t-T_{\alpha l})}{t-T_{\alpha l}} \neq 0$ and $\frac{f_{\alpha l}(t-T_{\beta l})}{t-T_{\beta l}} \neq 0$ hold for $t = T_{\alpha l}$ and $t = T_{\beta l}$, respectively, then they will still be valid for some time intervals containing $T_{\alpha l}$ or $T_{\beta l}$. We assume that the intervals can be extended to $[t_{\alpha min}, t_{\alpha max}]$ and $[t_{\beta min}, t_{\beta max}]$.

Assumption 4.2 Let, for all $l = 1, 2, \ldots, q$,

$$f_{\beta l}(t - T_{\alpha l}) \neq 0, \quad \forall t \in [t_{\alpha min}, t_{\alpha max}], \ t \neq T_{\alpha l}, \tag{4.49}$$
$$f_{\alpha l}(t - T_{\beta l}) \neq 0, \quad \forall t \in [t_{\beta min}, t_{\beta max}], \ t \neq T_{\beta l}. \tag{4.50}$$

The main result in this section is as follows.

Theorem 4.5.1. *Consider system Σ_τ and suppose Assumptions 2.1, 2.2 (for $\tau > 0$ only) and 4.2 hold. The limit cycle x^* is globally stable if for some $k = 1, 2, \ldots, 2q$, the following linear parameter-varying discrete system is asymptotically stable.*

$$\widehat{X}_{i+1} = \widehat{W}_{i,k}\widehat{X}_i, \quad i = 1, 2, \ldots \tag{4.51}$$

where

$$\widehat{W}_{i,k} = W(t_{i,k}, t_{i,k-1}, \ldots, t_{i,1}, t_{i,2q}, t_{i,2q-1}, \ldots, t_{i,k+1}) \tag{4.52}$$
$$= W(t_{i,k})W(t_{i,k-1}) \ldots W(t_{i,1})W(t_{i,2q})W(t_{i,2q-1}) \ldots W(t_{i,k+1})$$

and

$$W(t_{i,2l-1}) = \left(I - \frac{F_{\beta l}(t_{i,2l-1} - T_{\alpha l})c}{f_{\beta l}(t_{i,2l-1} - T_{\alpha l})}\right)e^{At_{i,2l-1}}, \quad t_{i,2l-1} \in [t_{\alpha min}, t_{\alpha max}],$$
$$W(t_{i,2l}) = \left(I - \frac{F_{\alpha l}(t_{i,2l} - T_{\beta l})c}{f_{\alpha l}(t_{i,2l} - T_{\beta l})}\right)e^{At_{i,2l}}, \quad t_{i,2l} \in [t_{\beta min}, t_{\beta max}],$$
$$l = 1, 2, \ldots, q. \tag{4.53}$$

Proof. The proof is similar to the process for attaining Theorem 4.4.1 or the result in Remark 4.4.2, and thus is omitted here.

For simplicity, let

$$T_k = \begin{cases} T_{\alpha l}, & \text{if } k = 2l-1, \\ T_{\beta l}, & \text{if } k = 2l, \end{cases} \quad k = 1, 2, \ldots, 2q, \quad l = 1, 2, \ldots, q. \tag{4.54}$$

Then, $W(T_k, T_{k-1}, \ldots, T_1, T_{2q}, T_{2q-1}, \ldots, T_{k+1})$ is well-defined and is given as

$$W(T_k, T_{k-1}, \ldots, T_1, T_{2q}, T_{2q-1}, \ldots, T_{k+1})$$
$$= W_k W_{k-1} \ldots W_1 W_{2q} W_{2q-1} \ldots W_{k+1} \tag{4.55}$$

where

$$W_{2l-1} = \left(I - \frac{(Ax^*_{\beta l} + bu_\alpha)c}{c(Ax^*_{\beta l} + bu_\alpha)}\right) e^{AT_{\alpha l}},$$

$$W_{2l} = \left(I - \frac{(Ax^*_{\alpha l} + bu_\beta)c}{c(Ax^*_{\alpha l} + bu_\beta)}\right) e^{AT_{\beta l}}, \quad l = 1, 2, \ldots, q. \tag{4.56}$$

If

$$\rho(W^N(T_k, T_{k-1}, \ldots, T_1, T_{2q}, T_{2q-1}, \ldots, T_{k+1})) < 1 \tag{4.57}$$

holds for some integer $N \geq 1$, then sufficient conditions for solving the asymptotic stability of system (4.51) can be obtained analogous to Propositions 4.4.1–4.4.4.

Remark 4.5.1. In Theorems 4.4.1 and 4.5.1, it is obvious that the sufficient conditions which guarantee the asymptotic stability of system (4.32) or (4.51) also imply uniqueness of the limit cycle x^*.

So far, we have established sufficient conditions, in Sections 4.4 and 4.5, for the global stability of a limit cycle. In practice, from our experience and extensive simulations, we notice that many relay feedback systems of the form Σ_τ have limit cycles. A rough explanation for the global convergence to a limit cycle is addressed below. The condition (4.57) ensures the local stability of a limit cycle. (This is related to the local stability of x^*, which can be verified using a similar method to that in Chapter 3.) If the limit cycle is unique, under Assumptions 2.1 and 2.2, the system trajectories will enter the region Ω, and after some switchings, they will tend to x^* under additional constraints.

The following gives a simple numerical example to illustrate the use of the global stability result.

Example 4.5.1. Consider Example 4.2.1 with A, b and c being unchanged and $u_\alpha = 3$, $u_\beta = -1$, $\alpha = 0$ and $\beta = 1$. It is easy to check that for $\tau = 0$, this system does not have a solution for all $t > 0$. However, for any $\tau > 0$, given any initial condition $x_0 \in \mathbb{R}^2$, there is a solution $x(t)$ (indeed unique) for all

76   4. Global Stability of Limit Cycles

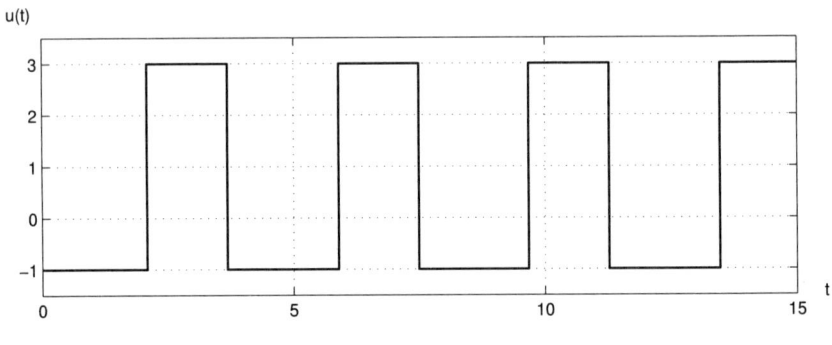

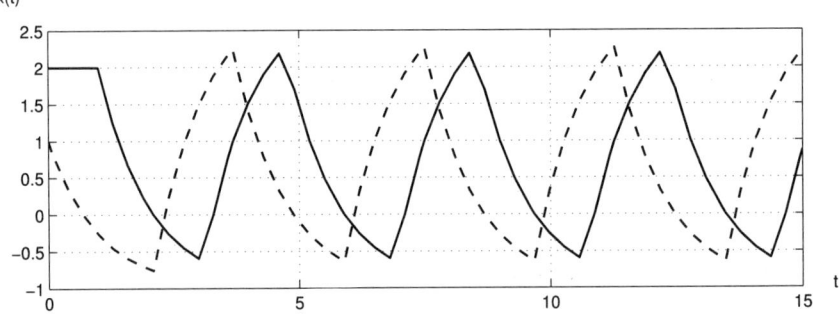

Fig. 4.3. The control $u(t)$ and the trajectory of $x(t) = [x_1(t)\ x_2(t)]^T$ starting from $[2\ 1]^T$ for Example 4.5.1 ($x_1(t)$ solid, $x_2(t)$ dashed)

$t > 0$. Also, we check that for any $\tau > 0$, there is a limit cycle of the form in Section 4.4. Figure 4.3 shows the trajectory starting from $[2\ 1]^T$ for $\tau = 1$.

Now, we check the global stability of a limit cycle x^* corresponding to a given $\tau > 0$. The validity of Assumptions 2.1 and 2.2 can easily be checked. The invariant region Ω in (2.2) is computed to be $\Omega = \{\xi \in \mathbb{R}^2 : \|\xi\| \leq 3\sqrt{2}\}$, and so $\omega_0 = 3\sqrt{2}$. We further compute that $\omega_{\tau\alpha}^{max} = \frac{1}{4\sqrt{2}}$ and $\omega_{\tau\beta}^{max} = \frac{1}{10\sqrt{2}}$ from (4.7); $t_{\alpha max} = t_{\tau\alpha}^{max} = \tau + 4\sqrt{2}$ and $t_{\beta max} = t_{\tau\beta}^{max} = \tau + 10\sqrt{2}$ from (4.13) and (4.18). Letting $x_\alpha^* = [\alpha_1\ \alpha_2]^T$ and $x_\beta^* = [\beta_1\ \beta_2]^T$, we see from $cx_\alpha^* = \alpha$ and $cx_\beta^* = \beta$ that $\alpha_1 = 0$ and $\beta_1 = 1$. The above gives that $f_\alpha(t) = -\frac{1}{2}f_\beta(t) = e^{-t} - 1 \neq 0$ for all $t \neq 0$, i.e., Assumption 4.1 is satisfied.

By using Theorem 4.4.1, to verify the global stability of the limit cycle, it is sufficient to check the asymptotic stability of the system (4.32), where $W(t_{\beta i}, t_{\alpha i})$ is computed to be

$$W(t_{\beta i}, t_{\alpha i}) = e^{-(t_{\alpha i} + t_{\beta i})} \begin{bmatrix} 0 & 0 \\ \frac{1}{2}(\beta_2 - 3) & 1 \end{bmatrix}. \tag{4.58}$$

It is easy to check (from Proposition 4.4.1 or 4.4.3) that such a system is asymptotically stable. This concludes that the system in Example 4.5.1 with any $\tau > 0$ has a unique globally stable limit cycle.

If we use system (4.45) in Remark 4.4.2 to check the global stability of the limit cycle, we have

$$W(t_{\alpha i}, t_{\beta i}) = e^{-(t_{\alpha i}+t_{\beta i})} \begin{bmatrix} 0 & 0 \\ -\alpha_2 - 1 & 1 \end{bmatrix}. \tag{4.59}$$

This gives the same conclusion.

4.6 Existence of Globally Stable Limit Cycles

In Sections 4.4 and 4.5, the given global stability results serve a wide class of relay feedback systems well. Indeed under our assumptions, it is hard to find a numerical example which conflicts with the global stability condition. This also implies that a wide class of systems has a unique globally stable limit cycle. We can thus have the following conjecture:

Conjecture: Suppose Assumptions 2.1 and 2.2 hold. If there is a unique limit cycle for system Σ_τ, then the limit cycle is globally stable.

It is known that the uniqueness of a limit cycle is necessary for its global stability. However, how to check the uniqueness of a limit cycle is difficult. The available method is to compute the period of a possible limit cycle through numerical procedures, and then confirm that only one corresponds to a limit cycle. Having obtained the period, the traversing points are computed straightforwardly. With this information, we are able to check the global stability of the limit cycle. Thus, how to determine that there does exist a globally stable limit cycle without assuming the existence of a limit cycle deserves study. This section deals with this problem.

4.6.1 Preliminaries

We introduce some concepts and prepare some lemmas in this subsection.

Definition 4.6.1. *Let (M, d) be a metric space and $D \subset M$ be a subset of M. A continuous mapping $T : D \to M$ is said to be a contraction mapping if there exists a constant $k \in (0, 1)$ such that*

$$d(T(x), T(y)) \leq k d(x, y), \quad \forall x, y \in M. \tag{4.60}$$

The following is a famous result regarding the unique fixed-point. For its proof, refer to Farkas (1994), Khamsi and Kirk (2001) and Smart (1974).

Lemma 4.6.1. *(Banach's Contraction Mapping Principle) Let $T : D \to M$ be a contraction mapping defined on the closed subset $D \subset M$ of the Banach space M. If $T(D) \subseteq D$, then T has a unique fixed-point x_0, and for each $x \in D$, $\lim_{n\to\infty} T^n(x) = x_0$.*

The above contraction mapping principle has the following extension.

Corollary 4.6.1. *Let $T : D \to D$ be a mapping defined on the closed subset $D \subset M$ of the Banach space M. If T^N is a contraction mapping for some positive integer N, then T has a unique fixed-point.*

Proof. By Lemma 4.6.1, T^N has a unique fixed-point x_0. Hence

$$T^{N+1}(x_0) = T(T^N(x_0)) = T(x_0), \tag{4.61}$$

which implies that $T(x_0)$ is also a fixed-point of T^N. By uniqueness, it must hold that $T(x_0) = x_0$. This shows x_0 is a fixed-point of T.

On the other hand, if $T(y_0) = y_0$, then $T^N(y_0) = y_0$. This, again by uniqueness, gives $y_0 = x_0$, proving that x_0 is the unique fixed-point of T.

4.6.2 Sufficient Conditions

It is known that under Assumptions 2.1 and 2.2 (for $\tau > 0$ only), any trajectory of system Σ_τ will eventually enter the region Ω. Moreover, for any trajectory in this invariant region, the time between any two successive traversing instants is within the interval $[t_{\alpha min}, t_{\alpha max}]$ (or $[t_{\beta min}, t_{\beta max}]$).

Let us consider two trajectories $\xi(t)$ and $\eta(t)$ for system Σ_τ. In the region Ω, let certain traversing points in Ω_α be $\xi_{\alpha 1}$ and $\eta_{\alpha 1}$, respectively, and the successive traversing points in Ω_β be $\xi_{\beta 1}$ and $\eta_{\beta 1}$, respectively. Let the consecutive traversing points in Ω_α and Ω_β be $\xi_{\alpha i}, \eta_{\alpha i}$ and $\xi_{\beta i}, \eta_{\beta i}$, $i = 2, 3, \ldots$, respectively. Denote the time taken for $\xi(t)$ (resp. $\eta(t)$) to go from $\xi_{\alpha i}$ (resp. $\eta_{\alpha i}$) to $\xi_{\beta i}$ (resp. $\eta_{\beta i}$) be $t_{\xi\alpha i}$ (resp. $t_{\eta\alpha i}$), and that from $\xi_{\beta i}$ (resp. $\eta_{\beta i}$) to $\xi_{\alpha(i+1)}$ (resp. $\xi_{\alpha(i+1)}$) be $t_{\xi\beta i}$ (resp. $t_{\eta\beta i}$). We now compare the distance between $\eta_{\alpha 2} - \xi_{\alpha 2}$ and $\eta_{\alpha 1} - \xi_{\alpha 1}$. Since

$$\xi_{\beta 1} = e^{At_{\xi\alpha 1}}(\xi_{\alpha 1} + A^{-1}bu_\beta) + e^{A(t_{\xi\alpha 1} - \tau)}A^{-1}b(u_\alpha - u_\beta) - A^{-1}bu_\alpha, \tag{4.62}$$

$$\eta_{\beta 1} = e^{At_{\eta\alpha 1}}(\eta_{\alpha 1} + A^{-1}bu_\beta) + e^{A(t_{\eta\alpha 1} - \tau)}A^{-1}b(u_\alpha - u_\beta) - A^{-1}bu_\alpha, \tag{4.63}$$

multiplying (4.62) by $e^{A(t_{\eta\alpha 1} - t_{\alpha\alpha 1})}$ and combining with (4.63), we get

4.6 Existence of Globally Stable Limit Cycles

$$\eta_{\beta 1} - \xi_{\beta 1} = e^{At_{\eta\alpha 1}}(\eta_{\alpha 1} - \xi_{\alpha 1}) + (e^{A(t_{\eta\alpha 1} - t_{\xi\alpha 1})} - I)(\xi_{\beta 1} + A^{-1}bu_\alpha). \quad (4.64)$$

To proceed, we assume the following.

Assumption 4.3 In the region Ω, any system trajectory is transversal to Ω_α and Ω_β at each of its traversing points.

Define

$$\begin{aligned} F(x_\beta, t) &:= (e^{At} - I)(x_\beta + A^{-1}bu_\alpha), \\ f(x_\beta, t) &:= cF(x_\beta, t), \\ F(x_\alpha, t) &:= (e^{At} - I)(x_\alpha + A^{-1}bu_\beta), \\ f(x_\alpha, t) &:= cF(x_\alpha, t). \end{aligned} \quad (4.65)$$

Under Assumption 4.3, we see that for $i = 1, 2, \ldots$,

$$\begin{aligned} t^{-1}f(\xi_{\beta i}, t)|_{t=0} &:= \lim_{t \to 0} t^{-1}f(\xi_{\beta i}, t) = cA\xi_{\beta i} + cbu_\alpha > 0, \\ t^{-1}f(\xi_{\alpha i}, t)|_{t=0} &:= \lim_{t \to 0} t^{-1}f(\xi_{\alpha i}, t) = cA\xi_{\alpha i} + cbu_\beta < 0, \\ t^{-1}F(\xi_{\beta i}, t)|_{t=0} &:= \lim_{t \to 0} t^{-1}F(\xi_{\beta i}, t) = A\xi_{\beta i} + bu_\alpha, \\ t^{-1}F(\xi_{\alpha i}, t)|_{t=0} &:= \lim_{t \to 0} t^{-1}F(\xi_{\alpha i}, t) = A\xi_{\alpha i} + bu_\beta, \end{aligned} \quad (4.66)$$

and similar arguments apply with respect to $\eta(t)$. Then, $t^{-1}f(\xi_{\beta i}, t), t^{-1}f(\xi_{\alpha i}, t), t^{-1}F(\xi_{\beta i}, t), t^{-1}F(\xi_{\beta i}, t), t^{-1}f(\eta_{\beta i}, t), t^{-1}f(\eta_{\alpha i}, t), t^{-1}F(\eta_{\beta i}, t)$ and $t^{-1}F(\eta_{\beta i}, t)$ are continuous on t. Since $t^{-1}f(\xi_{\beta i}, t) \neq 0, t^{-1}f(\xi_{\alpha i}, t) \neq 0, t^{-1}f(\eta_{\beta i}, t) \neq 0$ and $t^{-1}f(\eta_{\alpha i}, t) \neq 0$ hold for $t = 0$, then they will still be valid for some intervals containing $t = 0$. We assume that the intervals can be extended to $[t_{\alpha min} - t_{\alpha max}, t_{\alpha max} - t_{\alpha min}]$ and $[t_{\beta min} - t_{\beta max}, t_{\beta max} - t_{\beta min}]$.

Assumption 4.4 For $x_\alpha \in \Omega_\alpha$ and $x_\beta \in \Omega_\beta$, let

$$f(x_\beta, t) \neq 0, \quad \forall t \in [t_{\alpha min} - t_{\alpha max}, t_{\alpha max} - t_{\alpha min}], \ t \neq 0, \quad (4.67)$$

$$f(x_\alpha, t) \neq 0, \quad \forall t \in [t_{\beta min} - t_{\beta max}, t_{\beta max} - t_{\beta min}], \ t \neq 0. \quad (4.68)$$

Under Assumption 4.4, noting $c\xi_{\beta 1} = c\eta_{\beta 1} = \beta$, (4.64) gives

$$t_{\eta\alpha 1} - t_{\xi\alpha 1} = -ce^{At_{\eta\alpha 1}}(\eta_{\alpha 1} - \xi_{\alpha 1}) \frac{t_{\eta\alpha 1} - t_{\xi\alpha 1}}{f(\xi_{\beta 1}, t_{\eta\alpha 1} - t_{\xi\alpha 1})}. \quad (4.69)$$

Since $t^{-1}F(\xi_{\beta 1}, t)$ is well defined for $t \in [t_{\alpha min} - t_{\alpha max}, t_{\alpha max} - t_{\alpha min}]$, substituting (4.69) into (4.64) gives

$$\eta_{\beta 1} - \xi_{\beta 1} = e^{At_{\eta\alpha 1}}(\eta_{\alpha 1} - \xi_{\alpha 1}) + (t_{\eta\alpha 1} - t_{\xi\alpha 1})\frac{F(\xi_{\beta 1}, t_{\eta\alpha 1} - t_{\xi\alpha 1})}{t_{\eta\alpha 1} - t_{\xi\alpha 1}}$$

$$= e^{At_{\eta\alpha 1}}(\eta_{\alpha 1} - \xi_{\alpha 1}) - c e^{At_{\eta\alpha 1}}(\eta_{\alpha 1} - \xi_{\alpha 1})\frac{F(\xi_{\beta 1}, t_{\eta\alpha 1} - t_{\xi\alpha 1})}{f(\xi_{\beta 1}, t_{\eta\alpha 1} - t_{\xi\alpha 1})}$$

$$= \left(I - \frac{F(\xi_{\beta 1}, t_{\eta\alpha 1} - t_{\xi\alpha 1})c}{f(\xi_{\beta 1}, t_{\eta\alpha 1} - t_{\xi\alpha 1})}\right) e^{At_{\eta\alpha 1}}(\eta_{\alpha 1} - \xi_{\alpha 1}). \tag{4.70}$$

Similarly, from

$$\xi_{\alpha 2} = e^{At_{\xi\beta 1}}(\xi_{\beta 1} + A^{-1}bu_\alpha) + e^{A(t_{\xi\beta 1} - \tau)}A^{-1}b(u_\beta - u_\alpha) - A^{-1}bu_\beta,$$
$$\eta_{\alpha 2} = e^{At_{\eta\beta 1}}(\eta_{\beta 1} + A^{-1}bu_\alpha) + e^{A(t_{\eta\beta 1} - \tau)}A^{-1}b(u_\beta - u_\alpha) - A^{-1}bu_\beta,$$
$$c\xi_{\alpha 2} = c\eta_{\alpha 2} = \alpha, \tag{4.71}$$

we obtain

$$\eta_{\alpha 2} - \xi_{\alpha 2} \tag{4.72}$$

$$= e^{At_{\eta\beta 1}}(\eta_{\beta 1} - \xi_{\beta 1}) + (e^{A(t_{\eta\beta 1} - t_{\xi\beta 1})} - I)(\xi_{\alpha 2} + A^{-1}bu_\beta)$$

$$= \left(I - \frac{F(\xi_{\alpha 2}, t_{\eta\beta 1} - t_{\xi\beta 1})c}{f(\xi_{\alpha 2}, t_{\eta\beta 1} - t_{\xi\beta 1})}\right) e^{At_{\eta\beta 1}}(\eta_{\beta 1} - \xi_{\beta 1})$$

$$= \left(I - \frac{F(\xi_{\alpha 2}, t_{\eta\beta 1} - t_{\xi\beta 1})c}{f(\xi_{\alpha 2}, t_{\eta\beta 1} - t_{\xi\beta 1})}\right) e^{At_{\eta\beta 1}} \cdot$$

$$\left(I - \frac{F(\xi_{\beta 1}, t_{\eta\alpha 1} - t_{\xi\alpha 1})c}{f(\xi_{\beta 1}, t_{\eta\alpha 1} - t_{\xi\alpha 1})}\right) e^{At_{\eta\alpha 1}}(\eta_{\alpha 1} - \xi_{\alpha 1}).$$

The above equation shows the relationship between $\eta_{\alpha 2} - \xi_{\alpha 2}$ and $\eta_{\alpha 1} - \xi_{\alpha 1}$. So far, we have established the following result.

Theorem 4.6.1. *Suppose Assumptions 2.1, 2.2 (for $\tau > 0$ only), 4.3 and 4.4 hold. If there exists a constant $k \in (0,1)$ such that*

$$\left\|\left(I - \frac{F(x_\alpha, t_2 - t_{22})c}{f(x_\alpha, t_2 - t_{22})}\right) e^{At_2} \left(I - \frac{F(x_\beta, t_1 - t_{11})c}{f(x_\beta, t_1 - t_{11})}\right) e^{At_1}\right\| \leq k, \tag{4.73}$$

$\forall x_\beta \in \Omega_\beta, \quad x_\alpha \in \Omega_\alpha,$

$\forall t_1, t_{11} \in [t_{\alpha min}, t_{\alpha max}], \quad t_2, t_{22} \in [t_{\beta min}, t_{\beta max}],$

then there is a globally stable limit cycle for system Σ_τ.

Proof. View Ω_α as a closed subset of the Banach space $\mathbb{R}^n$ where the norm is defined as the spectral norm $\|\cdot\|$. For an arbitrary traversing point $x_\alpha \in \Omega_\alpha$, define the image $T(x_\alpha)$ of the mapping T as the successive traversing point in Ω_α. For the arbitrary trajectories of $\xi(t)$ and $\eta(t)$, let two specific traversing points in Ω_α be $\eta_{\alpha 1}, \xi_{\alpha 1}$. From the above analysis, the successive traversing points $\eta_{\alpha 2}, \xi_{\alpha 2} \in \Omega_\alpha$ are related by (4.72). In view of (4.73), we see that

$$\|\eta_{\alpha 2} - \xi_{\alpha 2}\| = \|T(\eta_{\alpha 1}) - T(\xi_{\alpha 1})\| \leq k\|\eta_{\alpha 1} - \xi_{\alpha 1}\|,$$

which, by using Lemma 4.6.1, suggests that there is a fixed-point, say $x_0 \in \Omega_\alpha$, of T. This implies that there is a trajectory which traverses Ω_α at x_0 and still traverses it next time. Hence, the theorem is proved.

Using Corollary 4.6.1, the following result is straightforward.

Corollary 4.6.2. *Suppose Assumptions 2.1, 2.2 (for $\tau > 0$ only), 4.3 and 4.4 hold. If there exist a positive integer N and a constant $k \in (0, 1)$ such that*

$$\left\| \prod_{i=1}^{N} \left(I - \frac{F(x_{\alpha i}, t_{2i} - t_{2i,2i})c}{f(x_{\alpha i}, t_{2i} - t_{2i,2i})}\right) e^{At_{2i}} \left(I - \frac{F(x_{\beta i}, t_{2i-1} - t_{2i-1,2i-1})c}{f(x_{\beta i}, t_{2i-1} - t_{2i-1,2i-1})}\right) e^{At_{2i-1}} \right\| \leq k,$$

$$\forall\, x_{\beta i} \in \Omega_\beta,\quad x_{\alpha i} \in \Omega_\alpha, \tag{4.74}$$

$$\forall\, t_{2i-1},\, t_{2i-1,2i-1} \in [t_{\alpha min},\, t_{\alpha max}],\quad t_{2i},\, t_{2i,2i} \in [t_{\beta min},\, t_{\beta max}],$$

then there is a globally stable limit cycle for system Σ_τ.

Indeed, the given result can be slightly generalized. To do this, we first introduce the concept of contractive mapping.

Definition 4.6.2. *A continuous mapping $T : M \to M$ is said to be contractive if*

$$d(T(x), T(y)) < d(x, y),\quad \forall x, y \in M. \tag{4.75}$$

The following result is proved in Khamsi and Kirk (2001).

Lemma 4.6.2. *Let (M, d) be a complete metric space and $D \subset M$ be a compact subset. If $T : D \to D$ is a contractive mapping, then T has a unique fixed-point x_0, and for each $x \in D$, $\lim_{n \to \infty} T^n(x) = x_0$.*

Using Lemma 4.6.2, the following is a direct corollary which is a slight generalization of Corollary 4.6.2.

Corollary 4.6.3. *Suppose Assumptions 2.1, 2.2 (for $\tau > 0$ only), 4.3 and 4.4 hold. If there exists a positive integer N such that*

$$\left\| \prod_{i=1}^{N} \left(I - \frac{F(x_{\alpha i}, t_{2i} - t_{2i,2i})c}{f(x_{\alpha i}, t_{2i} - t_{2i,2i})}\right) e^{At_{2i}} \left(I - \frac{F(x_{\beta i}, t_{2i-1} - t_{2i-1,2i-1})c}{f(x_{\beta i}, t_{2i-1} - t_{2i-1,2i-1})}\right) e^{At_{2i-1}} \right\| < 1,$$

$$\forall\, x_{\beta i} \in \Omega_\beta,\quad x_{\alpha i} \in \Omega_\alpha, \tag{4.76}$$

$$\forall\, t_{2i-1},\, t_{2i-1,2i-1} \in [t_{\alpha min},\, t_{\alpha max}],\quad t_{2i},\, t_{2i,2i} \in [t_{\beta min},\, t_{\beta max}],$$

then there is a globally stable limit cycle for system Σ_τ.

82   4. Global Stability of Limit Cycles

Until now, the above analysis and results are with respect to the traversing points in Ω_α. Analogous results can easily be established if consideration is restricted to traversing points in Ω_β. We list one below which is analogous to Corollary 4.6.3.

Corollary 4.6.4. *Suppose Assumptions 2.1, 2.2 (for $\tau > 0$ only), 4.3 and 4.4 hold. If there exists a positive integer N such that*

$$\left\| \prod_{i=1}^{N} \left(I - \frac{F(x_{\beta i}, t_{2i} - t_{2i,2i})c}{f(x_{\beta i}, t_{2i} - t_{2i,2i})} \right) e^{At_{2i}} \left(I - \frac{F(x_{\alpha i}, t_{2i-1} - t_{2i-1,2i-1})c}{f(x_{\alpha i}, t_{2i-1} - t_{2i-1,2i-1})} \right) e^{At_{2i-1}} \right\| < 1,$$

$$\forall\, x_{\beta i} \in \Omega_\beta,\quad x_{\alpha i} \in \Omega_\alpha, \tag{4.77}$$

$$\forall\, t_{2i-1},\, t_{2i-1,2i-1} \in [t_{\beta min}, t_{\beta max}],\quad t_{2i},\, t_{2i,2i} \in [t_{\alpha min}, t_{\alpha max}],$$

then there is a globally stable limit cycle for system Σ_τ.

Remark 4.6.1. If we already know that there is a globally stable limit cycle for system Σ_τ, then the location of the limit cycle can be specified through the limit points of all traversing points of a trajectory in Ω_α and Ω_β. These limit points are exactly the traversing points of the limit cycle. The limit cycle period can then be computed using the information of these limit points.

Finally, let us re-examine Example 4.5.1 with $\tau > 0$. The validity of Assumptions 2.1 and 2.2 has been checked. Similar to the checking of Assumption 4.1, we have that $f(x_\alpha, t) = -\frac{1}{2}f(x_\beta, t) = e^{-t} - 1 \neq 0$ for all $t \neq 0$, i.e., Assumption 4.4 is satisfied. Since $\tau > 0$, at any intersecting point x_β satisfying $cx_\beta = \beta = 1$, we get $\dot{y}_\beta = c(Ax_\beta + bu_\alpha) = 2$. Similarly, at any intersecting point x_α satisfying $cx_\alpha = \alpha = 0$, we have $\dot{y}_\alpha = c(Ax_\alpha + bu_\beta) = -1$. This verifies Assumption 4.3. Next, we use Corollary 4.6.3 to check whether or not there is a globally stable limit cycle.

For
$$x_{\beta i} = [x_{1\beta i}\ x_{2\beta i}]^T \in \Omega_\beta,\quad x_{\alpha i} = [x_{1\alpha i}\ x_{2\alpha i}]^T \in \Omega_\alpha,$$

and
$$t_{2i-1},\, t_{2i-1,2i-1} \in [t_{\alpha min}, t_{\alpha max}],\quad t_{2i},\, t_{2i,2i} \in [t_{\beta min}, t_{\beta max}],$$

where $t_{\alpha min} = t_{\beta min} = \tau$, $t_{\alpha max} = \tau + 4\sqrt{2}$ and $t_{\beta max} = \tau + 10\sqrt{2}$ (already computed in Example 4.5.1), we compute that

$$\left(I - \frac{F(x_{\alpha i}, t_{2i} - t_{2i,2i})c}{f(x_{\alpha i}, t_{2i} - t_{2i,2i})} \right) e^{At_{2i}} \left(I - \frac{F(x_{\beta i}, t_{2i-1} - t_{2i-1,2i-1})c}{f(x_{\beta i}, t_{2i-1} - t_{2i-1,2i-1})} \right) e^{At_{2i-1}}$$

$$= e^{-(t_{2i-1} + t_{2i})} \begin{bmatrix} 0 & 0 \\ \frac{1}{2}(x_{2\beta i} - 3) & 1 \end{bmatrix}.$$

4.6 Existence of Globally Stable Limit Cycles

Since $x_{2\beta i}$ is bounded by $\omega_0 = 3\sqrt{2}$, we see that there exists a positive integer N such that the condition in Corollary 4.6.3 holds. This concludes that the system in Example 4.5.1 with any $\tau > 0$ has a globally stable limit cycle.

If we use Corollary 4.6.4 to check whether or not there is a globally stable limit cycle, we have

$$\left(I - \frac{F(x_{\beta i}, t_{2i} - t_{2i,2i})c}{f(x_{\beta i}, t_{2i} - t_{2i,2i})}\right) e^{At_{2i}} \left(I - \frac{F(x_{\alpha i}, t_{2i-1} - t_{2i-1,2i-1})c}{f(x_{\alpha i}, t_{2i-1} - t_{2i-1,2i-1})}\right) e^{At_{2i-1}}$$

$$= e^{-(t_{2i-1}+t_{2i})} \begin{bmatrix} 0 & 0 \\ -x_{2\alpha i} - 1 & 1 \end{bmatrix},$$

where $t_{2i-1}, t_{2i-1,2i-1} \in [t_{\beta min}, t_{\beta max}]$, $t_{2i}, t_{2i,2i} \in [t_{\alpha min}, t_{\alpha max}]$. This gives the same conclusion.

Part II

Process Identification from Relay Feedback Test

Introduction to Part II

Among the many applications of relay feedback is process identification, pioneered by Åström and Writtenmark (1984). Identifying an unknown system from some test on it has been an active area of research for several decades, involves many engineering and science disciplines, and is a broad topic itself. Very interesting features of the relay test are that it is a closed-loop test which keeps the process variable under control and is usually preferred to an open-loop test, and that a linear stable process with relay feedback is likely to automatically reach a sustained stationary oscillation, a limit cycle. Furthermore, the relay feedback technique does not require prior information about the system time constants to ensure a careful choice of the sampling period. The choice of the sampling period has always been a tricky problem for traditional parameter estimation techniques. If the sampling interval is too long, the dynamics of the process will not be adequately captured in the data and the accuracy of the model subsequently obtained will be poor as a consequence. While a conservative safety-first approach to this decision may be to select the smallest sampling period supported by the data acquisition equipment, this would result in too much data collection with inconsequential information. Corrective action requires data decimation in the post-treatment phase, which for real-time parameter estimation may not be tolerable. Spared from these cumbersome and difficult decisions, the relay feedback method is therefore an attractive method for process identification.

From the amplitude and period of the oscillatory waveforms, information about the process critical point can easily be acquired. The Ziegler–Nichols-like rules may be used to tune simple controllers such as PID. Relay-based auto-tuning methods have been integrated into commercial controllers and are successful in many process control applications (Hägglund and Åström, 1991; Åström et al., 1993; Åström and Hägglund, 1995). Years of industrial practice have shown that the main problems with current relay auto-tuning technology are: (i) Due to the adoption of a describing function approximation, estimation of the critical point is not accurate and could be fairly inaccurate under some circumstance (Slotine and Li, 1991; Huang and Chen, 1996). (ii) Only crude PID controller settings can be obtained with this single point identified and the resulting performance may deteriorate for, say, oscillatory processes or processes with a long dead time.

Several modified relay feedback identification methods have been reported (Li et al., 1991; Leva, 1993; Palmor et al., 1993; Palmor and Blau, 1994; Palmor et al., 1995a; Tan et al., 1999; Yu, 1999). To identify two or more points on the process frequency response, additional linear components (or varying hysteresis width) have to be introduced and additional relay tests have to be performed. These methods are time consuming and the resulting estimation is still approximate in nature since these methods make repeated use of the standard method. Consequently, it is important to develop techniques for identifying multiple and accurate points on the process frequency response from a single relay test, which is necessary for the enhancement of PID control performance and for auto-tuning advanced controllers.

In our work (Wang et al., 1997b; Wang et al., 1997c; Bi et al., 1997; Bi et al., 1999; Wang et al., 1999a; Wang et al., 1999b; Wang et al., 1999c; Bi et al., 2000; Wang and Zhang, 2000; Wang et al., 2000a; Wang et al., 2001b; Wang et al., 2001d; Wang and Zhang, 2001a; Wang and Zhang, 2001b) to be included in this part, we improve relay identification in two directions: one is to modify the standard relay so as to have better excitation of the process at a number of important frequencies; the other is to devise new algorithms which make better use of the available information contained in the relay response. In the first category, biased, parasite and cascade relays are designed. In the second category, in addition to the steady-state response, the relay transient response, which has not been used by others, is employed with appropriate application of the fast Fourier transform. Improved identification is thus achieved. Other issues such as closed-loop testing, multivariable process identification and conversion from frequency response to transfer function are also addressed.

5. Relay Feedback and its Variations

Suppose that a relay feedback system is stable and eventually yields limit cycles. The information contained in the limit cycles can be used for process frequency response estimation, as pioneered by Åström and co-workers. Fundamentals are first provided in Section 5.1, followed by their refinements in the subsequent sections, which expand applicability to more scenarios. Note that in this chapter, only stationery oscillations are used for process identification while the following chapter involves relay transient response. This chapter focuses on non-parametric models while Chapter 7 addresses conversion from frequency responses to transfer function models.

5.1 Fundamentals

Consider a single-input single-output process described by

$$\dot{x}(t) = Ax(t) + bu(t - L),$$
$$y(t) = cx(t), \tag{5.1}$$

where $x(t) \in \mathbb{R}^n$, $y(t) \in \mathbb{R}$ and $u(t - \tau) \in \mathbb{R}$ are the state, output and control input, respectively; A, b, c are constant real matrices or vectors with appropriate dimensions; $L \geq 0$ indicates the time delay. More often, we will use the transfer function representation of the process:

$$Y(s) = G(s)U(s), \tag{5.2}$$

where

$$G(s) = G_0(s)e^{-Ls},$$

with $G_0(s) = c(sI - A)^{-1}b$ being a strictly proper rational function. Let $r(t)$ be the reference or the set-point for the process output $y(t)$ to track. The error between them is

$$e(t) = r(t) - y(t). \tag{5.3}$$

Assume that the process is controlled by relay feedback:

$$u(t) = \begin{cases} \mu_+, & \text{if } e(t) > \varepsilon_+, \text{ or } e(t) \geq \varepsilon_- \text{ and } u(t_-) = \mu_+, \\ \mu_-, & \text{if } e(t) < \varepsilon_-, \text{ or } e(t) \leq \varepsilon_+ \text{ and } u(t_-) = \mu_-, \end{cases} \quad (5.4)$$

where $\varepsilon_+, \varepsilon_- \in \mathbb{R}$ with $\varepsilon_- \leq \varepsilon_+$ indicating hysteresis; $\mu_-, \mu_+ \in \mathbb{R}$ and $\mu_- \neq \mu_+$. For easy reference, the relay function in (5.4), which maps $e(t)$ to $u(t)$, is denoted by $\rho(\epsilon_+, \epsilon_-, \mu_+, \mu_-)$. Its functionality is shown in Figure 5.1. A relay is said to have hysteresis if $\varepsilon_+ \neq 0$ or $\varepsilon_- \neq 0$; and to be symmetric if $\varepsilon_+ = \varepsilon$, $\varepsilon_- = -\varepsilon$, and $\mu_+ = \mu$, $\mu_- = -\mu$, denoted by $\rho(\varepsilon, \mu)$; otherwise, it is called a biased relay. The standard relay corresponds to a symmetric relay with no hysteresis and is denoted by $\rho(\mu)$.

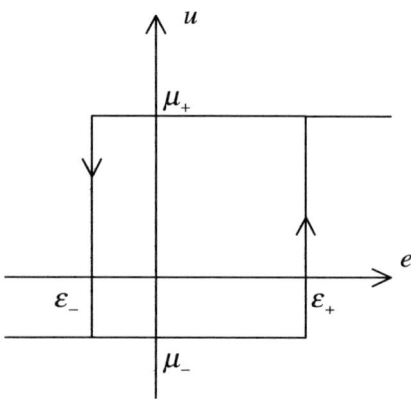

Fig. 5.1. General relay function $\rho(\varepsilon_+, \varepsilon_-, \mu_+, \mu_-)$

Due to time delay $L \geq 0$, we have to specify the initial function $u(\tilde{t})$ for $\tilde{t} \in [-L, 0]$. The most natural one, which is also used in practice, is

$$u(\tilde{t}) \equiv \begin{cases} \mu_+, & \text{if } e(0) > \varepsilon_+, \\ \mu_-, & \text{if } e(0) < \varepsilon_-, \\ u_0 \in \mathcal{U}, & \text{if } \varepsilon_- \leq e(0) \leq \varepsilon_+, \end{cases} \quad (5.5)$$

where

$$\mathcal{U} := \{\mu_-, \mu_+\}. \quad (5.6)$$

This completes the description of a linear process with relay feedback control.

We call (5.1)–(5.6) a relay feedback system (abbreviated as RFS), denote it by Σ_L, and depict it in Figure 5.2. Experience shows that a RFS is likely to have limit cycle oscillations as its steady state. Readers are referred to Part I of this book for a detailed analysis of the existence and stability of limit cycles in RFS and to the next section for the simple case of first-order systems to get a rough idea on them. In relay feedback experiments for analysis and identification, the set-point r is always kept unchanged, and we thus assume $r(t) = 0$ throughout this chapter. Assume that a limit cycle results from a RFS. Our task for this chapter is to extract the process dynamic information from such a limit cycle. We will start with the simplest method, the describing function approximation.

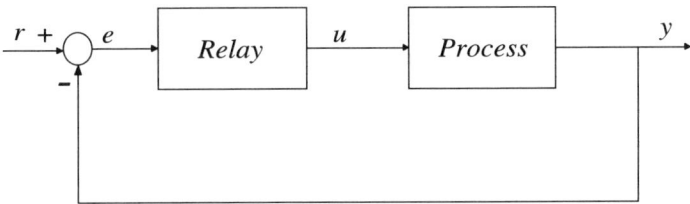

Fig. 5.2. Relay feedback system

Describing Function Method The describing function method approximates the relay with an "equivalent" linear time-invariant system. To this end, the input to the relay is assumed to be sinusoidal:

$$e(t) = a \sin \omega t,$$

and the resulting signals in the overall system are analyzed. Consider first the standard relay case. Then, the relay output $u(t)$ in response to $e(t)$ would be a square wave having a frequency ω and an amplitude equal to the relay output level μ. Using Fourier's series expansion, the periodic $u(t)$ can be written as

$$u(t) = \frac{4\mu}{\pi} \sum_{k=1}^{\infty} \frac{\sin(2k-1)\omega t}{2k-1}.$$

The describing function (DF) of the relay, $N(a)$, is simply the complex ratio of the fundamental component of $u(t)$ to the input sinusoid, i.e.

$$N(a) = \frac{4\mu}{\pi a}.$$

One sees that the DF analysis ignores harmonics beyond the fundamental component. The residual ϱ is the entire sinusoidally-forced relay output minus the fundamental component, i.e. the part of the output that is ignored in the DF development,

$$\varrho = \frac{4\mu}{\pi} \sum_{k=2}^{\infty} \frac{\sin(2k-1)\omega t}{2k-1}.$$

In the DF analysis of the relay feedback system, the relay is replaced with its quasi-linear equivalent DF, and a self-sustained oscillation of amplitude a and frequency ω_c is assumed. Then, for the process with the transfer function $G(s)$, it follows from Figure 5.2 that the variables in the loop satisfy the following relations,

$$E = -Y,$$
$$U = N(a)E,$$
$$Y = G(j\omega_c)U.$$

This implies

$$G(j\omega_c) = -\frac{1}{N(a)} = -\frac{\pi a}{4\mu}, \tag{5.7}$$

which gives an estimation of the process frequency response at one frequency, the RFS oscillation frequency.

The above DF analysis assumes that the Nyquist curve of $G(j\omega)$ intersects with the real axis at $-\frac{1}{N(a)}$ at ω_c in the complex plane. Recall that the intersection point of a process Nyquist curve with the real axis is called the critical point of the process and defines the critical or ultimate frequency, ω_u, of the process, for which

$$arg\{G(j\omega_u)\} = -\pi. \tag{5.8}$$

We can thus estimate the ultimate frequency and ultimate gain k_u by

$$\omega_u = \omega_c,$$

and

$$k_u := \frac{1}{|G(j\omega_u)|} = \frac{1}{|G(j\omega_c)|} = \frac{4\mu}{\pi a}.$$

For experiment design, the standard relay has only one parameter to tune, the relay output amplitude μ. Large μ will cause strong excitation of the process and thus better identification. On the other hand, a large signal will make the process output deviate further from its set-point, which is not desirable. The choice of μ is thus a trade-off between identification and control performance, as always in any identification problem, and much depends on measurement noise in the process output. If permissible, the relay output level should be adjusted such that the oscillation amplitude of the process output is about three times as large as its noise band. If the adjustment is not possible for a limited testing time, μ may be set to 3–10 % of the maximum range of the manipulated variable.

For preservation of regular relay switchings and estimating robustness against noise, it may be advantageous to replace a standard relay by a symmetric relay with suitable hysteresis so that the resultant system is less sensitive to measurement noise. The describing function of a (symmetric) hysteretic relay, $\rho(\epsilon, \mu)$, is

$$N(a) = \frac{4\mu}{\pi(\sqrt{a^2 - \epsilon^2} + j\epsilon)}.$$

The process frequency response, like the standard relay case, is estimated as the inverse negative describing function of the new relay

$$G(j\omega_c) = -\frac{1}{N(a)} = -\frac{\pi}{4\mu}\left(\sqrt{a^2 - \epsilon^2} + j\epsilon\right).$$

In this case, the oscillation corresponds to the point where the negative inverse describing function of the relay crosses the Nyquist curve of the process as shown in Figure 5.3.

Noise is always present in output measurements and a big concern for process identification since it uses noisy data from measurements. As mentioned, hysteresis in the relay is a simple way to reduce the influence of measurement noise. Before a relay test is performed, the noise bank in the process output measurements can be estimated using simple statistics. The hysteresis width, ϵ, should be greater than the noise band to avoid wrong switchings in the relay output and is usually chosen to be twice the noise band (Hang et al., 1993b) so that reliable stationery oscillations can be produced, maintained and observed easily. Filtering is another possible anti-noise measure (Åström and Wittenmark, 1984). Note that noise is usually of high frequency while most processes are of low-pass nature. A low-pass filter may be employed to pre-process noisy output and the pre-processed data are then used for model estimation. The filter bandwidth is usually set to be 3–5 times higher than the process critical frequency. Yet another anti-noise measure is to utilize multiple periods of

94   5. Relay Feedback and its Variations

limit cycles instead of a single period so as to filter out noise by the averaging method. A detailed and quantified analysis of this is given in Section 5.3, where the disturbance issue is also addressed.

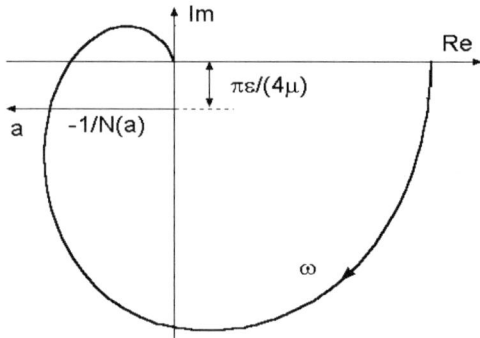

Fig. 5.3. Negative inverse describing function of hysteretic relay

Fourier Series Method   The oscillation waveform of the relay input signal $e(t)$ under a RFS is usually not precisely sinusoidal as assumed in the DF analysis above, which will thus cause estimation error. The error increases as the waveform differs from the sinusoidal function. For a linear process, this error can easily be removed by extracting the fundamental harmonics of both input and output of the linear process $G(s)$ using the Fourier series method. The resulting formula for estimating the process frequency response at the oscillation frequency is

$$G(j\omega_c) = \frac{Y(j\omega_c)}{U(j\omega_c)} = \frac{\int_{1period} y(t)e^{-j\omega_c t}dt}{\int_{1period} u(t)e^{-j\omega_c t}dt}, \qquad (5.9)$$

where $y(t)$ and $u(t)$ are one period of the process output and input stationery oscillations, respectively. This formula holds for a general relay and is precise if the system does not have any noise or disturbance.

Static Gain Estimation   If the relay used is symmetric, the resulting limit cycles will also be symmetric. No DC components are contained in such oscillations. This enables one-point estimation only, as shown above. To further estimate the DC gain or static gain of the process, we may introduce a bias, either in the relay input ($\epsilon_+ \neq \epsilon_-$), or relay output ($\mu_+ \neq \mu_-$), or both, so as to create an asymmetric relay and asymmetric limit cycles in the process output. If this is the case, the static gain can be obtained using the Fourier series expansion again as

$$G(0) = \frac{Y(0)}{U(0)} = \frac{\int_{1period} y(t)dt}{\int_{1period} u(t)dt}. \tag{5.10}$$

5.2 First-order Modelling

First-order plus dead time (FOPDT) transfer functions are often used in process modelling and control because of its simplicity although actual processes could be of high order. In general, relay feedback systems are a hard problem for theoretical analysis, see Part I of this book. For FOPDT processes, however, we are able to establish the complete results. They are presented in this section and give some idea of the existence of solutions, the existence of limit cycles, and the stability of limit cycles for relay feedback systems, without much mathematics. They also provide some feeling and insight into what will happen to relay feedback systems. The information on limit cycles is adequate to determine the FOPDT model and this is also covered in this section, after relay feedback theory.

Relay Feedback Theory Let the process be represented by the first-order plus dead time transfer function,

$$G(s) = \frac{K}{\tau s + 1} e^{-Ls}, \tag{5.11}$$

or in terms of a state space model,

$$\dot{x}(t) = ax(t) + bu(t - L),$$
$$y(t) = cx(t), \tag{5.12}$$

where for a non-integral process with $\tau \neq \infty$, we have

$$a = -\frac{1}{\tau}, \quad cb = \frac{K}{\tau}, \tag{5.13}$$

while for an integral process with $\tau = \infty$, (5.11) reduces to $G(s) = \frac{\kappa}{s} e^{-Ls}$ so that

$$a = 0, \quad cb = \kappa. \tag{5.14}$$

Suppose that the process is under general relay feedback control as described by (5.4). Since $r(t) = 0$, the relay is given by

$$u(t) = \begin{cases} \mu_-, & \text{if } y(t) > -\varepsilon_-, \text{ or } y(t) \geq -\varepsilon_+ \text{ and } u(t_-) = \mu_-, \\ \mu_+, & \text{if } y(t) < -\varepsilon_+, \text{ or } y(t) \leq -\varepsilon_- \text{ and } u(t_-) = \mu_+, \end{cases} \tag{5.15}$$

and the initial condition is

$$u(\tilde{t}) \equiv \begin{cases} \mu_-, & \text{if } y(0) > -\varepsilon_-, \\ \mu_+, & \text{if } y(0) < -\varepsilon_+, \\ u_0 \in \mathcal{U}, & \text{if } -\varepsilon_+ \leq y(0) \leq -\varepsilon_-. \end{cases} \quad (5.16)$$

The resulting relay feedback system is rather simple and enables us to easily analyze it completely. It turns out that the results depend on the nature of the parameter τ, whether it is positive, negative or zero, and have to be presented separately for these three different cases.

In what follows, define the switching planes (which are "switching lines" for the present case of the first-order system):

$$\mathcal{S}_+ := \{\xi \in R : c\xi = -\varepsilon_-\}, \quad (5.17)$$
$$\mathcal{S}_- := \{\xi \in R : c\xi = -\varepsilon_+\}. \quad (5.18)$$

Proposition 5.2.1. *Consider a RFS for the stable process in the form of (5.11) with $\tau > 0$ (i.e., $a < 0$ in (5.12)).*

(i) A unique solution exists for any given initial condition if and only if any of the following holds.

  (a) $L > 0$,
  (b) $L = 0$ and $-\varepsilon_+ > \max\{K\mu_-, K\mu_+\}$,
  (c) $L = 0$ and $-\varepsilon_- < \min\{K\mu_-, K\mu_+\}$,
  (d) $L = 0$ and $K\mu_+ \leq -\varepsilon_-$ and $K\mu_- \geq -\varepsilon_+$.

(ii) A limit cycle exists if and only if $L > 0$ and $K\mu_+ > -\varepsilon_- \geq -\varepsilon_+ > K\mu_-$. If this is the case, the limit cycle is unique with two switchings per period.

(iii) If a limit cycle exists, then the limit cycle is globally stable. Moreover, for a given process, the limit cycle is the common trajectory after the first switch, independent of the initial conditions.

Proof. We first show (i). For $L > 0$, it is easy to show that there exists a unique solution for any given initial condition. We now concentrate on the case for $L = 0$. Let the initial state x_0 satisfy $cx_0 \geq -\varepsilon_+$ and the relay start at μ_-. Then the trajectory of $x(t)$ will be governed by

$$x(t) = e^{at}(x_0 + ba^{-1}\mu_-) - ba^{-1}\mu_-. \quad (5.19)$$

It is easy to see that if $-\varepsilon_+ > K\mu_-$, then $x(t)$ will intersect $\mathcal{S}_-$ at some instant $t_1 > 0$. However, if $-\varepsilon_+ \leq K\mu_+$, after $t = t_1$, the trajectory $x(t)$ cannot evolve. Otherwise, for $t > 0$, we have

$$y(t_1 + t) = cx(t_1 + t) = \begin{cases} e^{at}(-\varepsilon_+ - K\mu_+) + K\mu_+ \geq -\varepsilon_+, & \text{for } u = \mu_+ \\ e^{at}(-\varepsilon_+ - K\mu_-) + K\mu_- < -\varepsilon_+, & \text{for } u = \mu_-, \end{cases}$$

which contradicts the control law (5.15). If $-\varepsilon_+ > K\mu_-$ and $-\varepsilon_+ > K\mu_+$ after the instant $t = t_1$, the trajectory will be governed by $x(t) = e^{at}(x_1 + ba^{-1}\mu_+) - ba^{-1}\mu_+$. Next, if $-\varepsilon_+ \leq K\mu_-$, we also check that if $-\varepsilon_- \geq K\mu_+$ holds, a unique solution exists for any initial condition. For $-\varepsilon_+ \leq K\mu_-$, if $-\varepsilon_- < K\mu_+$, then a similar analysis leads to a unique solution for any initial condition if $-\varepsilon_- < K\mu_-$ also holds. So far, (i) is proved.

Next we show (ii) and (iii). It is seen from the above that for $L = 0$, there is no limit cycle since the solution, if any, tends to $K\mu_-$ or $K\mu_+$. For $L > 0$, if and only if $K\mu_+ > -\varepsilon_- \geq -\varepsilon_+ > K\mu_-$, can the relay switch continuously. Moreover, any trajectory $x(t)$ will traverse $\mathcal{S}_-$ and $\mathcal{S}_+$ at fixed points $-\varepsilon_+/c$ and $-\varepsilon_-/c$, respectively. This proves (ii) and (iii).

Proposition 5.2.2. *Consider a RFS for the unstable process in the form* (5.11) *with* $\tau < 0$ *(i.e.,* $a > 0$ *in* (5.12) *).*

(i) A unique solution exists for any given initial condition if and only if any of the following holds.
  (a) $L > 0$,
  (b) $L = 0$ *and* $-\varepsilon_+ < \min\{K\mu_-, K\mu_+\}$,
  (c) $L = 0$ *and* $-\varepsilon_- > \max\{K\mu_-, K\mu_+\}$,
  (d) $L = 0$ *and* $K\mu_+ \geq -\varepsilon_- \geq -\varepsilon_+ \geq K\mu_-$.

(ii) A limit cycle exists if and only if $K\mu_+ < -\varepsilon_+ \leq -\varepsilon_- < K\mu_-$ *and*

$$0 < L < \min\left\{-\tau \ln\frac{K(\mu_+ - \mu_-)}{-\varepsilon_+ - K\mu_-}, \ -\tau \ln\frac{K(\mu_- - \mu_+)}{-\varepsilon_- - K\mu_+}\right\}.$$

If this is the case, the limit cycle is unique with two switchings per period.

(iii) If a limit cycle exists, then the limit cycle is locally stable, and the stability range is $K\mu_+ < cx(0) < K\mu_-$. *Moreover, for the given process, the limit cycle is the common trajectory after the first switch, independent of the initial conditions in the stability range.*

Proof. We first show (i). For $L > 0$, it is easy to show that there exists a unique solution for any given initial condition. We now concentrate on the case $L = 0$. Let the initial state x_0 satisfy $cx_0 \geq -\varepsilon_+$ and the relay start at μ_-. Then the trajectory of $x(t)$ will be governed by

$$x(t) = e^{at}(x_0 + ba^{-1}\mu_-) - ba^{-1}\mu_-. \tag{5.20}$$

Since $a > 0$, it is easy to see that if $-\varepsilon_+ < K\mu_-$, then for $cx_0 \geq K\mu_-$, the relay will remain μ_- for all $t \geq 0$; and for $cx_0 < K\mu_-$, $x(t)$ will intersect $\mathcal{S}_-$ at some instant $t_1 > 0$. However, if $-\varepsilon_+ \geq K\mu_+$, after $t = t_1$, the trajectory $x(t)$ cannot evolve. Otherwise, for $t > 0$, we have

$$y(t_1+t) = cx(t_1+t) = \begin{cases} e^{at}(-\varepsilon_+ - K\mu_+) + K\mu_+ \geq -\varepsilon_+, & \text{for } u = \mu_+ \\ e^{at}(-\varepsilon_+ - K\mu_-) + K\mu_- < -\varepsilon_+, & \text{for } u = \mu_-, \end{cases}$$

which contradicts the control law (5.15). If $-\varepsilon_+ < K\mu_-$ and $\varepsilon_- \leq K\mu_+$, after the instant $t = t_1$, the trajectory will be governed by $x(t) = e^{at}(x_1 + ba^{-1}\mu_+) - ba^{-1}\mu_+$. Next, if $-\varepsilon_+ \geq K\mu_-$, we check that if also $-\varepsilon_- \leq K\mu_+$ holds, a unique solution exists for any initial condition. For $-\varepsilon_+ \geq K\mu_-$, if $-\varepsilon_- > K\mu_+$, then a similar analysis leads to a unique solution for any initial condition if $-\varepsilon_- > K\mu_-$ also holds. So far, (i) is proved.

Next we show (ii) and (iii). It is seen from the above that for $L = 0$, there is no limit cycle since the solution, if any, tends to $+\infty$ or $-\infty$, or is equivalent to $K\mu_-/c$ or $K\mu_+/c$. We now concentrate on $L > 0$. Without loss of generality, assume that $cx_0 > -\varepsilon_-$. It is easy to see, as for the case $L = 0$, that if $cx_0 \geq K\mu_-$, then the trajectory $x(t)$ starting from the initial condition x_0 exists for all $t \geq 0$ while it does not make the relay switch (i.e., $u(t) \equiv \mu_-$). Let the initial state x_0 satisfy $K\mu_- > cx_0 > -\varepsilon_-$. Then the trajectory of $x(t)$ will be governed by

$$x(t) = e^{at}(x_0 + ba^{-1}\mu_-) - ba^{-1}\mu_-$$

until for some time $t_1 > 0$, it satisfies $cx(t_1) = -\varepsilon_+$. After $t = t_1$, due to the time delay $L > 0$, the trajectory will satisfy

$$x(t_1+t) = e^{at}(x(t_1) + ba^{-1}\mu_-) - ba^{-1}\mu_-, \quad 0 \leq t \leq L$$

before the switch occurs at $t = L$. It is easy to check that $cx(t_1+t) < -\varepsilon_+$ for all $t \in (0, L]$. After time $t_1 + L$, the trajectory of $x(t)$ will be governed by

$$x(t_1+L+t) = e^{at}(x(t_1+L) + ba^{-1}\mu_+) - ba^{-1}\mu_+. \tag{5.21}$$

Similarly, the switch will occur if and only if

$$cx(t_1+L) + cba^{-1}\mu_+ > 0. \tag{5.22}$$

With some simple manipulation, we see that (5.22) is equivalent to

$$0 < L < -\tau \ln \frac{K(\mu_+ - \mu_-)}{-\varepsilon_+ - K\mu_-}. \tag{5.23}$$

Under condition (5.23), for some time $t_2 > 0$, the trajectory in (5.21) satisfies $cx(t_1+L+t_2) = -\varepsilon_-$. After time $t_1 + L + t_2$, due to the time delay $L > 0$, the trajectory will satisfy

$$x(t_1+L+t_2+t) = e^{at}(x(t_1+L+t_2) + ba^{-1}\mu_+) - ba^{-1}\mu_+, \quad 0 \leq t \leq L$$

before the switch occurs at $t_1 + L + t_2 + L$. Again, we can check that the next switch will occur if and only if

$$cx(t_1 + L + t_2 + L) + cba^{-1}\mu_- < 0. \tag{5.24}$$

Also, with some simple manipulation, we see that (5.24) holds if and only if

$$0 < L < -\tau \ln \frac{K(\mu_- - \mu_+)}{-\varepsilon_- - K\mu_+}. \tag{5.25}$$

So, by combining (5.23) and (5.23), (ii) and (iii) are proved by noting that after time t_1, the trajectory $x(t)$ will be a limit cycle with two switchings per period. Moreover, any trajectory $x(t)$ starting from $K\mu_+ < cx(0) < K\mu_-$ will traverse $\mathcal{S}_-$ and $\mathcal{S}_+$ at fixed points $-\varepsilon_+/c$ and $-\varepsilon_-/c$, respectively.

Proposition 5.2.3. *Consider a RFS for the integral process in the form $G(s) = \frac{\kappa}{s} e^{-Ls}$ (i.e., (5.12) with (5.14)).*

(i) A unique solution exists for any given initial condition if and only if any of the following holds.

   (a) $L > 0$,
   (b) $L = 0$ and $0 > \max\{\kappa\mu_-, \kappa\mu_+\}$,
   (c) $L = 0$ and $0 < \min\{\kappa\mu_-, \kappa\mu_+\}$,
   (d) $L = 0$ and $\kappa\mu_- \geq 0 \geq \kappa\mu_+$.

(ii) A limit cycle exists if and only if $L > 0$ and $\kappa\mu_+ > 0 > \kappa\mu_-$. If this is the case, the limit cycle is unique with two switchings per period.

(iii) If a limit cycle exists, then the limit cycle is globally stable. Moreover, for the given process, the limit cycle is the common trajectory after the first switch, independent of the initial conditions.

Proof. Noting that in this case the trajectories of $x(t)$ and $y(t)$ will be governed by

$$x(t) = but + x_0,$$
$$y(t) = \kappa u t + cx_0,$$

the proof is similar to but simpler than those for the case $a \neq 0$, and thus is omitted here.

It can be checked that if there is a limit cycle for system (5.12), the relay switches at the instants when the trajectory $y(t) = cx(t)$ reaches the peak values; see Figure 5.4. Based on the information on this and other points, we can derive expressions for the limit cycle amplitudes and periods and use them to find the parameters in the FOPDT model of the process.

Parameter Estimation for Non-integral Processes Consider the case $a \neq 0$ (i.e., $\tau \neq \infty$). From (5.12) and Figure 5.4(a), we can see that

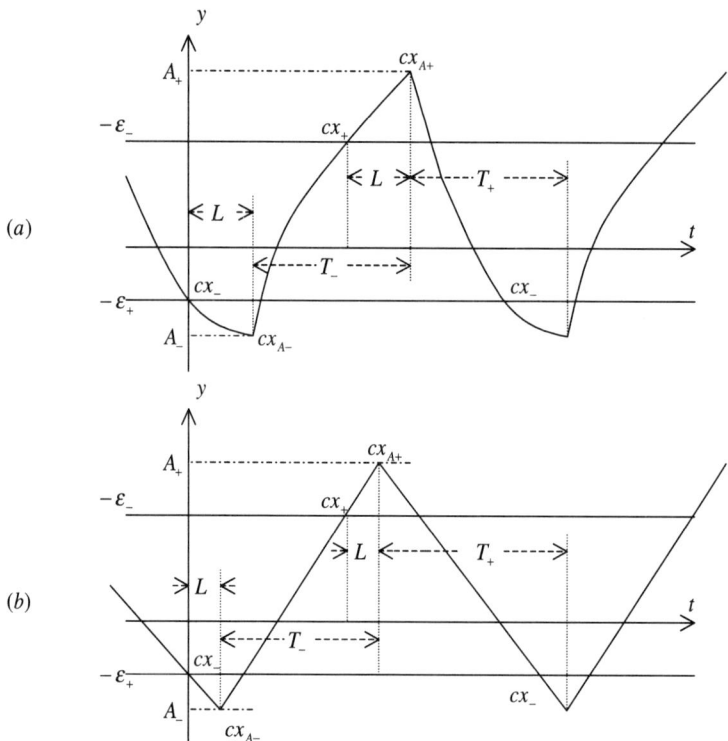

Fig. 5.4. Limit cycles for system (5.12) (a) $a \neq 0$; (b) $a = 0$

$$A_- = cx_{A_-},$$
$$x_{A_-} = e^{aL}(x_- + a^{-1}b\mu_-) - a^{-1}b\mu_-,$$
$$cx_- = -\varepsilon_+.$$

Taking into account (5.13), the above yields

$$A_- = e^{-L/\tau}(-\varepsilon_+ - K\mu_-) + K\mu_-. \tag{5.26}$$

Similarly, from

$$A_+ = cx_{A_+},$$
$$x_{A_+} = e^{aL}(x_+ + a^{-1}b\mu_+) - a^{-1}b\mu_+,$$
$$cx_+ = -\varepsilon_-,$$

we have

$$A_+ = e^{-L/\tau}(-\varepsilon_- - K\mu_+) + K\mu_+. \tag{5.27}$$

Let the time taken for the limit cycle to go from x_{A_-} (resp. x_{A_+}) to x_{A_+} (resp. x_{A_-}) be T_- (resp. T_+). Then, we get

$$x_{A_+} = e^{aT_-}(x_{A_-} + a^{-1}b\mu_+) - a^{-1}b\mu_+,$$

$$cx_{A_+} = A_+,$$

$$cx_{A_-} = A_-,$$

which gives, by taking into account (5.13),

$$T_- = -\tau \ln \frac{e^{-L/\tau}(-\varepsilon_- - K\mu_+)}{e^{-L/\tau}(-\varepsilon_+ - K\mu_-) + K(\mu_- - \mu_+)}. \tag{5.28}$$

Also, from

$$x_{A_-} = e^{aT_+}(x_{A_+} + a^{-1}b\mu_-) - a^{-1}b\mu_-,$$

$$cx_{A_+} = A_+,$$

$$cx_{A_-} = A_-,$$

we get

$$T_+ = -\tau \ln \frac{e^{-L/\tau}(-\varepsilon_+ - K\mu_-)}{e^{-L/\tau}(-\varepsilon_- - K\mu_+) + K(\mu_+ - \mu_-)}. \tag{5.29}$$

Luyben (1987) proposed the following method for first-, second- and third-order process modelling (called the ATV method): (i) The ultimate gain and ultimate frequency are obtained by using Åström's auto-tuning method. (ii) The dead time is read off from the initial response of the system to the auto-tuning test. (iii) The steady-state gain is obtained from a steady-state model of the process, or by using the step response method (Luyben, 1990). (iv) First-, second- and third-order transfer functions are fitted to the data at zero and the ultimate frequencies.

Table 5.1. Parameter estimation from biased relay

Case	Process			Biased relay				New method			ATV method		
	K	τ	L	T_+	T_-	A_+	A_-	K	τ	L	K	τ	L
1	1	2	2	2.79	3.91	0.859	-0.480	1.000	1.999	2.002	1	1.658	2
2	1	1	3	3.50	4.18	1.241	-0.670	1.000	0.999	3.006	1	1.042	3
3	1	5	2	3.44	5.46	0.497	-0.299	0.999	4.990	2.009	1	4.068	2
4	1	5	1	2.15	3.65	0.318	-0.209	1.001	5.003	1.004	1	4.055	1

In fact, the expressions (5.26)–(5.29) can be used to determine the three parameters in (5.11). However, they are coupled and nonlinear. Closed-form formulas for calculating the model parameters are not possible. Notice that for a biased relay, we can use (5.10) to find $G(0) = K$. Then, we obtain the normalized dead time, $\bar{L} = \frac{L}{\tau}$, from either (5.26) or (5.27), τ from (5.28) or (5.29), and finally obtain $L = \tau \bar{L}$. Simulation is carried out for processes with different normalized dead time to illustrate the accuracy of the above identification method. The outputs of the biased relay are set at 1.3 and -0.7 respectively, and the hysteresis of relay is made symmetrical and set at 0.1. The resultant limit cycles and model parameters are presented in Table 5.1. For comparison, the parameters obtained by the ATV (Luyben, 1987) are also given in Table 5.1, where it is assumed that the steady-state gain is known and the dead time is read exactly.

In practice, many high-order processes can be well approximated by first-order plus dead time models. The proposed method can do so effectively. The results for some typical processes are listed in Table 5.2. The Nyquist curves of the real processes and the models are shown in Figure 5.5, and they are very close to each other over the phase range 0 to $-\pi$.

Table 5.2. FOPDT models for the high-order processes

Case	Process	Model
1	$\frac{1}{(2s+1)^2}e^{-2s}$	$\frac{1.00}{4.072s+1}e^{-2.93s}$
2	$\frac{1}{(2s+1)^5}e^{-2s}$	$\frac{1.00}{6.809s+1}e^{-7.26s}$
3	$\frac{1}{(s+1)(s^2+s+1)}e^{-0.5s}$	$\frac{1.00}{1.152s+1}e^{-2.1s}$
4	$\frac{-s+1}{(s+1)^5}e^{-s}$	$\frac{1.00}{2.99s+1}e^{-4.24s}$

Integral Processes We now turn to the case $a = 0$ (i.e., $\tau = \infty$), which implies integral processes. We compute as follows. It is obvious from Figure 5.4 that

$$A_- = cb\mu_- L - \varepsilon_+, \tag{5.30}$$

$$A_+ = cb\mu_+ L - \varepsilon_-. \tag{5.31}$$

From

$$A_+ = cb\mu_+ T_- + A_-,$$

$$A_- = cb\mu_- T_+ + A_+$$

and taking into account (5.14), we have

$$T_- = \frac{\kappa L(\mu_+ - \mu_-) - \varepsilon_- + \varepsilon_+}{\kappa \mu_+}, \tag{5.32}$$

$$T_+ = \frac{\kappa L(\mu_- - \mu_+) - \varepsilon_+ + \varepsilon_-}{\kappa \mu_-}. \tag{5.33}$$

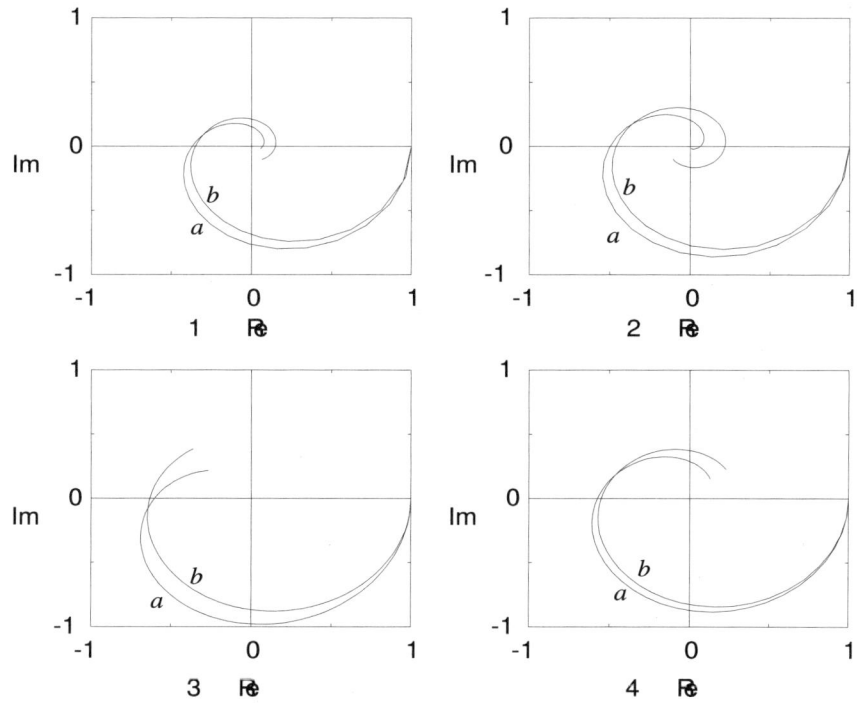

Fig. 5.5. Nyquist curves of processes and their FOPDT models
(a: real process; b: model)

5.3 Robustness Enhancement

It is well known that the difficulty in system identification is attributed to the existence of noise and disturbance. Noise is inevitable in practical situations and it contaminates the sampled data as picked up by the sensors. It is of so much a concern that once the samples are corrupted, there is basically no means by which they can be totally recovered. Distortions of the identification results are bound to arise when the samples used are subjected to random effects. Though it is true that recovery of the samples is not possible, many statistical methods, such as the stochastic least square algorithm employed in parameter estimation (Franklin *et al.*, 1994), have been devised to account for the effect of noise in the identification procedure. They essentially make use of the mean and variance characteristics of noise in their computations. In consideration of the growing popularity of the use of relays in performing system identification, effective ways to minimize the effect of noise will be of much importance. This is especially so in situations where the amount of noise has become so large that it can no longer be ignored.

Before we can work on the relevant design, we need first to study the nature and characteristics of noise. In most applications, noise is modelled as white noise with zero mean. White noise is the most random type of signal possible, so that any samples taken at different instants are totally uncorrelated. If the white noise concerned has zero mean, then it is likely that the noise can apparently be rejected by use of averaging, although it cannot be removed from the associated signals, and this is the main idea behind the method introduced in this section. To test the validity of the concept, the accuracy of the process gains at the critical point and the static frequency are evaluated by taking different numbers of limit cycles in the relay experiment. It is found that as samples are averaged over a larger number of limit cycles, the relative error in the process gains at the two frequencies drops, and the results conform to expectations. Apart from noise, disturbances are another common source of error in many identification problems. They can appear in many different forms depending on their sources. Some of the types of disturbances are load disturbances, measurement errors and parameter variations (Åström and Writtenmark, 1984). In this section, disturbances due to offsets in measurement and load disturbances, which are typically modelled as steps acting in the loop, are considered.

Suppose that a SISO linear plant is subjected to a relay test as shown in Figure 5.6, where white noise $n(t)$ of zero mean acts in the loop through the sensor at the output of the plant, and $w(t)$ is a constant disturbance. Consider first the disturbance-free case, i.e., $w(t) = 0$. With noise present in the system, the contaminated plant output $y(t)$ is measured instead of the actual value $\tilde{y}(t)$. If $y(t)$ and $u(t)$ are employed directly for identification, impaired results will be obtained. Since noise samples do not follow a traceable pattern and can assume any random value at different instants of time when they are picked up at different locations in the loop, they cannot be calculated or predicted from previous records and thus, the actual output is not recoverable from the corrupted samples.

Despite the fact that signal corruption by noise is an irreversible process, the behaviour of noise can still be described by statistical measurements. Therefore, estimation of the actual signal $\tilde{y}(t)$ from the infected $y(t)$ to achieve more accurate identification results is always possible by examining the statistical properties of noise. We start by noting that the noise involved has zero mean.

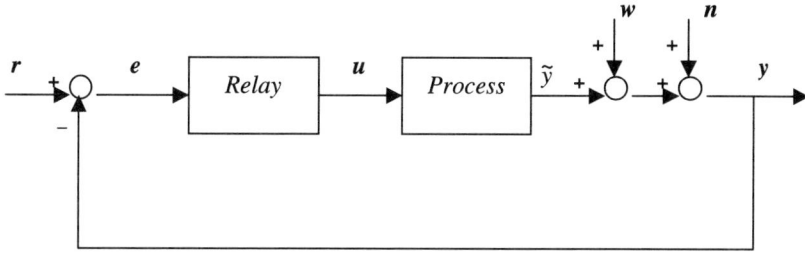

Fig. 5.6. Relay feedback system with noise and disturbance

This suggests that if a large number of samples extending over a sufficient number of periods (limit cycles) are collected and averaged out, it is possible to obtain a period of processed samples in which noise is significantly reduced. This is because noise attached to the different periods cancel each other and in the limit when the number of periods taken approaches infinity, the noise can be completely cancelled in the ideal situation. However, this is practically unattainable and it is usually sufficient to take up to a certain number of periods depending on the noise-to-signal ratio.

Let the relay feedback system give rise to limit cycle oscillations and the oscillation frequency be ω_c. Let $\hat{y}(t)$ be an estimate of one period of $\tilde{y}(t)$ from $y(t)$ using the averaging method. The associated frequency response $G(j\omega_c)$ can be obtained from $\hat{y}(t)$ and $u(t)$ using the equation

$$\hat{G}(j\omega_c) = \frac{\hat{Y}(j\omega_c)}{U(j\omega_c)} = \frac{\int_{1period} \hat{y}(t) e^{-j\omega_c t} dt}{\int_{1period} u(t) e^{-j\omega_c t} dt}. \tag{5.34}$$

Since the static gain of the process can be attained at the same time in the same relay test with no extra effort if biased relays are used in place of symmetrical relays, the second frequency point is conveniently chosen to correspond to the d.c. component of the process. The relevant frequency response $G(0)$ is obtained from the equation

$$\hat{G}(0) = \frac{\hat{Y}(0)}{U(0)} = \frac{\int_{1period} \hat{y}(t) dt}{\int_{1period} u(t) dt}. \tag{5.35}$$

It is, in essence, the ratio of the area of $\hat{y}(t)$ over one period to that of $u(t)$.

Two observations can be made in performing the relay experiment described above. First, if the noise power is high, the relay will be easily switched by noise on top of the switchings by the actual process output $\tilde{y}(t)$. To overcome this problem, a higher hysteresis level can be used for situations with higher noise power. A convenient indicator of the noise power P_n is its standard deviation σ_n and the two are related by $P_n = \sigma_n^2$. It is found from simulation studies that correct switchings can be achieved if the hysteresis level is set at twice the standard deviation of the noise together with an upper limit equal to 0.95 times the minimum of the on and off relay output level.

The second observation concerns the reliability of judging the period of the limit cycles from the separation between two switching points of the relay output at high noise power. To derive a more accurate value for the period, samples taken during the transient are excluded from the entire span of $u(t)$. The total number of switching points N is then counted. If N is even, the last point is retrenched so as to keep N odd. This is to make sure that a window size of an integral number of periods is used since an odd value of N implies an even number of separations and each separation denotes a half-period. The total time covering these switching points, T, is then recorded and the quotient $2T/N$ is taken as the final answer for the period of the limit cycle. Since this method uses the average of the periods of all limit cycles, it is intuitively more credible in the presence of noise and is justified to be so in the simulation results to be presented later.

The noise power and signal power are defined as

$$P_n = \frac{\int_{T_i}^{T_f} n^2(t)dt}{T_f},$$

$$P_y = \frac{\int_{T_i}^{T_f} y^2(t)dt}{T_f - T_i},$$

respectively, where T_i and T_f are the corresponding initial and final time instants between which the samples $n(t)$ and $y(t)$ are taken for integration in the respective equations. The noise-to-signal ratio (NSR) may be measured by

$$NSR = \frac{P_n}{P_y}.$$

The relative estimation errors at the two frequency points are defined as

$$\text{for the point } s = j\omega_c, \quad \text{relative error} = \left| \frac{\hat{G}(j\omega_c) - G(j\omega_c)}{G(j\omega_c)} \right|,$$

$$\text{for the point } s = 0, \quad \text{relative error} = \left| \frac{\hat{G}(0) - G(0)}{G(0)} \right|.$$

The mean relative error (MRE) represents the average value of the errors in $G(0)$ and $G(j\omega_c)$. By varying the noise power and hence the NSR, the number of limit cycles used to produce a pre-specified mean relative error is calculated. In such a way, the relationship between NSR and MRE is established. In practice, this can be used to decide how many limit cycle periods are required to reach the pre-specified estimation accuracy in terms of MRE, i.e. how long the testing should last.

Example 5.3.1. The proposed method is applied to a first-order plus delay process

$$G(s) = \frac{1}{s+1} e^{-3s}.$$

The noise power is gradually increased to raise the noise-to-signal ratio. The results of the computations are shown in Table 5.3. It can be concluded from the results that

- For the same NSR, as the number of limit cycles used increases, the relative errors decrease.
- As the NSR increases, the same relative errors can be achieved if more limit cycles are used.

It must be emphasized that the identification method used here serves only as a tool to show and verify the effectiveness of the proposed averaging method. The choice of method is entirely optional but the results found on the relationship between the number of limit cycles adopted, the relative error in the frequency response and the noise-to-signal ratio are applicable in general to all relay-based identification methods.

Consider next the situation where both noise and disturbance are present as shown in Figure 5.6. Apart from the white noise model $n(t)$ encountered earlier, also appearing in the sensor is a constant value disturbance represented by $w(t) = w$. Owing to the extra d.c. term introduced into the output signal by the disturbance and the fact that its value is uncertain and may change with time, the static component of the output due solely to the relay bias cannot be separated from $y(t)$ and hence the method described above for the estimation of $G(0)$ will fail. Nevertheless, the calculation of $G(j\omega_c)$ for the process is not affected since the disturbance contains no frequency content other than d.c. and therefore (5.34) remains applicable despite the unknown w. This can be verified mathematically by noting that

$$\frac{\int_{1period} y(t)e^{-j\omega_c t} dt}{\int_{1period} u(t)e^{-j\omega_c t} dt} = \frac{\int_{1period} y(t)e^{-j\omega_c t} dt - \int_{1period} w e^{-j\omega_c t} dt}{\int_{1period} u(t)e^{-j\omega_c t} dt}$$

108   5. Relay Feedback and its Variations

Table 5.3. Required limit cycles vs NSR and MRE without disturbance

		Mean Relative Error				
		10%	8%	6%	4%	2%
	0.0744	3	3	3	3	11
	0.1382	3	3	3	3	13
	0.1938	3	3	3	11	11
NSR	0.2510	3	3	6	10	22
	0.3025	4	4	4	6	16
	0.3502	5	5	12	80	80
	0.3968	10	22	22	74	80
	0.4520	33	33	39	80	80
		Minimum number of limit cycles required				

$$\frac{\int_{1period}(y(t)-w)e^{-j\omega_c t}dt}{\int_{1period}u(t)e^{-j\omega_c t}dt} = G(j\omega_c). \tag{5.36}$$

In the above equation, $y(t)$ can be replaced by its average over a number of periods, in much the same way as in the averaging technique illustrated earlier. Therefore, although a measurement of the magnitude of the disturbance is not available, $G(j\omega_c)$ can still be deduced by using the output samples $y(t)$.

Example 5.3.2. Simulations are performed using the same model with transfer function $G(s) = \frac{1}{s+1}e^{-3s}$ as in the previous example and details of the adjustment of the hysteresis level, the determination of the period of limit cycles, and the calculation of noise power and signal power follow precisely those stipulated there. The final results are given in Table 5.4. The same conclusions as in the previous example are obtained.

5.4 Parasitic Relay

To get more information on a process, identifying multiple points on the process frequency response from one relay test is more appealing than the use of several relay tests. A standard or symmetric relay can excite the process at the limit cycle oscillation frequency ω_c, as well as $3\omega_c, 5\omega_c, \ldots$, where ω_c is usually very close to the process critical frequency for which the process phase lag is π. Due

Table 5.4. Required limit cycles vs NSR and MRE with disturbance

		Mean Relative Error				
		10%	8%	6%	4%	2%
	0.0819	3	3	3	3	3
	0.1499	3	3	3	3	4
	0.2144	3	3	3	3	7
NSR	0.2665	3	3	3	3	9
	0.3165	3	3	6	9	16
	0.3649	8	8	8	18	18
	0.4076	18	21	27	27	42
	0.4489	28	28	28	28	28
		Minimum number of limit circle required				

to the low-pass nature of most practical processes, the signal-to-noise ratios at $3\omega_c$, $5\omega_c$, ..., are too low to enable meaningful estimation of the process frequency response at these points. Effectively, we can only get the critical point information from such a relay test. By adding a bias to the relay, we may obtain the process static gain as well. The frequency response information between zero and ω_c is most important for an understanding of the process dynamics and its use in controller design. To estimate more points around this region in one relay test, a modified relay is proposed in this section.

Our modified relay consists of a standard relay and a parasitic relay as shown in Figure 5.7. The standard relay operates as usual with amplitude of the sampled output $u_1(k)$ being μ_1, where $u_1(k)$ is the kth sample of $u_1(t)$. It is well known that this relay excites process mainly at frequency ω_c. In order to provide additionally effective excitation to the process at frequencies other than ω_c while maintaining the process output oscillation under such an arrangement, a parasitic relay with output amplitude $\alpha\mu_1$ and twice the period of $u_1(k)$ is introduced and superimposed on $u_1(k)$. This implies that the output $u_2(k)$ of the parasitic relay flip-flops immediately when every period of oscillations in $u_1(k)$ is reached. The parasitic relay is realized by

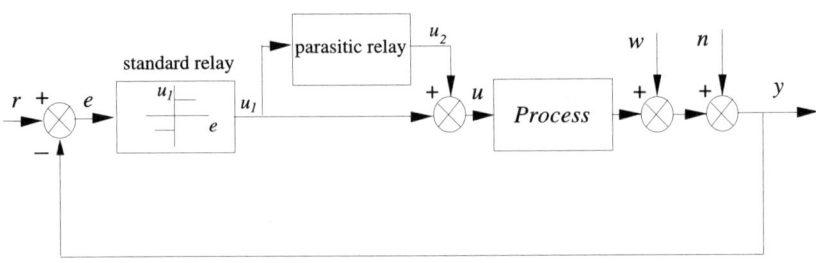

Fig. 5.7. Modified relay feedback system

$$\begin{cases} u_2(0) = \alpha\mu_1; \\ u_2(k) = -\alpha\mu_1 \cdot sign(u_2(k-1)), \text{ if } u_1(k-1) > 0 \text{ and } u_1(k) < 0; \\ u_2(k) = u_2(k-1), \text{ otherwise.} \end{cases}$$
(5.37)

The constant α should be large enough to sufficiently stimulate the process while it should also be small enough that the parasitic relay will not change the period of oscillation generated by the main relay by too much. According to extensive simulations, α is recommended to be $0.1 \sim 0.3$. The output of the modified relay test is thus given by

$$u(k) = u_1(k) + u_2(k),$$

and is sent to the process. In this way, the process is stimulated by two different excitations whose periods are T_c and $2T_c$. The resultant process output y from the modified relay test is shown in Figure 5.8 and reaches a stationary oscillation of period $2T_c$. Due to the two excitations in u, y consists of frequency components at $\frac{2\pi}{T_c}$, $\frac{\pi}{T_c}$ and their odd harmonics at $\frac{6\pi}{T_c}$, $\frac{10\pi}{T_c}$, ..., and $\frac{3\pi}{T_c}$, $\frac{5\pi}{T_c}$, ..., respectively. Let y_s and u_s be a period ($2T_c$) of the stationary oscillations of $u(k)$ and $y(k)$ respectively. For a linear process, the process frequency response can be obtained by

$$G(j\omega_i) = \frac{\int_0^{2T_c} y_s(t)e^{-j\omega_i t}dt}{\int_0^{2T_c} u_s(t)e^{-j\omega_i t}dt}, \quad i = 1, 2, \ldots, \qquad (5.38)$$

where

$$\omega_i = \frac{(2i-1)2\pi}{2^l T_c}, \quad l = 0, 1,$$

are the basic and odd harmonic frequencies in u_s and y_s. Equation (5.38) can be implemented using the FFT algorithm as

$$G(j\omega_i) = \frac{FFT(y_s)}{FFT(u_s)}. \tag{5.39}$$

Since the method adopts spectrum analysis instead of the describing function, it will lead to accurate process frequency response estimation. The proposed method employs the FFT only and the required computation burden is light. It can identify multiple points on the frequency response from a single relay test. Moreover, the method can easily be extended to find other points on the frequency response. You may flip-flop the parasitic relay every three or four periods of the main oscillations generated by the standard relay to get other frequency points. You may also use more than one parasitic relay in a relay test and find more points on the frequency response in one relay test. As discussed earlier, to estimate the static gain of a process, a bias has to be introduced to the relay input or output.

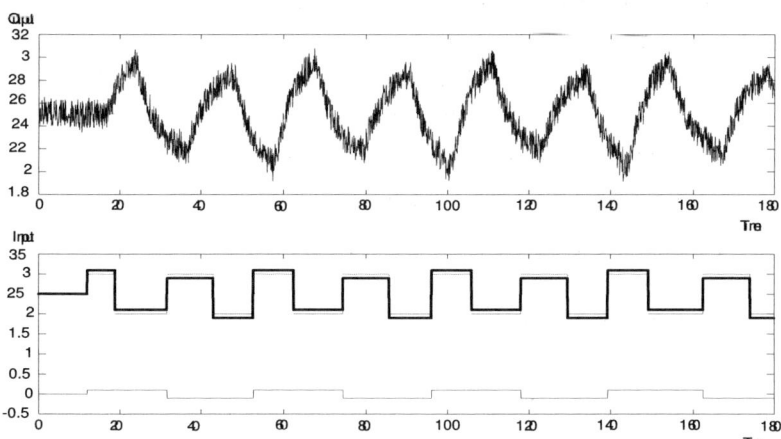

Fig. 5.8. Process input and output in the modified relay test
($\text{———}\ u,\ \text{······}\ u_1,\ \text{———}\ u_2$)

Reduction of Noise and Disturbance Effects on Estimation In a realistic environment, the major concerns for any identification method are disturbance and noise. As in Section 5.3, it can be shown that our new identification is also unaffected by a step-like load disturbance w as shown in Figure 5.7. As to measurement noise in the relay test, the same anti-noise measures as presented in Section 5.1 for simple relays, such as the introduction of hysteresis and a low-pass filter, can be used for the master relay in the new scheme. To reduce the noise effect further, especially in the case of large noise-to-signal ratio, we can average several periods of the stationary oscillations to enhance estimation robustness; see Section 5.3 for details. With these anti-noise measures, the proposed method can reject noise very effectively, and provide accurate frequency response estimation at frequencies $0.5\omega_c$, ω_c and $1.5\omega_c$. It should also be noted that a nonzero initial condition of the process at the start of a relay test has no effect on our estimation because only stationary oscillations u_s and y_s after the transient are used in the estimation, where u_s and y_s are independent of the initial condition.

For assessment of identification accuracy, the identification error is measured here by the worst-case error

$$ERR = \max_i \left\{ \left| \frac{\hat{G}(j\omega_i) - G(j\omega_i)}{G(j\omega_i)} \right| \times 100\%, \ i = 1, 2, 3 \right\}, \tag{5.40}$$

where $G(j\omega_i)$ and $\hat{G}(j\omega_i)$ are the actual and estimated process frequency responses respectively. The process frequency responses at $\frac{\pi}{T_c}$, $\frac{2\pi}{T_c}$ and $\frac{3\pi}{T_c}$ are considered since the frequency response in these region is especially important to controller design. To test estimation robustness against noise, the process output may be corrupted by some noise and the corrupted output used for identification. The noise level is judged, in the context of system identification, by the noise-to-signal ratio, which is usually defined as

$$N_1 = Noise\text{-}to\text{-}Signal\ Power\ Spectrum\ Ratio$$
$$= \frac{mean\ power\ spectrum\ density\ of\ noise}{mean\ power\ spectrum\ density\ of\ signal}, \tag{5.41}$$

or

$$N_2 = Noise\text{-}Signal\ Mean\ Ratio$$
$$= \frac{mean(abs(noise))}{mean(abs(signal))}. \tag{5.42}$$

In order to test our method in a realistic environment, real-time relay tests were performed using the *Dual Process Simulator KI 100* from KentRidge Instruments, Singapore. The simulator is an analogue process simulator and can be configured to simulate a wide range of industrial processes with different levels of noise and disturbance. The simulator is connected to a PC via an A/D and

D/A board. The window-based *DT VEE 3.0* from *DataTranslation* is used as the system control platform, on which the relay control code is written in C++. The fastest sampling time of the *VEE* system is 0.06 second. A few examples of real-time testing are presented below.

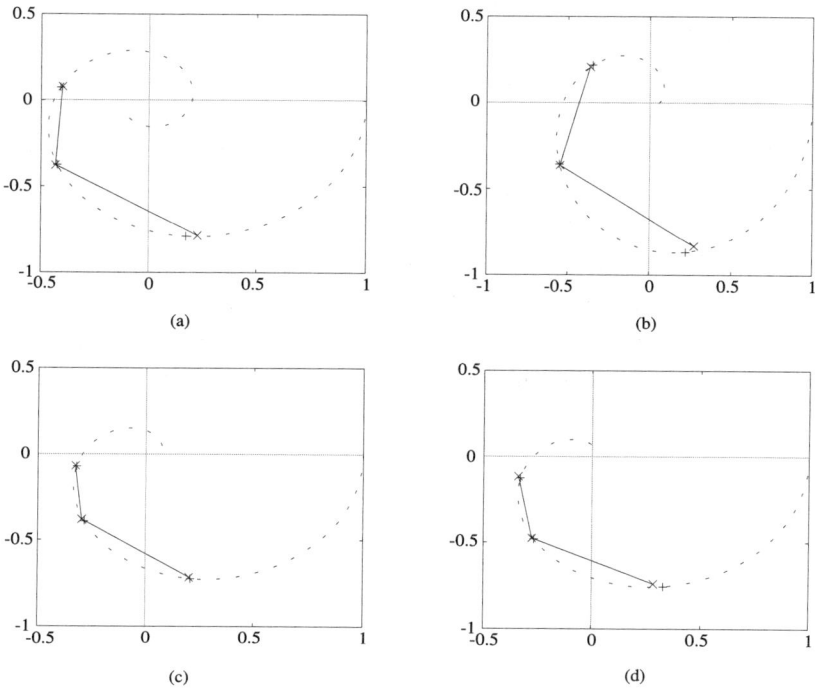

Fig. 5.9. Estimation of frequency response with $N_1 = 10\%$ noise
(+ actual, x estimated)

Example 5.4.1.   Consider a first-order plus dead time process:

$$G(s) = \frac{1}{5s+1}e^{-5s}.$$

In our relay test, the standard relay amplitude is chosen as 0.5 and the parasitic relay height is set to $20\% \times 0.5$. Without additional noise, the noise-to-signal ratio N_1 of the inherent noise in our test environment is 0.025% ($N_2 = 4\%$). The identification error ERR is 2.57%. To see noise effects, extra noise is introduced with the noise source in the Simulator. Time sequences of $y(t)$ and $u(t)$ in a relay test under $N_1 = 10\%$ ($N_2 = 31\%$) are shown in Figure 5.8, where $t = 0 \sim 12$ is the "listening period", in which the noise bands of $y(t)$ and $u(t)$ at steady state are measured. At this noise level, the hysteresis is chosen as 0.3. With averaging of four periods of stationary oscillations, the estimated frequency response points at this noise level are shown in Figure 5.9(a). The result is pretty good. At noise level $N_1 = 10\%(N_2 = 31\%)$, real-time testing of the proposed method was also performed on other typical processes whose transfer functions are listed in Table 5.5. Figure 5.9 compares frequency responses of the actual processes and their respective estimates. The identification results are shown to be satisfactory.

To ensure estimation accuracy at different noise levels, the number of stationary oscillation periods adopted in average calculation should be different. The estimation error ERR vs the number of stationary oscillation periods adopted in the averaging is plotted in Figure 5.10, which can be used as a guide in deciding how many periods are needed to achieve a given estimation accuracy at a given noise level. Table 5.5 shows the identification accuracy of four real-time examples at different noise and disturbance levels.

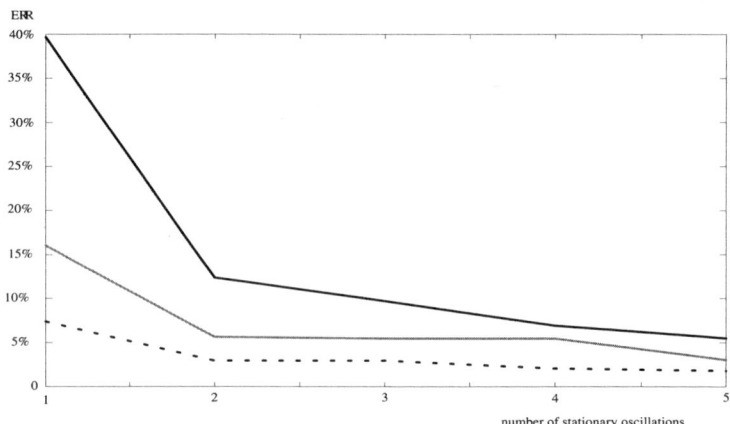

Fig. 5.10. ERR vs number of stationary oscillations adopted
(- - - $N_1 = 0\%$, ... $N_1 = 1\%$, —— $N_1 = 10\%$)

In this section, a new method for process frequency response identification has been developed in the context of the relay feedback test. The method has several unique features. Firstly, it can estimate multiple points on the process frequency response simultaneously with one single relay experiment and this reduces testing time significantly. Secondly, the method is accurate since no approximation is made. The computations involved are simple so that it can be easily implemented on microprocessors. Thirdly, the method is insensitive to noise and step-like load disturbances, and nonzero initial condition. Various processes have been employed to demonstrate the effectiveness of the method in real time. The identified process frequency response is useful for process analysis and controller design.

Table 5.5. Identification error (ERR)

Disturbance		d=0			d=0.5
Noise		N_1=0%	1%	10%	10%
		N_2=4%	11%	31%	31%
Processes		ERR			
(a)	$\frac{1}{5s+1}e^{-5s}$	2.57%	5.02%	6.83%	7.17%
(b)	$\frac{1}{(s+1)^8}$	2.93%	5.46%	6.90%	6.35%
(c)	$\frac{1}{(s+1)(5s+1)}e^{-2.5s}$	5.01%	5.08%	5.41%	5.16%
(d)	$\frac{1-s}{(2s+1)^2(5s+1)}e^{-0.5s}$	3.55%	5.17%	6.38%	5.13%

5.5 Cascade Relay

In the preceding section, one notes that the amplitude of the parasite relay cannot be chosen freely. It should be large enough to sufficiently stimulate the process while it should also be small enough that the parasite relay will not change the period of oscillation generated by the main relay by too much. Since the recommended value for it is small, the resultant estimation at $0.5\omega_c$ might be sensitive to measurement noise due to small signal-to-noise ratio there. In this section, cascade relay feedback is proposed as an alternative to parasitic feedback. The former can achieve almost the same objectives and results as the latter while the generation of limit cycles is less restrictive in the former than the latter.

The proposed cascade relay feedback consists of a master relay in the outer loop and a slave relay in the inner loop, as shown in Figure 5.11. The slave relay

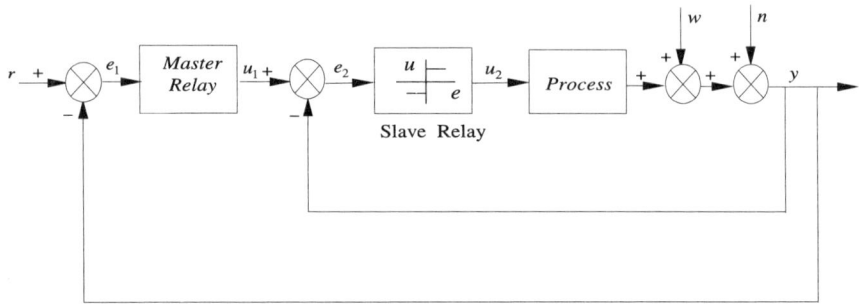

Fig. 5.11. Cascade relay feedback system

is just a standard relay with amplitudes of the sampled output $u_2(k)$ being d_2. With the inner loop closed, this relay can excite the process at the frequency ω_c sufficiently. In order to provide additional effective excitations to the process at other frequencies while maintaining the process output oscillation, the master relay in the outer loop is introduced with its output amplitude of d_1 and bias of μ_1, and operated at the frequency $0.5\omega_c$. It is realized by

$$u_1(k) = \begin{cases} -u_1(k-1) + 2\mu_1, & \text{if } e_1(k-1) < 0 \text{ and } e_1(k) > 0; \\ u_1(k-1) - 2d_1, & \text{if } e_1(k-1) > 0 \text{ and } e_1(k) < 0; \\ u_1(k-1), & \text{otherwise.} \end{cases} \quad (5.43)$$

The sampled output $u_1(k)$ from the master relay is a periodic stair wave with three amplitudes, $2d_1 + \mu_1$, μ_1 and $-2d_1 + \mu_1$, respectively. This relay is introduced to obtain persistent excitation at frequencies of $0.5\omega_c$ and $1.5\omega_c$, in addition to ω_c. In this way, the process is stimulated by two different excitations whose periods are T_c and $2T_c$. The waveforms for u_1, u_2, and the resultant output response are shown in Figure 5.12. The output reaches a stationary oscillation with period $2T_c$. The bias μ_1 is introduced to reduce possible unnecessary switchings due to noise and disturbances. One can see that the difference between the master relay output and the process output determines the switchings in the slave relay. Since the output of the master relay has three possible values, $2d_1 + \mu_1$, μ_1 and $-2d_1 + \mu_1$, load disturbance noise will not cause any relay switching unless its amplitude is larger than μ_1. Hence, a suitable μ_1 helps to establish robust oscillations in the process output at two fundamental frequencies.

Due to the two excitations in the input, y consists of frequency components at $\frac{2\pi}{T_c}$ and $\frac{\pi}{T_c}$ and their odd harmonics at $\frac{6\pi}{T_c}$, $\frac{10\pi}{T_c}$, ..., and $\frac{3\pi}{T_c}$, $\frac{5\pi}{T_c}$, ..., respectively. This enables process frequency response estimation at these points.

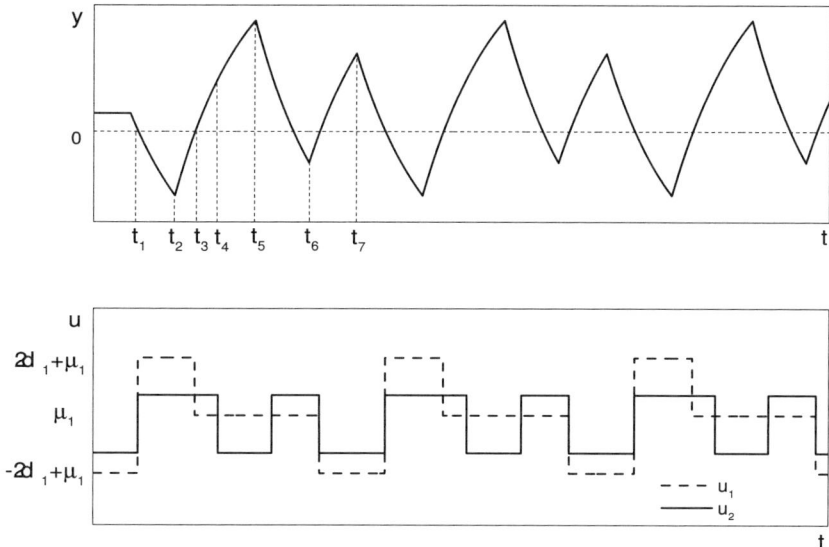

Fig. 5.12. Signals from cascade relay test

The estimation formula and its implementation are given by (5.38) and (5.39), respectively. Moreover, the cascade relay shares the same properties of estimation robustness against noise and constant disturbances as in the master and parasite relay case, and can use some anti-noise measures as discussed in Section 5.4.

It should be pointed out that in principle, the proposed method can be extended to find other points on the frequency response. One can realize the master relay and generate u_1 as a periodic stair wave which can enable y to have frequency components other than ω_c, $\omega_c/2$, and their odd harmonics. Another possible way is to use more than one cascade outer loop in a relay test and find more points on the frequency response in one relay test. Practically, the information about three points on the process frequency response available from the proposed cascade relay method is usually adequate to represent the process dynamics and to tune a good controller. Although more points can be identified from the extension as mentioned above, the structure will inevitably become more complicated, leading to implementation problems.

Guidelines for Relay Parameter Selection Most processes in industry are open-loop stable, and it is conjectured (Åström and Writtenmark, 1984) that most of them will exhibit a stable limit cycle with standard relay feedback. This is also true when the proposed cascade relay feedback is used. Extensive simulation shows that stationary oscillation is obtained for most processes if the parameters are chosen properly. We are thus motivated to consider the parameter selection problem for the generation of stationary oscillation in the cascade relay test.

To simplify the problem and gain insight into the solution, let us consider, as in Section 5.2, a first-order plus dead time (FOPDT) model,

$$G(s) = \frac{K}{\tau s + 1} e^{-Ls}. \tag{5.44}$$

Lemma 5.5.1. *For the cascade relay feedback system of Figure 5.11 with the process given by (5.44), if stationary oscillations at two different fundamental frequencies exist, then $d_2 K > 0$ and*

$$\mu_1 < \min\left\{d_2 K,\ d_2 K(e^{\frac{L}{\tau}} - 1)\right\}. \tag{5.45}$$

Proof. Refer to Figure 5.12 for the definition of t_i. Consider for illustration that the output initial condition is $y_0 > 0$ and that the slave relay switches to $u_2 = -d_2$ at $t = 0$. Obviously, the other case can be proven similarly. y will decrease while e_1 increases monotonically after a delay L. The output response for $t \geq L$ is described by

$$y(t) = -d_2 K(1 - e^{-(t-L)/\tau}) + y_0 e^{(t-L)/\tau}.$$

At $t = t_1$, y becomes 0, which causes u_1 to switch to $2d_1 + \mu_1$ and u_2 to d_2. Before $u_2 = d_2$ takes its effect on y, the output after t_1 can be described by

$$y(t) = -d_2 K(1 - e^{-(t-t_1)/\tau}), \tag{5.46}$$

where t_1 can be calculated from $y(t_1) = 0$, i.e.,

$$-d_2 K(1 - e^{-(t_1-L)/\tau}) + y_0 e^{(t_1-L)/\tau} = 0,$$

as

$$t_1 = L + \tau \ln(1 + \frac{y_0}{d_2 K}).$$

However, $y(t)$ will not respond to the positive switching $u_2 = d_2$ until it has continued its monotonic downward trend for a time L, as seen from (5.46). At $t_2 = t_1 + L$, $y(t)$ reaches its first peak in the cycle, and its value can be calculated from (5.46) as

$$y(t_2) = y(t_1 + L) = -d_2 K(1 - e^{-L/\tau}). \tag{5.47}$$

For $t > t_2$, the output response can be described by

$$y(t) = d_2 K(1 - e^{-(t-t_2)/\tau}) + y(t_2) e^{-(t-t_2)/\tau}. \tag{5.48}$$

The output increases monotonically. At time $t = t_3$, y becomes 0 and u_1 switches to μ_1. For $t_3 < t < t_4$, y can still be described by (5.48) and it keeps increasing with decreasing e_2. At $t = t_4$, e_2 becomes 0, which leads to the switching of u_2 to $-d_2$. t_4 can be calculated from the following equation:

$$y(t_4) = \mu_1,$$

or

$$d_2 K(1 - e^{-(t_4-t_2)/\tau}) + y(t_2) e^{-(t_4-t_2)/\tau} = \mu_1. \tag{5.49}$$

However, $y(t)$ will not respond to the negative switching of u_2 until it has continued for a time L. Therefore, at $t = t_5 = t_4 + L$, y reaches its second peak point as

$$\begin{aligned} y(t_5) &= y(t_4 + L) \\ &= d_2 K(1 - e^{-(t_4+L-t_2)/\tau}) + y(t_2) e^{-(t_4+L-t_2)/\tau} \\ &= d_2 K - (d_2 K - \mu_1) e^{-L/\tau}. \end{aligned} \tag{5.50}$$

The other two peaks in one cycle can be found similarly. Denote these four points as

$$\begin{aligned} y(t_2) &= -d_2 K(1 - e^{-L/\tau}), \\ y(t_5) &= d_2 K - (d_2 K - \mu_1) e^{-L/\tau}, \\ y(t_6) &= -d_2 K + (d_2 K + \mu_1) e^{-L/\tau}, \\ y(t_7) &= d_2 K(1 - e^{-L/\tau}). \end{aligned} \tag{5.51}$$

To enable the slave relay in the inner loop to switch between two output levels while the master relay in the outer loop switches between the three output levels with two different fundamental frequencies indefinitely, the following condition holds:

$$y(t_6) < 0,$$

or

$$\mu_1 < d_2 K(e^{L/\tau} - 1).$$

It also follows from (5.49) that

$$e^{-(t_4-t_1)/\tau} = \frac{d_2 K - \mu_1}{d_2 K(2 - e^{-L/\tau})} > 0,$$

i.e.,

$$\mu_1 < d_2 K.$$

This completes the proof of the lemma.

Several remarks are now made regarding this lemma.

- The condition therein provides a guideline for choosing the bias μ_1 to obtain limit cycles. Other parameters can be chosen similarly to the standard relay case. In practice, the relay amplitude is adjusted so that the oscillation at the process output is about three times the amplitude of the noise.
- Though Lemma 5.5.1 is derived for a hysteresis-free relay, it can easily be extended to relays with hysteresis.
- For a process with negative steady-state gain, u_1 and u_2 have to change their signs to attain stationary oscillations.
- The assumption $y_0 > 0$ is made only for simplicity of illustration, and can be removed without affecting the derivation. In fact, as can be seen from (5.51), y_0 has no influence on the four peaks.

Simulation A few simulation examples are presented below, and a comparison is made with the standard relay in Section 5.1 and parasite relay in Section 5.4. The relay parameters in these three cases are chosen such that the resultant output oscillations have almost the same amplitudes. Performance is measured by the worst-case error, ERR, as defined in (5.40), and the noise-to-signal ratio in the form of (5.41) and (5.42) is also adopted here.

Example 5.5.1. (**Simple Dynamics**) Consider a FOPDT process:

$$G(s) = \frac{1}{5s+1} e^{-5s}.$$

In our relay test, the slave relay amplitude d_1 is chosen as 1, the master relay height d_2 is set to 1, and its bias μ_1 is 0.5. The responses are shown in Figure 5.13, where u is the input to the process. For multiple-point estimation evaluation, the parasite relay test sets its standard relay amplitude to 0.5 and the parasitic relay height to 0.2 × 0.5. For the standard relay test, its height is set to 1 and only one point ω_c on the process frequency response is available and then used for calculating its *ERR*. In the noise-free case, the identification error *ERR* is 0.30% for the cascade relay test, 0.31% for the parasite relay test and 11.19% for the standard relay test, respectively. Afterwards, noise is introduced using the band-limited white noise module in Matlab. Under this noise condition, hysteresis is set to 0.3 for all three relays. To reduce the noise effect, especially in the case of large noise-to-signal ratio, we use the average of the last two–four periods of oscillation as the stationary oscillation period, depending on the noise level. Further, the accuracy of the relay test depends on the reliability of judging the period of the limit cycles. To derive a more accurate value of the period, the averaging technique in Section 5.3 is adopted. With these noise rejection techniques, for $N_1 = 10\%$, ERR is 1.87% for the cascade relay test, 6.83% for the parasite relay and 10.01% for the relay test, respectively. By averaging four periods of stationary oscillations, the estimated frequency responses for noise levels of $N_1 = 10\%$ and 20%, are shown in (a) of Figure 5.14 and Figure 5.15, respectively.

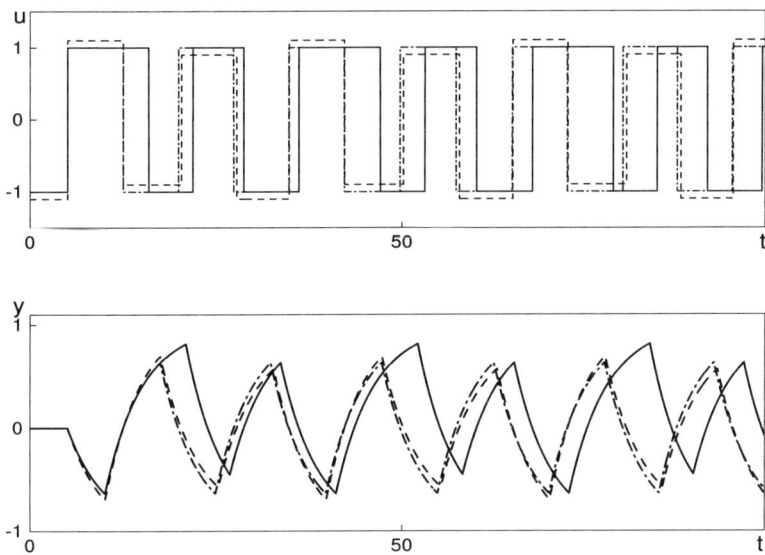

Fig. 5.13. Responses obtained during various relay tests
(- · · --: standard; - - - : parasite; ——— : cascade)

Example 5.5.2. (**Complex Dynamics**) Consider now three processes having different dynamics:

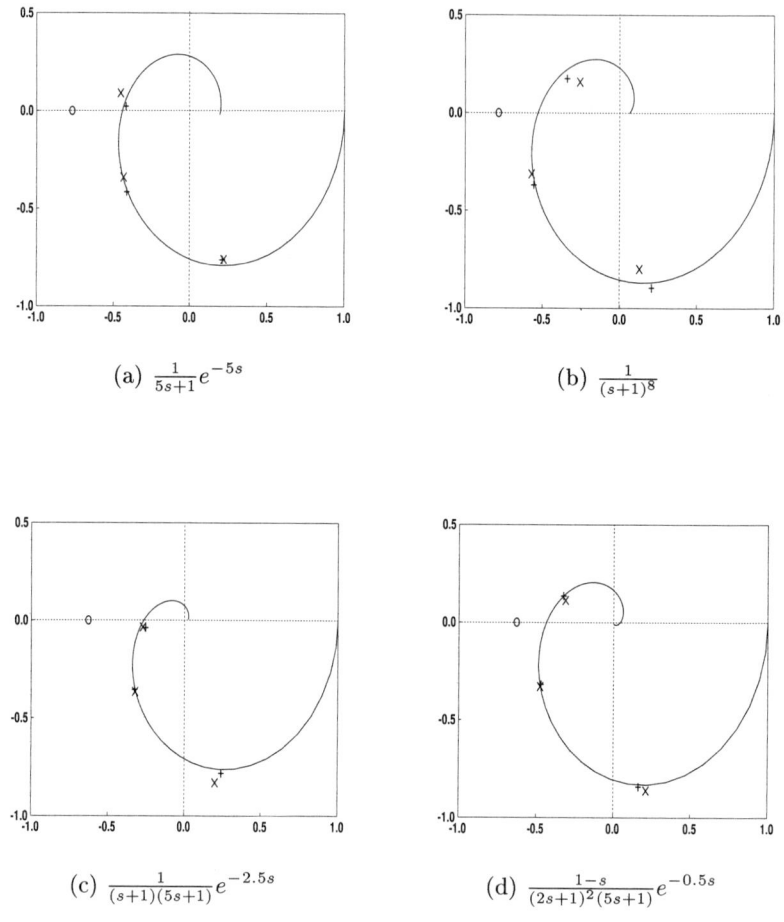

Fig. 5.14. Evaluation of $G(j\omega)$ for $N_1 = 10\%$
('+': cascade, 'x': parasite, 'o': standard)

$$G(s) = \frac{1}{(s+1)^8}; \qquad (5.52)$$

with a multi-lag high order,

$$G(s) = \frac{1}{(s+1)(5s+1)}e^{-2.5s};$$

with different poles, and

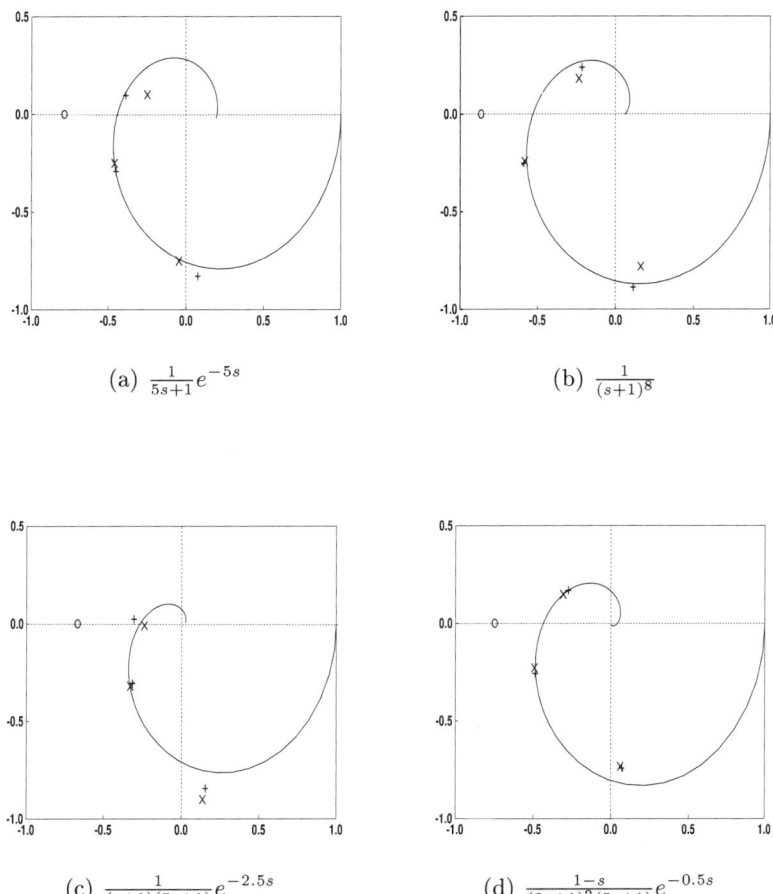

Fig. 5.15. Evaluation of $G(j\omega)$ for $N_1 = 20\%$
('+': cascade, 'x': parasite, 'o': standard)

$$G(s) = \frac{1-s}{(2s+1)^2(5s+1)}e^{-0.5s};$$

with non-minimum phase plus dead time. The actual and estimated frequency responses are shown in (b), (c) and (d) of Figure 5.14 and Figure 5.15, respectively. Table 5.6 shows the identification accuracy obtained under different relay tests with/without noise. The identification results are seen to be satisfactory for both parasite and cascade relays. For the noisy case, one can see from

Table 5.6 that significant improvement is achieved by the cascade relay over the parasite relay for most processes. The processes used here for simulation are exactly the same as those in Section 5.4 and thus the results achieved with the cascade relay are typical and representative for the method.

Table 5.6. Identification errors (ERR)

Noise levels	Different relays	Processes			
		$\frac{e^{-5s}}{5s+1}$	$\frac{1}{(s+1)^8}$	$\frac{e^{-2.5s}}{(s+1)(5s+1)}$	$\frac{(1-s)e^{-0.5s}}{(2s+1)^2(5s+1)}$
$N_1 = 0\%$	Cascade	0.30%	0.62%	0.41%	0.31%
($N_2 = 0\%$)	Parasite	0.31%	0.62%	0.42%	0.32%
w= 0	Standard	11.19%	0.53%	7.06%	3.71%
$N_1 = 10\%$	Cascade	1.87%	2.76%	5.38%	4.39%
($N_2 = 29\%$)	Parasite	6.83%	6.90%	5.41%	6.38%
w= 0	Standard	10.01%	3.70%	10.73%	9.37%
$N_1 = 20\%$	Cascade	5.80%	3.91%	12.61%	9.62%
($N_2 = 41\%$)	Parasite	14.52%	6.12%	14.20%	16.96%
w= 0	Standard	15.35%	7.41%	17.20%	25.31%
$N_1 = 10\%$	Cascade	2.01%	4.52%	4.93%	4.14%
($N_2 = 29\%$)	Parasite	7.17%	6.35%	5.16%	5.13%
w= 0.5	Standard	17.28%	10.08%	36.91%	15.31%

All real processes have some nonlinearity. If the nonlinearity is associated with operating point change (which is the usual case), then the proposed method may be applied to each operating point with a linearized model and gain scheduling can be used to handle this change. When the nonlinearity is modest, our method can be applied without any gain adaptation.

Example 5.5.3. (**Nonlinearity**) Introduce a nonlinearity into a linear model such that the process is described by

$$y = \frac{1}{(s+1)^8} v,$$

where $v = k(u)$ and

$$k(u) = \begin{cases} u, & \text{if } |u| > 0.2, \\ 0, & \text{if } |u| \leqslant 0.2, \end{cases}$$

and u is the process input. Its input and output responses $u(t)$ and $y(t)$ under the cascade relay test are processed as usual with the proposed method to give $G(j\omega)$. Since the frequency response of a nonlinear process is not defined, the effectiveness of the proposed identification method is judged from the control performance. For illustration, a multiple-point fitting method (Wang et al., 1998a) for PID tuning is designed with the resultant $\hat{G}(j\omega)$. The refined gain and phase method (Zhuang and Atherton, 1993), which uses only the critical point on the process frequency response available from the standard relay feedback test, is also applied for comparison. The resulting closed-loop response is shown in Figure 5.16, where the solid line is for the proposed method, and the dashed line is for the standard relay feedback test with Zhuang's tuning method. The effectiveness of the proposed method for nonlinear processes is verified.

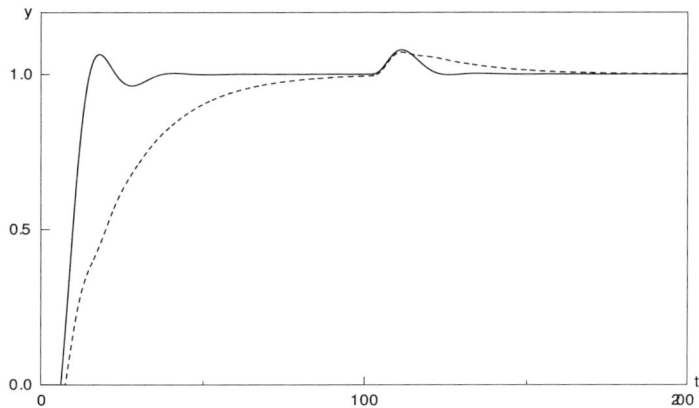

Fig. 5.16. Control performance for $\frac{1}{(s+1)^8}k(u)$

In this section, a new relay, the cascade relay, has been proposed for robust estimation of process frequency response. It shows some improvement over the master-and-parasite relay in terms of estimation results and the likehood of limit cycle generation.

5.6 Extension to MIMO Case

We have so far dealt with SISO processes only. We are looking for extension to the MIMO case. Obviously, it is too tedious and also unnecessary to consider all the types of relays covered so far. Instead, we will illustrate such an extension for the simple relays of Section 5.1 only, and extract the process frequency response matrix at the critical and zero frequencies. When the relay technique is extended to an MIMO system, there are three possible relay feedback schemes.

- *Independent Single Relay Feedback (IRF)*: Only one loop at a time is subjected to relay feedback while all others are kept open.
- *Sequential Relay Feedback (SRF)*: A loop is closed with a simple controller once a relay test has been made on that loop. This is repeated until all the loops have been tested.
- *Decentralized Relay Feedback (DRF)*: All loops are subjected to relay feedback simultaneously, as shown in Figure 5.17.

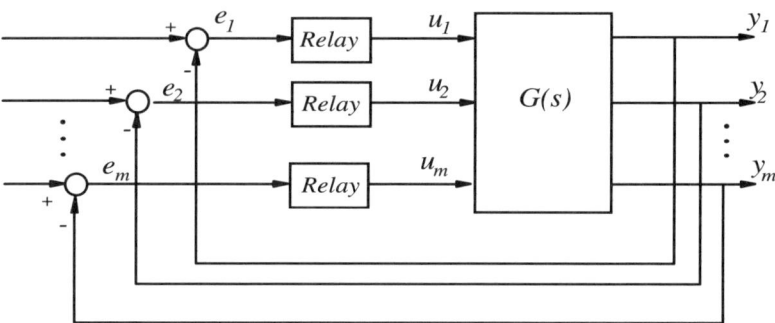

Fig. 5.17. Decentralized relay test

Among the three relay feedback schemes, decentralized relay feedback is the most desirable and will be used as our test for process frequency response matrix estimation in this section. Note that DRF is a *complete* closed-loop test, meaning that for an $m \times m$ plant at any instant during a test, all the m outputs are simultaneously under feedback control, while IRF and SRF are only partial closed-loop tests. For IRF, only one loop is closed, with $(m-1)$ open. For SRF, at the ith test, i loops are closed with $(m-i)$ loops open. Closed-loop testing is preferred to open-loop testing since a closed-loop test keeps outputs close to the set-points so that it causes less perturbation to the process and makes the linear model assumption (such as frequency response or transfer function) valid.

Analysis of Decentralized Relay Feedback If a $m \times m$ process is controlled by decentralized relay feedback, its outputs usually oscillate in the form of limit cycles after an initial transient. Each output has its own oscillation frequency, denoted ω_{ic}, $i = 1, 2, \ldots, m$, and they are, in general, different. For instance, a 2×2 process consisting of two independent (or very lightly coupled) loops has different output oscillation frequencies. However, it was found in Atherton (1975) that for typical coupled multivariable processes, m outputs normally have the same oscillation frequencies, that is, $\omega_{1c} = \omega_{2c} = \cdots = \omega_{mc}$, but different phases. For ease of reference later, we call this kind of multivariable oscillation *oscillations of a common frequency* and the frequency as a *process critical frequency*, and denote it by ω_c.

The describing function method is extended in Loh and Vasnani (1994) to analyze multivariable oscillations under decentralized relay feedback control. In this context, it is assumed that the m-input and m-output process has low-pass characteristics in each element of its transfer function matrix and one of its characteristic loci has at least $180°$ phase lag. Analysis of decentralized relay feedback based on the describing function provides a basic understanding of the behaviour of the resulting system and shows the effects of relay parameters on the behaviour so that insight and guidelines can be gained for the design of such relay tests. It is not intended to be comprehensive, but just to capture the major features of the system, as the analysis is approximate in nature. Therefore, for simplicity, suppose that each relay in the DRF is standard. Let the output amplitudes of standard relays be μ_i, and the inputs to the relays have amplitudes a_i. Then, the describing function matrix of such a decentralized relay controller is

$$N(a, \mu) = diag\left\{\frac{4\mu_i}{\pi a_i}\right\}.$$

Lemma 5.6.1. *(Loh and Vasnani, 1994). If the decentralized relay feedback system oscillates at a common frequency, then at least one of the characteristic loci of $N(a,\mu)G(j\omega)$ crosses the $(-1+j0)$ point on the complex plane, and the oscillation frequency corresponds to the frequency at which the crossing occurs. Further, if the process is stable, then the limit cycle oscillation is stable, the outermost characteristic locus of $N(a,\mu)G(j\omega)$ passes through the $(-1+j0)$*

point and the process critical frequency is the same as the critical frequency of the outermost characteristic locus.

It is noted that the crossing condition and the oscillation frequency in Lemma 5.6.1 are related to $N(a,\mu)$, which cannot be calculated until the oscillations are observed and the amplitudes a_i of relay inputs measured from the oscillation waveforms. It would be useful if the frequency could be given in terms of the information on the process only but independent of the relay controller. To this end, consider an $m \times m$ multivariable process $G(s)$ with row Gershgorin bands as shown in Figure 5.18. For each band, let $c_{i1} = g_{ii}(\omega_{i1})$ and $c_{i2} = g_{ii}(\omega_{i2})$ be the centres of the circles which are tangential to the negative real axis, and $(-\beta_{i1}+j0)$ and $(-\beta_{i2}+j0)$ be the points at which the outer-rim and inter-rim of the ith Gershgorin band intersect the negative real axis respectively. If the ith Gershgorin band does not intersect the negative real axis, $[\omega_{i1}, \omega_{i2}]$ is defined to be empty. The following result gives an estimate for ω_c in terms of ω_{i1} and ω_{i2}.

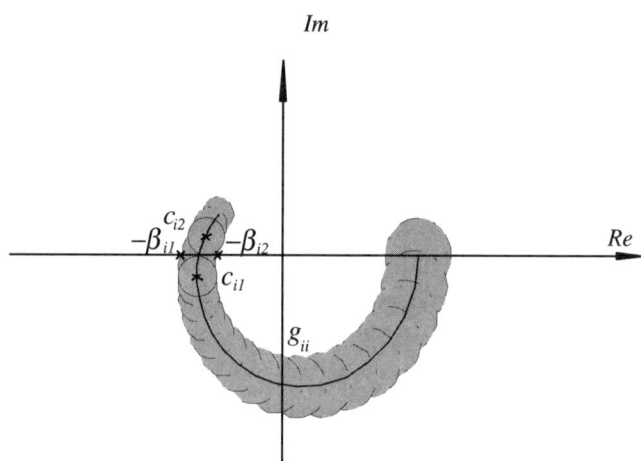

Fig. 5.18. Gershgorin bands

Proposition 5.6.1. *If the decentralized relay feedback system oscillates at a common frequency ω_c, then there exists a $k \in \{1, 2, \ldots, m\}$ such that $\omega_c \in [\omega_{k1}, \omega_{k2}]$.*

Proof. By the Gershgorin theorem (Maciejowski, 1989), we know that the characteristic loci of $G(s)$ lie in the union of its Gershgorin bands. The point at which the i-th characteristic locus $\lambda_i(j\omega)$ of G crosses the negative real axis if it exists can only lie in the union of circles with centres from c_{i1} to c_{i2}. It follows that for $\lambda_i(j\omega)$, the critical frequency ω_{ic} at which the crossing occurs is in the range $[\omega_{i1}, \omega_{i2}]$. Suppose now that the transfer function matrix $G(s)$ is multiplied by a diagonal constant matrix $K = diag\{k_i\}$ as

$$Q = KG = \begin{bmatrix} k_1 g_1 \\ \vdots \\ k_i g_i \\ \vdots \\ k_m g_m \end{bmatrix}$$

where g_i, $i = 1, \ldots, m$ are the row vectors of $G(s)$. The centres of the circles for the ith Gershgorin band of Q have now been shifted to $k_i g_{ii}$ with the radii of the circles magnified k_i times as shown in Figure 5.19. Since k_i is constant, the centre $k_i g_{ii}(\omega_{i1})$ has the same phase as that of $c_{i1} = g_{ii}(\omega_{i1})$ and is on the straight line drawn through the origin and $g_{ii}(\omega_{i1})$. Further, the magnitude $|k_i g_{ii}(\omega_{i1})|$ differs from $|g_{ii}(\omega_{i1})|$ by a factor $|k_i|$. Therefore, the distance between the point $k_i g_{ii}(\omega_{i1})$ and the negative real axis is $|k_i|$ times as large as that between the point $g_{ii}(\omega_{i1})$ and the axis, which is exactly the radius of the circle with centre $k_i g_{ii}(\omega_{i1})$. This implies that this circle is still tangential to the negative real axis and thus $\tilde{\omega}_{i1}$ for Q is equal to ω_{i1} for G. The same can be said for c_{i2} and $\tilde{\omega}_{i2} = \omega_{i2}$. It follows that the critical frequency $\tilde{\omega}_{ic}$ for the ith characteristic locus of $Q(s)$ is still in $[\omega_{i1}, \omega_{i2}]$. Since the describing matrix $N(a, \mu)$ is also a constant diagonal matrix, the critical frequency for the ith characteristic locus of $N(a,\mu)G(s)$ is thus in $[\omega_{i1}, \omega_{i2}]$. By Lemma 5.6.1, the limit cycle oscillation frequency must be in one of $[\omega_{i1}, \omega_{i2}]$, $i = 1, 2, ..., m$ and our result follows.

In view of Lemma 5.6.1 and Proposition 5.6.1, the oscillation frequency ω_c for a stable process depends on which characteristic locus of $G(s)$ is moved to the outermost by the multiplication of the corresponding relay element describing function $N_i = \frac{4\mu_i}{\pi a_i}$. In general, one can enlarge the gain N_i by increasing the ratios of the relay amplitudes in the ith loop to those in other loops. We call this outermost loop the *dominant* loop. It is noted that the dominant loop remains dominant and the critical frequency varies very little with a fairly large

130   5. Relay Feedback and its Variations

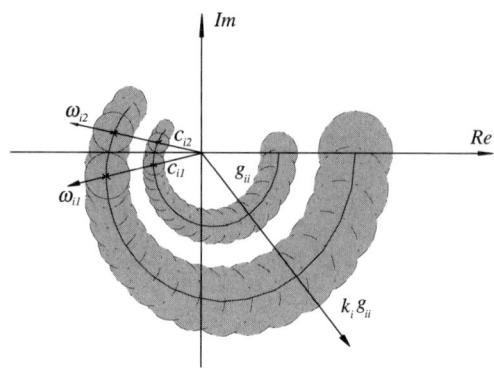

Fig. 5.19. Expansion of a Gershgorin band

change of relay amplitude ratios unless an inner characteristic locus becomes a new outermost. As an example, consider the following typical process (Wood and Berry, 1973):

$$G(s) = \begin{bmatrix} \frac{12.8e^{-s}}{1+16.7s} & \frac{-18.9e^{-3s}}{1+21s} \\ \\ \frac{6.6e^{-7s}}{1+10.9s} & \frac{-19.4e^{-3s}}{1+14.4s} \end{bmatrix}.$$

Let μ_1 and μ_2 be the relay amplitudes in loop 1 and loop 2 respectively. When $r := \frac{\mu_1}{\mu_2}$ varies from 1 to 2 by 100%, the process always exhibits oscillations of a common frequency, and the process critical frequency ω_c changes from 0.494 to 0.496, i.e., by 0.4%. This feature is addressed in the following proposition.

Proposition 5.6.2. *If the decentralized relay feedback system for a stable process oscillates at a common frequency and for some k, $N_k > \frac{N_i \beta_{i1}}{\beta_{k2}}$, $i = 1, 2, \ldots, m$, $i \neq k$, then only the kth characteristic locus of $N(a, \mu)G(j\omega)$ crosses the $(-1 + j0)$ point and the oscillation frequency satisfies $\omega_c \in [\omega_{k1}, \omega_{k2}]$.*

Proof. The conditions, $N_k > \frac{N_i \beta_{i1}}{\beta_{k2}}$, $i = 1, 2, \ldots, m$, $i \neq k$, guarantee that the kth Gershgorin band of $N(a, \mu)G(j\omega)$ is the outermost among all the m Gershgorin bands. Since the kth characteristic locus of $N(a, \mu)G(j\omega)$ is in this band, it is the outermost locus of $N(a, \mu)G(j\omega)$. It follows from Lemma 5.6.1 that the kth characteristic locus of $N(a, \mu)G(j\omega)$ crosses the $(-1 + j0)$ point and the oscillation frequency ω_c is equal to ω_{kc}, which is in $[\omega_{k1}, \omega_{k2}]$.

By Proposition 5.6.2, if we vary the relay amplitudes such that the resulting describing function gain matrix $N'(a,\mu)$ still satisfies $N'_k > \frac{N'_i \beta_{i1}}{\beta_{k2}}$, $i = 1, 2, \ldots, m$, $i \neq k$, then the resulting limit cycle oscillation frequency is expected to be in the range $[\omega_{k1}, \omega_{k2}]$ and thus close to the previous value if the interval $[\omega_{k1}, \omega_{k2}]$ is small. In general, the condition $N_k > \frac{N_i \beta_{i1}}{\beta_{k2}}$, $i = 1, 2, \ldots, m$, $i \neq k$, remains true if one increases the relay amplitude of the dominant loop or decreases one or more relay amplitudes in the other loops.

Estimation of Process Frequency Response An $m \times m$ multivariable process can be described in the frequency domain as

$$\begin{bmatrix} y_1(j\omega) \\ \vdots \\ y_m(j\omega) \end{bmatrix} = \begin{bmatrix} g_{11}(j\omega) & \cdots & g_{1m}(j\omega) \\ \vdots & \ddots & \vdots \\ g_{m1}(j\omega) & \cdots & g_{mm}(j\omega) \end{bmatrix} \begin{bmatrix} u_1(j\omega) \\ \vdots \\ u_m(j\omega) \end{bmatrix}. \tag{5.53}$$

We want to estimate the process frequency response $G(j\omega)$ at the critical oscillation frequency ω_c. In order to additionally identify the steady-state gain matrix of the process, a biased relay instead of a standard relay should be used in the dominant loop to make the process inputs and outputs have nonzero means. Thus, a test as shown in Figure 5.17 with a biased relay in the dominant loop and symmetric relays in the other loops is applied to the process. When the process becomes stationary, the process stationary inputs $u_i(t)$ and outputs $y_i(t)$, $i = 1, 2, \ldots, m$, are all periodic, and can be expanded into Fourier series. If the oscillations in m loops have a common frequency ω_c, then the direct-current components and the first harmonics of these periodic waves are extracted as

$$U^1(0) := \begin{bmatrix} \int_0^{T_c} u_1(t) dt \\ \vdots \\ \int_0^{T_c} u_m(t) dt \end{bmatrix}, \quad Y^1(0) := \begin{bmatrix} \int_0^{T_c} y_1(t) dt \\ \vdots \\ \int_0^{T_c} y_m(t) dt \end{bmatrix} \tag{5.54}$$

and

$$U^1(j\omega_c) := \begin{bmatrix} \int_0^{T_c} u_1(t) e^{-j\omega_c t} dt \\ \vdots \\ \int_0^{T_c} u_m(t) e^{-j\omega_c t} dt \end{bmatrix}, \quad Y^1(j\omega_c) := \begin{bmatrix} \int_0^{T_c} y_1(t) e^{-j\omega_c t} dt \\ \vdots \\ \int_0^{T_c} y_m(t) e^{-j\omega_c t} dt \end{bmatrix}. \tag{5.55}$$

Then

$$Y^1(0) = G(0) U^1(0), \tag{5.56}$$

and

$$Y^1(j\omega_c) = G(j\omega_c)U^1(j\omega_c). \tag{5.57}$$

Since (5.56) and (5.57) are vector equations, they are not sufficient to determine $G(j\omega_c)$ and $G(0)$ from Y^1 and U^1 only. Next, we slightly increase the relay amplitude of the dominant loop or decrease that of another loop, and repeat the above procedure until m tests have been completed. According to Proposition 5.6.2, the process is likely to have all the oscillation frequencies close to each other for the m tests. $Y^2(0), U^2(0), Y^2(j\omega_c), U^2(j\omega_c), \ldots,$ $Y^m(0), U^m(0), Y^m(j\omega_c), U^m(j\omega_c)$ are obtained subsequently. We have

$$[Y^1(0) \quad \ldots \quad Y^m(0)] = G(0)[U^1(0) \quad \ldots \quad U^m(0)], \tag{5.58}$$

and

$$[Y^1(j\omega_c) \quad \ldots \quad Y^m(j\omega_c)] = G(j\omega_c)[U^1(j\omega_c) \quad \ldots \quad U^m(j\omega_c)]. \tag{5.59}$$

While (5.58) is accurate for any decentralized relay test, (5.59) is only approximate since ω_c is not exactly the same for all m tests. U^i, $i = 1, 2, \ldots, m$, are linearly independent since there is always a relay amplitude change for each test. It follows from (5.58) and (5.59) that the steady-state gain matrix $G(0)$ and frequency response matrix $G(j\omega_c)$ are determined, respectively, as

$$G(0) = [Y^1(0) \quad \ldots \quad Y^m(0)][U^1(0) \quad \ldots \quad U^m(0)]^{-1}, \tag{5.60}$$

and

$$G(j\omega_c) = [Y^1(j\omega_c) \quad \ldots \quad Y^m(j\omega_c)][U^1(j\omega_c) \quad \ldots \quad U^m(j\omega_c)]^{-1}. \tag{5.61}$$

Our relay experiment thus consists of m decentralized relay tests and continues from one to another without any stop in between. To design this experiment, one needs to specify relay amplitudes for each test. The following design parameters are recommended for use and are obtained through our extensive case studies. For the first test, the relay amplitude for each loop is set as in the single-variable case (see Section 1). In most circumstances, stationary oscillations of a common frequency will result in the system. For subsequent tests, either the relay amplitude in the dominant loop is increased or the relay amplitude in one of the other loops is decreased by 5–20%. This usually leads to oscillations with frequencies close to the previous ones.

It should be pointed out that m decentralized relays in our test scheme are reasonable and even necessary to identify an $m \times m$ system. Out test scheme may actually need *less* time than those for IRF and SRF. To see this, our scheme uses m *non-stop* relays, while both IRF and SRF also contain m relays for a $m \times m$ system. Furthermore, between their m relays, there are additional $(m-1)$ control transients to bring outputs back to the set-points before the next relays can be performed. In the context of resonance approximations, the number of relays should be at least m in order to identify an $m \times m$ frequency response matrix $G(j\omega)$, as explained above. In our opinion, the main shortcoming of the decentralized relay test is that it may cause complicated multivariable oscillations (Atherton, 1975; Loh et al., 1993; Zhuang and Atherton, 1993), where three modes of multivariable oscillations have been observed. If there are no oscillations or the oscillations have different frequencies at different outputs, our method cannot be used and this is a restriction on it. However, oscillations with a common frequency is the mode most likely to occur (Atherton, 1975) when the process has significant interaction, which is the case considered in this section.

Noise is an important issue in the identification problem. Like the SISO case, anti-noise measures such as hysteresis, low-pass filtering and multiple oscillation periods can also be used in the present case of a DRF for each relay. No further discussion is required.

Example 5.6.1. Consider the well-known Wood/Berry binary distillation column plant (Wood and Berry, 1973):

$$G(s) = \begin{bmatrix} \frac{12.8e^{-s}}{1+16.7s} & \frac{-18.9e^{-3s}}{1+21s} \\ \\ \frac{6.6e^{-7s}}{1+10.9s} & \frac{-19.4e^{-3s}}{1+14.4s} \end{bmatrix}.$$

It is a typical MIMO plant with strong interaction and significant time delays. For a tuning test, the relay in loop 1 is set as a symmetric relay with output switching levels of 1.00 and -1.00, and a relay with bias in its output giving switching levels 1.50 and -1.00 is used in loop 2. The system exhibits limit cycle oscillations having a common frequency with frequency $\omega_c^1 = 0.485$. The switching levels of the relay in loop 2 is then changed to 1.80 and -1.20. The system exhibits limit cycle oscillations having a common frequency with $\omega_c^2 = 0.484$ in this case. The steady-state gain matrix $\hat{G}(0)$ and frequency response matrix $\hat{G}(\omega_c)$ are computed from (5.60) and (5.61) as

$$\hat{G}(0) = \begin{bmatrix} 12.8 & -18.9 \\ 6.60 & -19.4 \end{bmatrix}, \text{ and } \hat{G}(j\omega_c) = \begin{bmatrix} 1.56e^{-1.92j} & 18.6e^{0.221j} \\ 1.21e^{1.46j} & 2.79e^{0.260j} \end{bmatrix},$$

where $w_c=\frac{1}{2}(w_c^1+w_c^2)=0.485$. They are very accurate, compared with their true values:

$$G(0) = \begin{bmatrix} 12.8 & -18.9 \\ 6.6 & -19.4 \end{bmatrix}, \text{ and } G(0.485j) = \begin{bmatrix} 1.57e^{-1.93j} & 18.5e^{0.21j} \\ 1.23e^{1.50j} & 2.75e^{0.26j} \end{bmatrix}.$$

Example 5.6.2. Consider the process in Palmor et al (1993):

$$G(s) = \begin{bmatrix} \frac{0.5}{(0.1s+1)^2(0.2s+1)^2} & \frac{-1}{(0.1s+1)(0.2s+1)^2} \\ \frac{1}{(0.1s+1)(0.2s+1)^2} & \frac{2.4}{(0.1s+1)(0.2s+1)^2(0.5s+1)} \end{bmatrix}.$$

There are large interactions in this process. Two decentralized relay tests are performed on it. The relay in loop 1 is symmetric with unit switching levels, and the switching levels of the relay in loop 2 are 1.40 and -0.933 in the first test and changed to 1.50 and -1.00 in the second. Both tests result in limit cycle oscillations with the same frequency $w_c = 4.29$. The estimated steady-state gain matrix $\hat{G}(0)$ and frequency response matrix $\hat{G}(jw_c)$ are

$$\hat{G}(0) = \begin{bmatrix} 0.500 & -1.00 \\ 1.00 & 2.40 \end{bmatrix}, \quad \hat{G}(4.29j) = \begin{bmatrix} 0.243e^{-2.25j} & 0.529e^{1.31j} \\ 0.529e^{-1.84j} & 0.537e^{-2.98j} \end{bmatrix}$$

while the true values are

$$G(0) = \begin{bmatrix} 0.5 & -1 \\ 1 & 2.4 \end{bmatrix}, \quad G(4.29j) = \begin{bmatrix} 0.24e^{-2.23j} & 0.53e^{1.32j} \\ 0.53e^{-1.82j} & 0.54e^{-2.96j} \end{bmatrix}.$$

In this section, multivariable oscillations under decentralized relay feedback control have been investigated. In particular, it is shown that for a stable $m \times m$ process, the oscillation frequencies remain almost unchanged under relatively large relay amplitude variations. Therefore, if m decentralized relay feedback tests are performed on the process, their oscillation frequencies are close to each other so that the process frequency response matrix can be estimated at that frequency. A bias may be introduced into the relay to additionally obtain the process steady-state matrix.

6. Use of Relay Transient Responses

In the preceding chapter, relay feedback has been employed to identify the process frequency response. It is, however, noted that the frequency response is obtained only at a few discrete frequency points. Essentially, only one point, the critical point on the process frequency response, is estimated from a symmetric relay test, and a few other points can also be estimated at the expense of possibly a longer testing time and a more complicated relay, such as biased, parasite or cascade relay. Fundamentally, this is because only the steady-state response, i.e., the limit cycle, is exploited for identification and this contains spectral information on the oscillation frequency and its higher harmonics only, and harmonics at high frequencies are practically not identifiable due to the low-pass nature of most process dynamics. Note that the transient responses of the process input and output are ignored although they are also available from a relay test and contain rich spectral information of the process. As a result, one can expect that use of the relay transient response in identification will produce more process information than an approach based solely on the steady-state response. To do so, time domain based least squares techniques may be utilized and good books on this topic are readily available (Ljung and Söderström, 1983), which is thus not pursued here. Instead, the goal of the present chapter is to show how to estimate the process frequency response from the system transient responses to a *single and simple* relay test. The main tool used is the fast Fourier transform (FFT). We demonstrate that direct application of the FFT to relay feedback system signals gives false results, and signal decomposition and weighting are necessary to yield correct results.

6.1 Signal Analysis

With a computer monitoring and control system in place, the signals collected from a process test such as a relay test are all discrete time. For spectrum analysis of discrete time signals, the discrete Fourier transform (DFT) may be used and the fast Fourier transform (FFT) is an efficient and reliable method for computing the DFT. Since the FFT can be applied to arbitrarily windowed signals, it seems possible to obtain the process frequency response by taking the FFT of the process input and output directly. However, this idea is shown in this section to be false, and the remedy to this failure is highlighted.

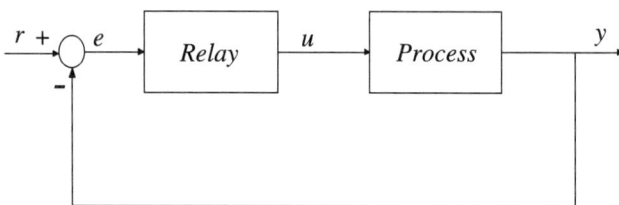

Fig. 6.1. Relay feedback system

Consider a SISO relay feedback system, depicted in Figure 6.1, where the process is a LTI system with the transfer function $G(s)$, and the relay can be any type, but preferred for simplicity to be a symmetric one with nonzero hysteresis as an anti-noise measure and with bias if static gain estimation is extremely important. Suppose that the process is at rest, i.e., has zero initial conditions, and then a chosen relay feedback is applied to the process at $t = 0$. Assume that the relay feedback system is stable and yields limit cycle oscillations as its steady state. The resultant process output response $y(t)$ and input response $u(t)$ are sampled as $y(kT)$ and $u(kT)$ with the sampling interval T, respectively. Our task is to identify the process frequency response, $G(j\omega)$, from the recorded samples, $y(kT)$ and $u(kT)$, $k = 1, 2, \cdots$.

Let $\mathcal{F}\{x(t)\}$ represent the Fourier transform of $x(t)$. Intuitively, estimation of the process frequency response, $G(j\omega)$, seems straightforward, and could be

$$G(j\omega) = \frac{Y(j\omega)}{U(j\omega)} = \frac{\mathcal{F}\{y(t)\}}{\mathcal{F}\{u(t)\}}.$$

This may be approximated by applying the FFT to $y(kT)$ and $u(kT)$, $k = 1, 2, \cdots$,

$$G(j\omega) = \frac{Y(j\omega)}{U(j\omega)} \approx \frac{T \cdot \mathcal{FFT}\{y(kT)\}}{T \cdot \mathcal{FFT}\{u(kT)\}}. \tag{6.1}$$

Unfortunately, it is wrong.

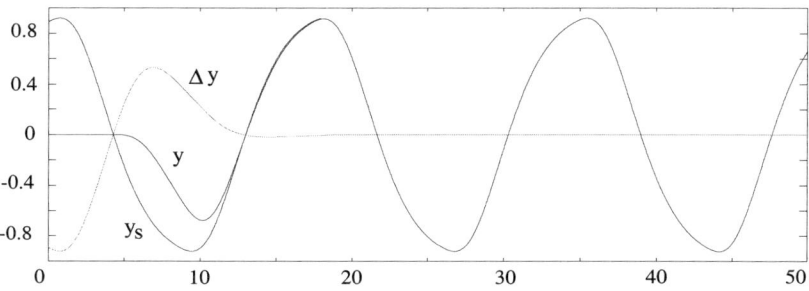

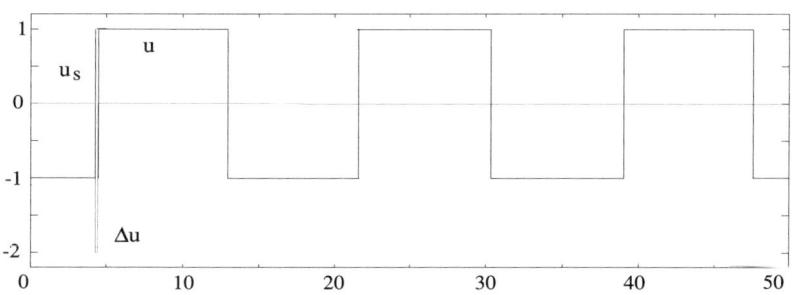

Fig. 6.2. Signals under relay feedback control

To determine the bugs in the formula, mathematically, $y(t)$ and $u(t)$ can be decomposed into the periodic stationary cycle parts $y_s(t)$ or $u_s(t)$ and the transient parts $\Delta y(t)$ or $\Delta u(t)$ as

$$y(t) = \Delta y(t) + y_s(t) \tag{6.2}$$

and

$$u(t) = \Delta u(t) + u_s(t), \tag{6.3}$$

as exhibited in Figure 6.2. Practically, when noise is not significant, the last stationary oscillation period of $y(t)$ and $u(t)$ are copied backwards to generate periodic stationary cycle parts $y_s(t)$ or $u_s(t)$, and $\Delta y(t)$ and $\Delta u(t)$ are obtained by taking respective differences between the overall and stationary responses. Obviously, the transients parts, $\Delta y(t)$ and $\Delta u(t)$, decay to zero at steady state.

Thus, they are absolutely integrable and Fourier transformable. We obtain their Fourier transforms

$$\Delta Y(j\omega) = \mathcal{F}\{\Delta y(t)\}, \text{ and } \Delta U(j\omega) = \mathcal{F}\{\Delta u(t)\}. \tag{6.4}$$

The FFT of a discrete signal, $x(kT)$, $k = 0, 1, \cdots, N - 1$, where N is an even integer, is defined by

$$\mathcal{FFT}\{x(kT)\} := \sum_{k=0}^{N-1} x(kT)e^{-j\omega_l kT}, \quad l = 1, 2, \cdots, M, \tag{6.5}$$

where $M = \frac{N}{2}$ and $\omega_l = 2\pi l/(NT)$. Suppose that the signal $x(t)$ decays to zero as time approaches infinity and it is a one-side signal, i.e., $x(t) = 0$, if $t < 0$; and $x(t) \approx 0$ if $t > T_f$. If the FFT window is bigger than T_f, the FFT of the corresponding discrete decaying signal will give the scaled Fourier transform of $x(t)$ at discrete frequencies as

$$X(j\omega) = \int_0^\infty x(t)e^{-j\omega t}dt \approx \int_0^{T_f} x(kT)e^{-j\omega kT}$$

$$\approx T\sum_{k=0}^{N-1} x(kT)e^{-j\omega kT} = T \cdot \mathcal{FFT}\{x(kT)\}, \tag{6.6}$$

where N is an integer such that $(N-1)T = T_f$. Equation (6.6) when applied to (6.4) implies

$$\Delta Y(j\omega) \approx T \cdot \mathcal{FFT}\{\Delta y(kT)\}, \tag{6.7}$$

and

$$\Delta U(j\omega) \approx T \cdot \mathcal{FFT}\{\Delta u(kT)\}, \tag{6.8}$$

that is, the FFT of the transient's part Δy or Δu from the relay test is approximate to its scaled Fourier transform. This fact is very important in FFT applications.

Strictly speaking, the Fourier transform of the periodic parts $y_s(t)$ and $u_s(t)$ actually does not exist since they are not decaying signals and thus not absolutely integrable, and are unable to yield the corresponding frequency response, $Y_s(j\omega)$ and $U_s(j\omega)$ as required in (6.1) together with (6.4). Nevertheless, taking the FFT of a periodic signal is a very common operation in signal processing. For a double-sided periodic signal $x(t)$ with a period T_c, its Fourier transform (Morrison, 1994) is

$$X(\omega) = 2\pi \sum_{n=-\infty}^{\infty} C_n \delta(\omega - n\omega_c), \tag{6.9}$$

where $\omega_c = \frac{2\pi}{T_c}$, $\delta(\omega - n\omega_c)$ is the Dirac delta or unit impulse, and C_n are the coefficients of the Fourier series of $x(t)$ given by

$$C_n = \frac{1}{T_c} \int_0^{T_c} x(t) e^{-jn\omega_c t} dt \approx \frac{T}{T_c} \mathcal{FFT}\{x(kT)\}, \tag{6.10}$$

where the FFT is taken over a period of $x(t)$, indicating that the FFT of one period of the sampled $x(t)$ gives $\frac{T_c C_n}{T}$, but not $X(j\omega)$ as desired.

To see what is wrong with (6.1), we apply (9.9) and (6.10) to the periodic parts $y_s(t)$ and $u_s(t)$ and compare the results with (6.7) and (6.8). It is true that the FFT of the transient parts Δy and Δu do arrive at their respective scaled Fourier transforms. However, the FFT of one period (or multiple periods) of the periodic part y_s or u_s will actually give the *scaled amplitudes* of impulses in its Fourier transform, or the coefficients of their Fourier series. Two FFT results have different meanings. Summing them together does not have any practical sense and will not produce the correct process frequency response. Therefore, directly applying the FFT to the process output and input and calculating the process frequency response by (6.1) is meaningless and will definitely lead to a wrong result.

Since the FFT of the transient part Δy or Δu gives its scaled Fourier transform and the FFT of one period of the periodic part y_s or u_s will actually give the scaled *amplitudes* of impulses in its Fourier transform, one may think that the process frequency response might be obtained by

$$G(j\omega) = \frac{Y(j\omega)}{U(j\omega)} = \frac{T \cdot \mathcal{FFT}[\Delta y(kT)] + 2\pi \sum_{n=-\infty}^{\infty} Y_{sn} \delta(\omega - n\omega_c)}{T \cdot \mathcal{FFT}[\Delta u(kT)] + 2\pi \sum_{n=-\infty}^{\infty} U_{sn} \delta(\omega - n\omega_c)} \tag{6.11}$$

where Y_{sn} and U_{sn} are the coefficients of the Fourier series of y_s and u_s respectively. However, this makes use of (6.9), which assumes that the signal is two-sided, while y_s and u_s from a relay feedback test are one-side signals. For a one-side periodic signal which we have at hand, its Fourier transform usually contains two parts. One is obtained by simply setting $s = j\omega$ in its Laplace transform while the other is a series of impulses at the singularity frequencies of the Laplace transform (McGillem and Cooper, 1984). If what one is interested in is the Fourier transform at frequencies other than these singularity frequencies, the Fourier transform is determined by the first part only and can be obtained from its Laplace transform with the simple substitution $s = j\omega$. For this purpose, we have to make use of the following lemma.

Lemma 6.1.1. *(Kuhfitting, 1978). Suppose that $x(t)$ is a periodic function with period T_c for $t \geq 0$, i.e.,*

140   6. Use of Relay Transient Responses

$$x(t) = \begin{cases} x(t+T_c), & t \in [0, +\infty), \\ 0, & t \in (-\infty, 0). \end{cases}$$

Assume that the Laplace transform, $L\{x(t)\} = X(s)$, exists. Then,

$$X(s) = \frac{1}{1 - e^{-sT_c}} \int_0^{T_c} x(t) e^{-st} dt. \tag{6.12}$$

It follows that the corresponding Fourier transform at frequencies other than the singularity frequencies, $\omega = m\omega_c = m\frac{2\pi}{T_c}, m = 0, 1, 2, \cdots$, can be obtained as

$$X(j\omega) = \frac{1}{1 - e^{-j\omega T_c}} \int_0^{T_c} x(t) e^{-j\omega t} dt. \tag{6.13}$$

Based on the above observations, two numerical techniques for multiple-point frequency response estimation from a single relay feedback test will be presented in the subsequent two sections.

6.2 Decomposition Method

The transfer function of a linear process is given by

$$G(s) = \frac{Y(s)}{U(s)}.$$

Laplace-transforming (6.2) and (6.3) gives

$$G(s) = \frac{Y(s)}{U(s)} = \frac{\Delta Y(s) + Y_s(s)}{\Delta U(s) + U_s(s)}, \tag{6.14}$$

where $\Delta Y(s)$ and $\Delta U(s)$ are the Laplace transforms of the transient parts $\Delta y(t)$ and $\Delta u(t)$ respectively, $Y_s(s)$ and $U_s(s)$ are the Laplace transforms of the periodic parts $y_s(t)$ and $u_s(t)$ respectively. By Lemma 6.1, (6.14) becomes

$$G(s) = \frac{\Delta Y(s) + \frac{1}{1-e^{-sT_c}} \int_0^{T_c} y_s(t) e^{-st} dt}{\Delta U(s) + \frac{1}{1-e^{-sT_c}} \int_0^{T_c} u_s(t) e^{-st} dt}, \tag{6.15}$$

where T_c is the period of the process output stationary oscillation from the relay feedback test. Setting $s = j\omega$, (6.15) becomes

$$G(j\omega) = \frac{\Delta Y(j\omega) + \frac{1}{1-e^{-j\omega T_c}} \int_0^{T_c} y_s(t) e^{-j\omega t} dt}{\Delta U(j\omega) + \frac{1}{1-e^{-j\omega T_c}} \int_0^{T_c} u_s(t) e^{-j\omega t} dt}. \tag{6.16}$$

We wish to have a numerical scheme for computing (6.15) for practical applications. Suppose that at $t = T_f$, $y(t)$ and $u(t)$ have nearly entered stationary oscillation status and after $t = T_f$, both $\Delta y(t)$ and $\Delta u(t)$ are approximately zero. Let N be an integer such that $(N-1)T \approx T_f$. As discussed before, the Fourier transform of $\Delta y(t)$ is approximated by

$$\Delta Y(j\omega) = \int_0^\infty \Delta y(t) e^{-j\omega t} dt \approx \int_0^{T_f} \Delta y(t) e^{-j\omega t} dt$$

$$\approx T \sum_{k=0}^{N-1} \Delta y(kT) e^{-j\omega_l T} = T \cdot \mathcal{FFT}\{\Delta y(kT)\}, \tag{6.17}$$

where $l = 1, 2, \cdots, M$, $M = \frac{N}{2}$ and $\omega = \omega_l = 2\pi l/(NT)$.

$Y_s(j\omega)$ at discrete frequencies ω_l is computed using the digital integral:

$$Y_s(j\omega_l) = \frac{T}{1 - e^{-j\omega_l T_c}} \sum_{k=0}^{N_c} y_s(kT) e^{-j\omega_l kT}, \quad l = 1, 2, \cdots, M, \tag{6.18}$$

where M and ω_l are defined as before, and the integer $N_c \approx (T_c - T)/T$, i.e., the summation is taken over one period. $\Delta U(j\omega_l)$ and $U_s(j\omega)$ can be calculated in the same way. Consequently, the process frequency response is obtained as

$$G(j\omega_l) = \frac{\mathcal{FFT}\{\Delta y(kT)\} + \frac{1}{1-e^{-j\omega_l T_c}} \sum_{k=0}^{N_c} y_s(kT) e^{-j\omega_l kT}}{\mathcal{FFT}\{\Delta u(kT)\} + \frac{1}{1-e^{-j\omega_l T_c}} \sum_{k=0}^{N_c} u_s(kT) e^{-j\omega_l kT}},$$
$$l = 1, 2, \cdots, M. \tag{6.19}$$

This method employs the FFT and digital integration only. The required computation burden is thus modest. It can identify multiple points on a frequency response from a single relay test. It is accurate in the absence of noise and disturbance.

For real implementation of (6.19), we must give values for the time span $T_f = (N-1)T$ required in the FFT computations (6.17). It turns out that T_f or N is related to the number of the frequency response points to be identified. Suppose that the number of the frequency response points to be identified from zero frequency to the process phase-crossover frequency ω_c is M, where M is specified by the user. It follows from (6.17) that the M frequency response points recovered by the FFT method are at the discrete frequencies $0, \Delta\omega, 2\Delta\omega, \cdots, (M-1)\Delta\omega$, where $\Delta\omega = \omega_{l+1} - \omega_l = 2\pi/NT$ is the frequency increment. The definition of M means that $\omega_c \approx (M-1)\Delta\omega$ and

$$\omega_c \approx (M-1)\frac{2\pi}{NT}. \tag{6.20}$$

On the other hand, one can measure the oscillation period T_c on line when a relay test is performed and ω_c can be estimated as

$$\omega_c \approx \frac{2\pi}{T_c}. \tag{6.21}$$

Equations (6.20) and (6.21) yield

$$N \approx (M-1)\frac{T_c}{T} \tag{6.22}$$

where M should be specified *a priori* and be large enough to ensure that the stationary oscillation is reached. Then, N is determined from (6.22) and the corresponding time span is obtained as $T_f = (N-1)T$.

Different identification methods are more or less sensitive to noise. It is apparent (Ljung and Söderström, 1983) that in almost all identification methods a low noise-to-signal ratio is required. Regarding measurement noise in the relay test, Åström and Hägglund (1984b) pointed out that hysteresis in the relay is a simple way of reducing the influence of measurement noise. The hysteresis width should be bigger than the noise band (Åström and Hägglund, 1988) and is usually chosen as twice the noise band (Hang et al., 1993b). Filtering is another possibility (Åström and Hägglund, 1984b). Measurement noise is usually of high frequency, while the process frequency response of interest for control analysis and design is usually in the low-frequency region. In particular, the process frequency response in $[0, \omega_c]$ is critical for controller design. We found in our experiments that the measurement noise is in a fairly high frequency region. Therefore, a low-pass filter can be employed to reduce the measurement noise. The cut-off frequency of the filter is determined with respect to the process frequency region of interest. A possible choice is $3\omega_c \sim 5\omega_c$. To reduce the effect of noise further, especially in the case of large noise-to-signal ratio, one may also use the average of the last four or five periods of oscillation as the stationary oscillation period and copy this average period backwards to obtain y_s. With the above three anti-noise measures, the proposed method can reject noise quite effectively.

When a relay feedback test is to be conducted, the process is first brought to the operating point before relay feedback tests start. After the process reaches a steady state, a 'listening period' (10–20 samples) begins. During this period, the steady state of the process input and output are measured. To prevent relay switching at the wrong instants, relay hysteresis is usually adopted. In the 'listening period', the peak value of noise is monitored, from which the hysteresis of the relay is determined. Usually, the hysteresis is chosen to be twice the peak value of the noise. After the 'listening period', the relay feedback test is launched. The process input and output are sampled until the system reaches a stationary oscillation. The stationary oscillation can be detected by monitoring the relay output oscillation periods.

For the assessment of identification accuracy, identification error is here measured by the worst-case error

$$ERR = \max_l \left\{ \left| \frac{\hat{G}(j\omega_l) - G(j\omega_l)}{G(j\omega_l)} \right| \times 100\%, \quad l = 1, 2, \cdots, M \right\}, \qquad (6.23)$$

where $G(j\omega_l)$ and $\hat{G}(j\omega_l)$ are the actual and the estimated process frequency responses respectively. The Nyquist curve for phase ranging from 0 to $-\pi$ is considered since this part is most significant for control design. To test estimation robustness against noise, the process output may be corrupted by noise and the corrupted output used for identification. The noise level is judged, in the context of system identification, by the noise-to-signal ratio, which is usually defined as

$$N_1 = \text{Noise-to-Signal Power Spectrum Ratio}$$
$$= \frac{\text{mean power spectrum density of noise}}{\text{mean power spectrum density of signal}}, \qquad (6.24)$$

or

$$N_2 = \text{Noise-Signal Mean Ratio} = \frac{\text{mean(abs(noise))}}{\text{mean(abs(signal))}}. \qquad (6.25)$$

Example 6.2.1. Consider a high-order plus dead time process:

$$G(s) = \frac{e^{-4s}}{(s+1)^5}.$$

Relay feedback is applied to it and $y(t)$ and $u(t)$ are logged and then decomposed into $y = \Delta y + y_s$ and $u = \Delta u + u_s$ as in (6.2) and (6.3). With 1000 samples per stationary cycle period and $M = 15$, the process frequency response is identified using (6.19) and is shown in Figure 6.3. The ERR is 0.32%.

In order to test our method in a realistic environment, real-time relay tests were performed using the *KI 100 Dual Process Simulator* from *KentRidge Instruments, Singapore*, as shown in Figure 6.4. The simulator is an analogue process simulator and can be configured to simulate a wide range of industrial processes with different levels of noise. The simulator is connected to a PC via an A/D and D/A board. The window-based *DT VEE* 3.0 is used as the system control platform, on which the relay control code is written in C++. The fastest sampling time of the *VEE* system is 0.06 s. It follows from our extensive

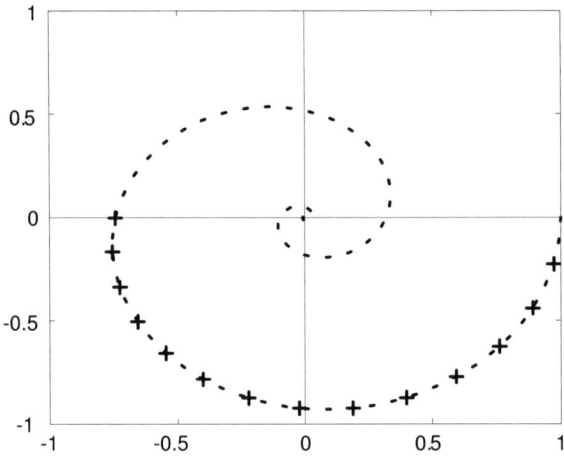

Fig. 6.3. Process Nyquist plot ($---$actual, +estimated)

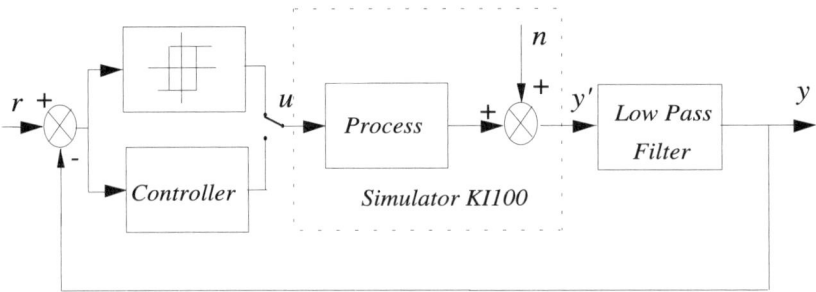

Fig. 6.4. Relay feedback system in the presence of noise

tests that 200–300 samples per stationary cycle during the relay test are good enough for identification and control design.

The filter in Figure 6.4 is selected as a Butterworth low-pass filter with cut-off frequency $3\omega_c \sim 5\omega_c$. Other anti-noise actions discussed earlier are also

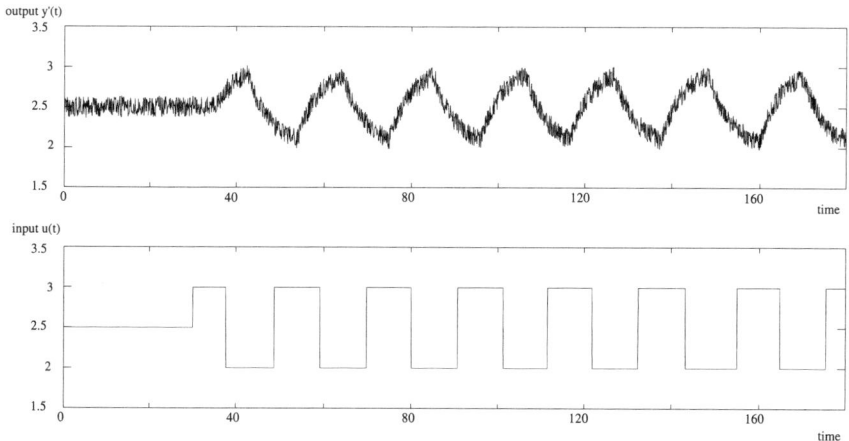

Fig. 6.5. Real-time signals under relay feedback control

taken. Without additional noise, the noise-to-signal ratio N_1 of the inherent noise in our test environment is 0.06% ($N_2 = 2\%$).

Example 6.2.2. Consider a first-order plus dead time process:

$$G(s) = \frac{1}{5s+1}e^{-5s}.$$

The identification error ERR is 5.41%. To see the effects of noise, extra noise is introduced with the noise source in the simulator. The time sequences of $y(t)$ and $u(t)$ in a relay test for $N_1 = 10\%(N_2 = 35\%)$ are shown in Figure 6.5. The first part of the test in Figure 6.5 ($t = 0 \sim 30$) is the "listening period", in which the noise bands of $y(t)$ and $u(t)$ are measured at steady state.

The testing was also carried out for a variety of dynamics available on the simulator. The typical results are summarized in Table 6.1 with different noise levels. Estimation of process frequency response is generally good. It can be seen that our decomposition method can obtain multiple points on the process frequency response points simultaneously with one single relay experiment, and this greatly saves testing time. The method is accurate, especially in the critical frequency range $[0, \omega_c]$, provided that the process is linear and no disturbance exists. The computations are simple, and easily implemented on microprocessors.

Table 6.1. Identification error (ERR) for decomposition method

Processes	Noise Levels		
	N_1=1%	6%	10%
	N_2=13%	25%	35%
$\frac{1}{5s+1}e^{-5s}$	10.14%	12.79%	13.57%
$\frac{1}{(s+1)^8}$	9.48%	12.46%	15.18%
$\frac{1}{(s+1)(5s+1)^2}e^{-2.5s}$	6.74%	7.98%	8.17%
$\frac{1-s}{(2s+1)^4(5s+1)}e^{-s}$	2.32%	6.00%	8.62%

6.3 Weighting Method

As revealed in Section 6.1, the signals in a relay feedback system are not directly Fourier transformable to obtain the process frequency response. The decomposition method in the preceding section decomposes the signals into the transient part and the stationary cycle parts. These parts are then transformed to their frequency responses using the DFT and digital integration respectively. An alternative way is to put a decay weighting on the signals such that the weighted signals do die off as time passes.

Let $y(t)$ and $u(t)$ be the process output and input responses in a relay feedback system, respectively. A decay exponential $e^{-\alpha t}$ ($\alpha > 0$) is introduced to moderate them:

$$\tilde{y}(t) = y(t)e^{-\alpha t}, \tag{6.26}$$

and

$$\tilde{u}(t) = u(t)e^{-\alpha t}, \tag{6.27}$$

so that $\tilde{y}(t)$ and $\tilde{u}(t)$ decay to zero exponentially as t approaches infinity and become Fourier transformable. Applying the Fourier transform to (6.26) and (6.27) yields

$$\tilde{Y}(j\omega) = \int_0^\infty \tilde{y}(t)e^{-j\omega t}dt = \int_0^\infty y(t)e^{-\alpha t}e^{-j\omega t}dt = Y(j\omega+\alpha), \tag{6.28}$$

and

$$\tilde{U}(j\omega) = \int_0^\infty \tilde{u}(t)e^{-j\omega t}dt = \int_0^\infty u(t)e^{-\alpha t}e^{-j\omega t}dt = U(j\omega+\alpha). \tag{6.29}$$

The integral intervals in (6.28) and (6.29) are infinite, and digital computation of infinite interval integration is not trivial. However, due to the introduction of the decay exponential $e^{-\alpha t}$, $\tilde{y}(t)$ and $\tilde{u}(t)$ are approximately zero after a certain time. The infinite interval integration problem then becomes a finite integration one. Thus, $\tilde{Y}(j\omega)$ can be computed at discrete frequencies, using the standard FFT technique. Suppose that $y(kT), k = 0, 1, 2, \cdots, N-1$, are samples of $y(t)$, where T is the sampling interval. N is chosen such that $y((N-1)T)$ has reached a stationary oscillation, and the decay coefficient α is selected such that $\tilde{y}((N-1)T)$ formed from (6.26) has approximately decayed to zero. Then, we have

$$\tilde{Y}(j\omega_l) \approx \int_0^{T_f} \tilde{y}(t)e^{-j\omega_l t} \approx T \sum_{k=0}^{N-1} \tilde{y}(kT)e^{-j\omega_l kT} = T \cdot \mathcal{FFT}\{\tilde{y}(kT)\},$$

$$l = 1, 2, \cdots, M, \quad (6.30)$$

where T_f is the time when $\tilde{y}(T_f) = \tilde{y}(NT-T) \approx 0$, $M = \frac{N}{2}$ and $\omega_l = 2\pi l/(NT)$. $\tilde{U}(j\omega_l)$ can be similarly computed by taking the FFT of $\tilde{u}(kT)$. Therefore, the shifted process frequency response $G(j\omega_l + \alpha)$ is obtained as

$$G(j\omega_l+\alpha) = \frac{Y(j\omega_l+\alpha)}{U(j\omega_l+\alpha)} = \frac{\mathcal{FFT}\{\tilde{y}(kT)\}}{\mathcal{FFT}\{\tilde{u}(kT)\}}, \quad l = 1, 2, \cdots, M. \quad (6.31)$$

The identified shifted process frequency response may be directly used for process modelling and controller design. However, if $G(j\omega)$ rather than $G(j\omega + \alpha)$ is needed, one can first take the inverse FFT of $G(j\omega_l + \alpha)$ as

$$\tilde{g}(kT) := \mathcal{FFT}^{-1}(G(j\omega_l+\alpha)) = g(kT)e^{-\alpha kT}, \quad l = 1, 2, \cdots, M. \quad (6.32)$$

It then follows that the process impulse response $g(kT)$ is

$$g(kT) = \tilde{g}(kT)e^{\alpha kT}. \quad (6.33)$$

Applying the FFT again to $g(kT)$ would result in the original process frequency response

$$G(j\omega_l) = \mathcal{FFT}(g(kT)). \quad (6.34)$$

For implementation of (6.31), values must be assigned to the decay factor α and the time span $T_f = (N-1)T$ required in the FFT computations (6.30). Consider first the selection of T_f. Following a similar argument to that in the preceding section on T_f, we also reach (6.22) and for easy reference we repeat

$$N \approx (M-1)\frac{T_c}{T} \quad (6.22)$$

where M should be specified *a priori*, and should be large enough to ensure that the stationary oscillation is reached. Then, N is determined from (6.22) and the corresponding time span is obtained as $T_f = (N-1)T$. Practically, if M is large, the computational load increases, but the points in a small frequency range may not necessarily give additional information on the process dynamics. The number of frequency response points M is usually selected to be around 10–20. To cover the entire transients of $\tilde{y}(t)$ and $\tilde{u}(t)$ with such an M, the process inputs and outputs samples may be screened, or one taken from a few, equivalently making a bigger T.

The value of the decay coefficient α should be such that $\tilde{y}(t)$ and $\tilde{u}(t)$ decay nearly to zero when the time approaches T_f, regardless of the nonzero $y(t)$ and $u(t)$. It is this decay coefficient α that enables the infinite integral of the Fourier transform to be replaced by the finite digital integral of the FFT. To make (6.30) a good approximation, the decay coefficient α should satisfy

$$e^{-\alpha T_f} \leq \varepsilon \tag{6.35}$$

or

$$\alpha \geq -\frac{\ln \varepsilon}{T_f}, \tag{6.36}$$

where ε is the specified threshold, and usually takes a value of $10^{-4} \sim 10^{-6}$. On the other hand, α should not be too large, better kept α less than 0.1–0.2 to prevent losing too much information when forming $\tilde{y}(t)$ and $\tilde{u}(t)$.

For noisy data, all the anti-noise measures discussed in Section 6.2 are also applicable to this weighting method. The identification error EER as defined in (6.23) is used for assessment of the estimation accuracy.

Example 6.3.1. This example is adopted from Li et al. (1991):

$$G(s) = \frac{e^{-2s}}{10s+1}.$$

The model estimated by Li *et al.* (1991) is

$$\hat{G}(s) = \frac{0.988 e^{-2s}}{8.02s+1},$$

and its identification error is $ERR = 22.32\%$. In Li's method, the dead time is assumed to be known. For our weighting method, relay feedback is applied to the process. The process output and input are logged. $y(t)$ and $\tilde{y}(t)$ of the relay test are shown in Figure 6.6(a), while $u(t)$ and $\tilde{u}(t)$ are given in Figure 6.6(b). The frequency response identified is shown in Figure 6.7. The ERR is 0.26%. This indicates that the proposed method provides a much more accurate result.

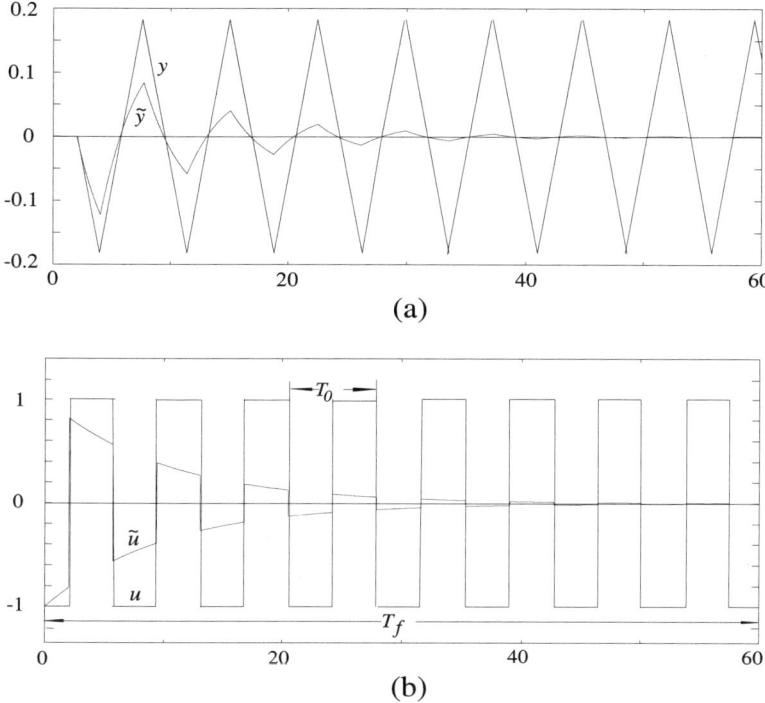

Fig. 6.6. Signals under relay feedback control

Example 6.3.2. In order to test the method in a realistic environment, realtime relay tests were also performed on the KI 100 Dual Process Simulator. The description of this process simulator and its experimental set-up for identification has been given in the preceding section. The same low-pass filter and other anti-noise actions as discussed in Example 6.2 were also taken. Table 6.2 presents the identification results of four typical examples at different noise levels. The accuracy is satisfactory. This weighting method yields slightly better estimation results, compared with the decomposition method.

6.4 Testing on Pilot Plants

In the preceding two sections, we have presented two methods for identifying *multiple* points on the process frequency response with *one single* relay feedback

150   6. Use of Relay Transient Responses

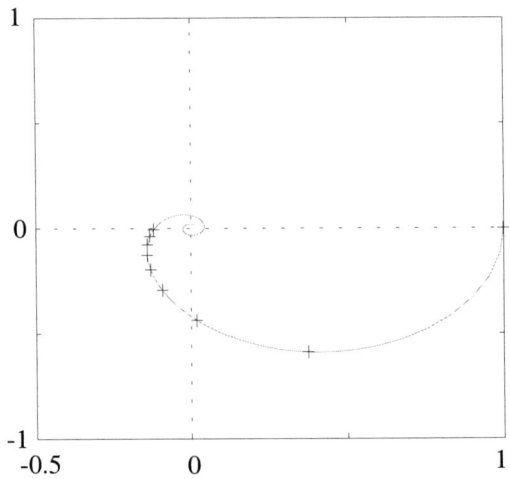

Fig. 6.7. Nyquist plot of $G(j\omega)$ (—actual, +estimated)

Table 6.2. Identification error (ERR) for weighting method

Processes	Noise levels		
	N_1=1%	6%	10%
	N_2=13%	25%	35%
$\frac{1}{5s+1}e^{-5s}$	6.14%	7.96%	9.67%
$\frac{1}{(s+1)^8}$	7.42%	8.97%	9.85%
$\frac{1}{(s+1)(5s+1)^2}e^{-2.5s}$	3.75%	6.52%	8.65%
$\frac{1-s}{(2s+1)^4(5s+1)}e^{-s}$	2.93%	4.48%	8.69%

experiment, and demonstrated their effectiveness by numerical simulation and real-time implementation on an analogue simulator. In this section, we test them on real plants. Since the two methods produce similar results, we will show only one method, the weighting method, for illustration. An essential difference between this physical test and the previous ones is that for a real plant no true frequency response is available to assess our identification results. In the context of control engineering, process identification is ultimately for process control. Control performance can easily be measured. Thus, we shall, in this section, combine our identification method with some simple PID controller tuning methods to form an auto-tuner and then compare the resultant control performance with the standard relay auto-tuner for PID controllers, to see if an improvement can be achieved with the new technique.

These multiple points on the process Nyquist curve can be used to tune a better PID controller, compared with those using the critical information only. Systematic techniques for the design of various controllers are covered in detail in Part III of this book. For this section, we present only a simple PID tuning method.

Consider a conventional negative feedback control system with a given process $G(s)$ and controller $K(s)$. The controller is to be designed such that the actual closed-loop transfer function is best fitted to the objective closed-loop transfer function H from r to y. The objective open-loop frequency response profile can be properly selected according to the user's requirements, such as gain and phase margins, bandwidth, robustness, etc. One simple solution is to choose H as

$$H = \frac{\omega_n^2}{s^2 + 2\zeta\omega_n s + \omega_n^2} e^{-Ls}, \tag{6.37}$$

where L is the apparent dead time of the process that can be read off from the relay test. ω_n and ζ dominate the behaviour of the desired closed-loop response.

We must give values for the objective damping rate ζ and natural frequency ω_n in (6.37). If the control specifications are given in terms of the phase margin φ_m and gain margin A_m, the ζ and ω_n in H are approximately determined (Wang et al., 1996) by

$$\zeta = \sqrt{\frac{1 - \cos^2 \varphi_m}{4 \cos \varphi_m}},$$

and

$$\omega_n = \frac{\tan^{-1}(\frac{2\zeta p}{p^2 - 1})}{pL},$$

where p is the positive root of equation

$$(A_m - 1)^2 = 4\zeta^2 p^2 + (1 - p^2)^2.$$

In general, we choose relatively large gain and phase margins for fast loops and choose small ones for slow loops. The default settings for the parameter ζ and $\omega_n L$ are $\zeta = 0.707$ and $\omega_n L = 2$, where the corresponding phase margin is $60°$ and gain margin is 2.2.

With H determined, the open-loop transfer function Q corresponding to the desired closed-loop transfer function H,

$$H = \frac{Q}{1+Q},$$

can be obtained as

$$Q = \frac{H}{1-H}.$$

The controller K is designed such that the actual GK is fitted to the desired transfer function Q in the frequency domain as well as possible. This is achieved by minimizing the cost function:

$$\min_{K} \sum_{l=1}^{M} |G(j\omega_l)K(j\omega_l) - Q(j\omega_l)|^2, \qquad (6.38)$$

with respect to K, where M is chosen such that ω_M is slightly greater than the critical frequency of Q. With G and Q given, a solution for K is obtained by its structure selection and parameter calculation. As the majority of regulators used in industry are of PID type, a controller having PID structure is considered here. For a PID controller in the form

$$K(s) = K_c(1 + \frac{1}{T_i s} + T_d s), \qquad (6.39)$$

the least squares solution to (6.38) is

$$X^* = (A^T A)^{-1} A^T B, \qquad (6.40)$$

where

$$A = \begin{bmatrix} Re(\Psi) \\ Im(\Psi) \end{bmatrix}, \quad B = \begin{bmatrix} Re(\Omega) \\ Im(\Omega) \end{bmatrix},$$

$$\Psi = \begin{bmatrix} G(j\omega_1) & \frac{G(j\omega_1)}{j\omega_1} & j\omega_1 G(j\omega_1) \\ G(j\omega_2) & \frac{G(j\omega_2)}{j\omega_2} & j\omega_2 G(j\omega_2) \\ \vdots & \vdots & \vdots \\ G(j\omega_M) & \frac{G(j\omega_M)}{j\omega_M} & j\omega_M G(j\omega_M) \end{bmatrix}, \quad \Omega = \begin{bmatrix} Q(j\omega_1) \\ Q(j\omega_2) \\ \vdots \\ Q(j\omega_M) \end{bmatrix},$$

and

$$X^* = [x_1^* \; x_2^* \; x_3^*]^T := [K_c^* \; \frac{K_c^*}{T_i^*} \; K_c^* T_d^*]^T.$$

Note that the above development computes PID settings from $G(j\omega)$. However, by replacing $j\omega$ with $j\omega + \alpha$ in (6.38), a PID controller can also be calculated using the same frequency response fitting idea from $G(j\omega + \alpha)$, which is the early result of the weighting method in Section 6.3, so that the conversion $G(j\omega + \alpha)$ to $G(j\omega)$ is unnecessary for this PID tuning.

The multiple-point identification and multiple-point PID fitting design have thus been married together to form a new relay auto-tuner. The implemented tuner was tested on a number of physical plants and its performance (solid line in the figures to come) is compared with the standard relay tuner (dotted line) (Åström and Hägglund, 1995).

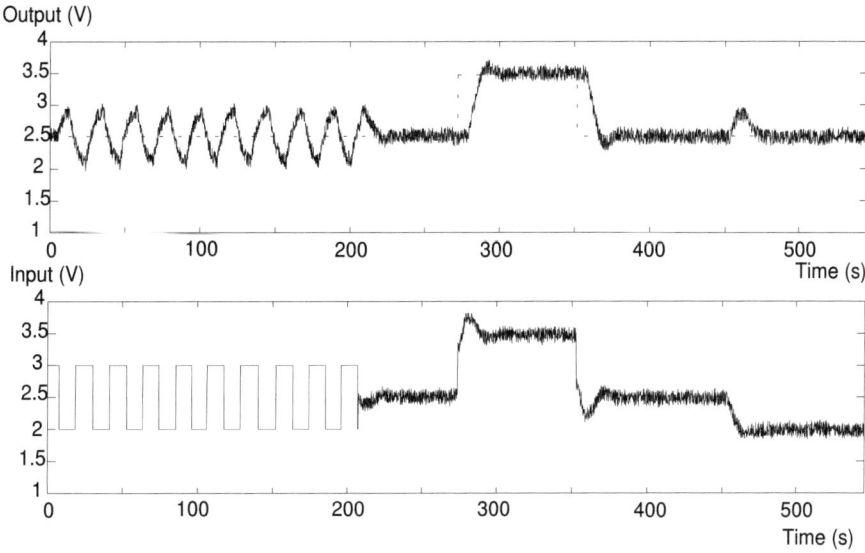

Fig. 6.8. Process and subsequent performance

Example 6.4.1. (**Dual Simulator**) It is interesting and useful to test our tuner on the Dual Simulator as described in Section 6.2 for simple verification of the results, as we know everything about it. Besides the test environment noise, extra noise was introduced with the noise source in the Simulator to see the effect of noise. This was in form of pseudo-white noise generated using the random number generator, and was inserted in the process output to act as

154   6. Use of Relay Transient Responses

measurement noise. The noise peak value was set on the simulator to be 50% of the relay signal amplitude, which is $0.5V$. This is equivalent to a noise-to-signal power spectrum ratio about 10–20%. For illustration, the simulator is configured into a first-order plus dead time process

$$G(s) = \frac{1}{5s+1} e^{-5s}.$$

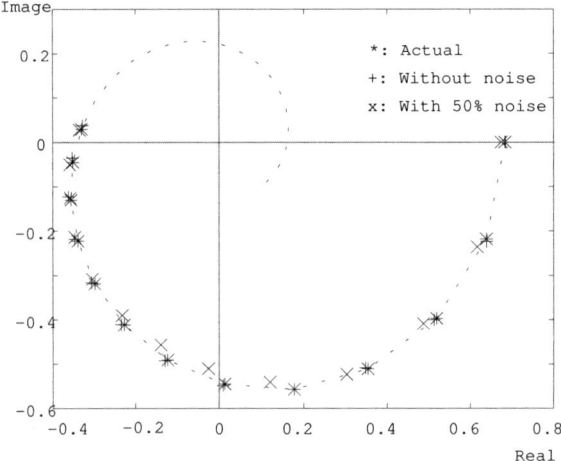

Fig. 6.9. Process frequency response $G(j\omega + \alpha)$

The complete auto-tuning process and subsequent control performance are shown in Figure 6.8. The process was in the steady state of $2.5V$ before the relay feedback test began. The resultant input and output transients were sampled and the multiple points on the process frequency response estimated by the identification method in Section 6.3. Figure 6.9 shows the estimated process frequency responses $G(j\omega_i + \alpha)$ from real-time tests with and without additional noise.

The apparent dead time L in (6.37) was estimated as 4.877, which was quite close to the real value of 5. With defaults of $\zeta = 0.707$ and $\omega_n L = 2$, the PID controller in (6.39) was tuned to $K_c = 0.6738$, $T_i = 6.1099$ and $T_d = 0.3941$ using (6.40). The PID control was then commissioned and the system

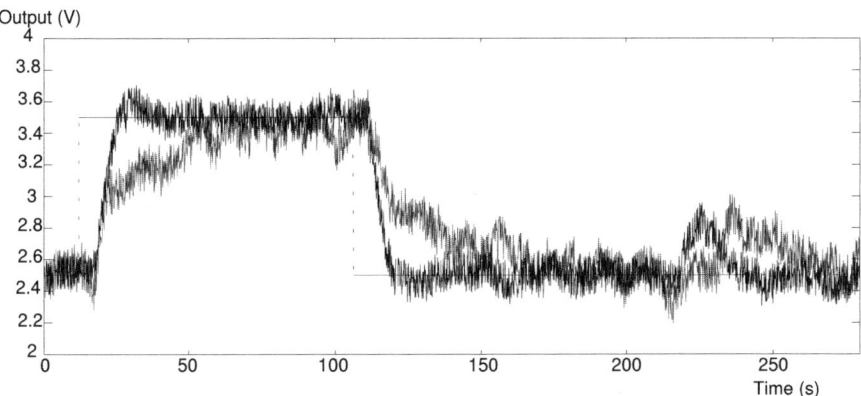

Fig. 6.10. Auto-tuner performance for $G_p(s) = \frac{1}{5s+1}e^{-5s}$

settled down. The auto-tuning process was thus completed. The standard relay tuner gave PID controller parameters $K_c = 0.5699$, $T_i = 11.6287$ and $T_d = 2.9072$. To see the closed-loop performance, a set-point change of 1V and a load disturbance of 0.5V were introduced afterwards. The control performance of the new auto-tuner (faster) and standard relay tuner (slower) are shown in Figure 6.10. It is observed that the response of the proposed auto-tuner is excellent. The performance under standard relay control is acceptable but shows a sluggish response.

Example 6.4.2. (**Coupled Tank**) Fluid level control is popular in the process industry. It aims at keeping the fluid level at a user-defined value in storage tanks, chemical blending and reaction vessels. Normally, the dynamics of fluid level systems is relatively long. For a single tank, although nonlinearity exists in this process, it can be approximated by a first-order linear process around the operating point. However, things become complex when multiple tanks are linked together. The water level in multiple storage tanks, in which one tank's output is another tank's input, shows a high-order dynamics. The proposed relay tuner was tested using the coupled tank system shown in Figure 6.11. The schematic diagram of the coupled tank is shown in Figure 6.12. The inflow (control input) is supplied by a variable speed pump which pumps water into Tank 1 through a long tube. Tank 1 and Tank 2 are coupled to each other

156   6. Use of Relay Transient Responses

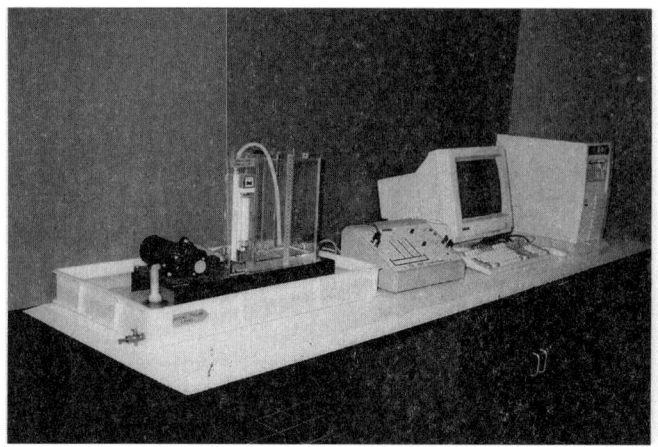

Fig. 6.11. Coupled-tank water level control system

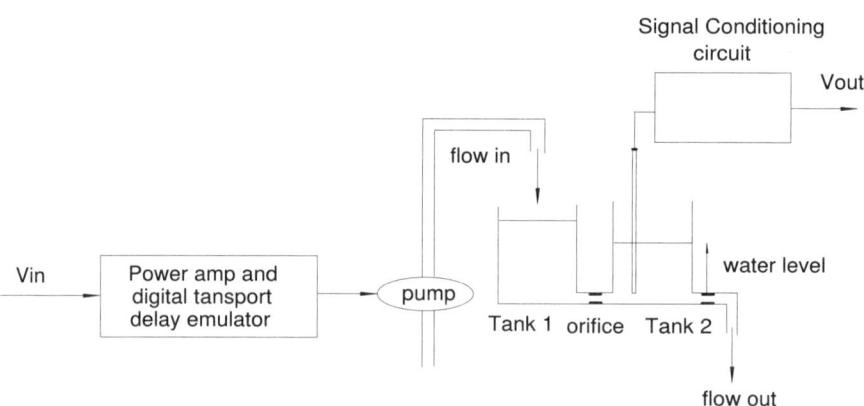

Fig. 6.12. Schematic diagram of coupled tank

through an orifice at the bottom of the tanks. In the test, the water level in Tank 2 was controlled by regulating the drive voltage to the pump. The performance of the coupled tank under the proposed tuner control is shown in

Figure 6.13, in which the process response under standard relay tuner control is also presented. The proposed tuner gives better closed-loop responses.

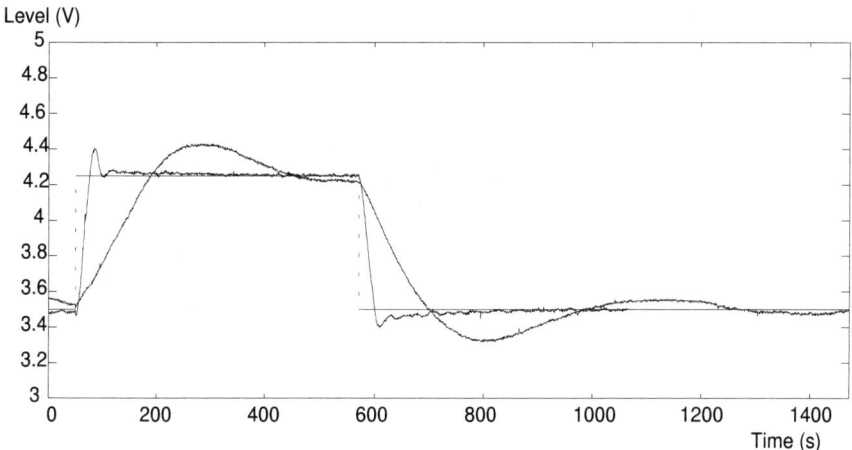

Fig. 6.13. Auto-tuner performance for coupled-tank level control

Example 6.4.3. (**Heat Exchanger**) A heat exchanger is a common installation in industry, where heat generated from one process is used in other part of the plant. Heat exchanger applications can be found in oil refineries, chemical processing plants, and in heating and cooling systems for buildings. Many different types of heat exchanger are available and the shell-and-tube heat exchanger is one of the most common. The pilot-scale shell-and-tube heat exchanger, shown in Figure 6.14 is used as a test bed. Figure 6.15 shows the schematic diagram of the heat exchanger. The heat exchanger is a steam-heated system in which water flows inside the tubes surrounded by steam flowing through the shell. Steam is generated by a relay-controlled steam generator. The outlet hot water is subsequently collected in a level-controlled tank. The flow of cold water into the heat exchanger is controlled by pneumatic valve PV1. The flow of steam into the heat exchanger is controlled by pneumatic valve PV2. The flow rate and pressure of the steam is measured by the differential pressure flow transducer (FT) and pressure sensor (PT). At the outlet flow of the

158   6. Use of Relay Transient Responses

Fig. 6.14. A pilot scale shell-and-tube heat exchanger

heat exchanger another platinum resistance thermometer (TT) measures the temperature of the water going into the hot water tank.

For the test, the opening of cold water supply valve PV1 was kept constant, while the opening of the steam supply valve PV2 was selected as the manipulated variable to regulate the temperature of the outlet water. The cold water supply valve was 40% open and the steam supply valve was 50% open with the system under open-loop control. At steady state, the temperature of the outlet water was 72^oC with a voltage value of 5.5V. The proposed auto-tuner was then applied to the heat exchanger. The sampling interval T_s was 0.5 second. The relay test was performed and the PID parameters obtained. A set-point change of 0.5V was then made to the closed-loop system. Figure 6.16 shows the control performances of the proposed tuning method and the standard relay tuning method. The proposed method works quite well. Because the capacity of the boiler is not sufficient, when cold water refills the boiler under relay control, the steam pressure drops. This periodic drop acts as a disturbance which can be seen clearly in the plot.

6.4 Testing on Pilot Plants

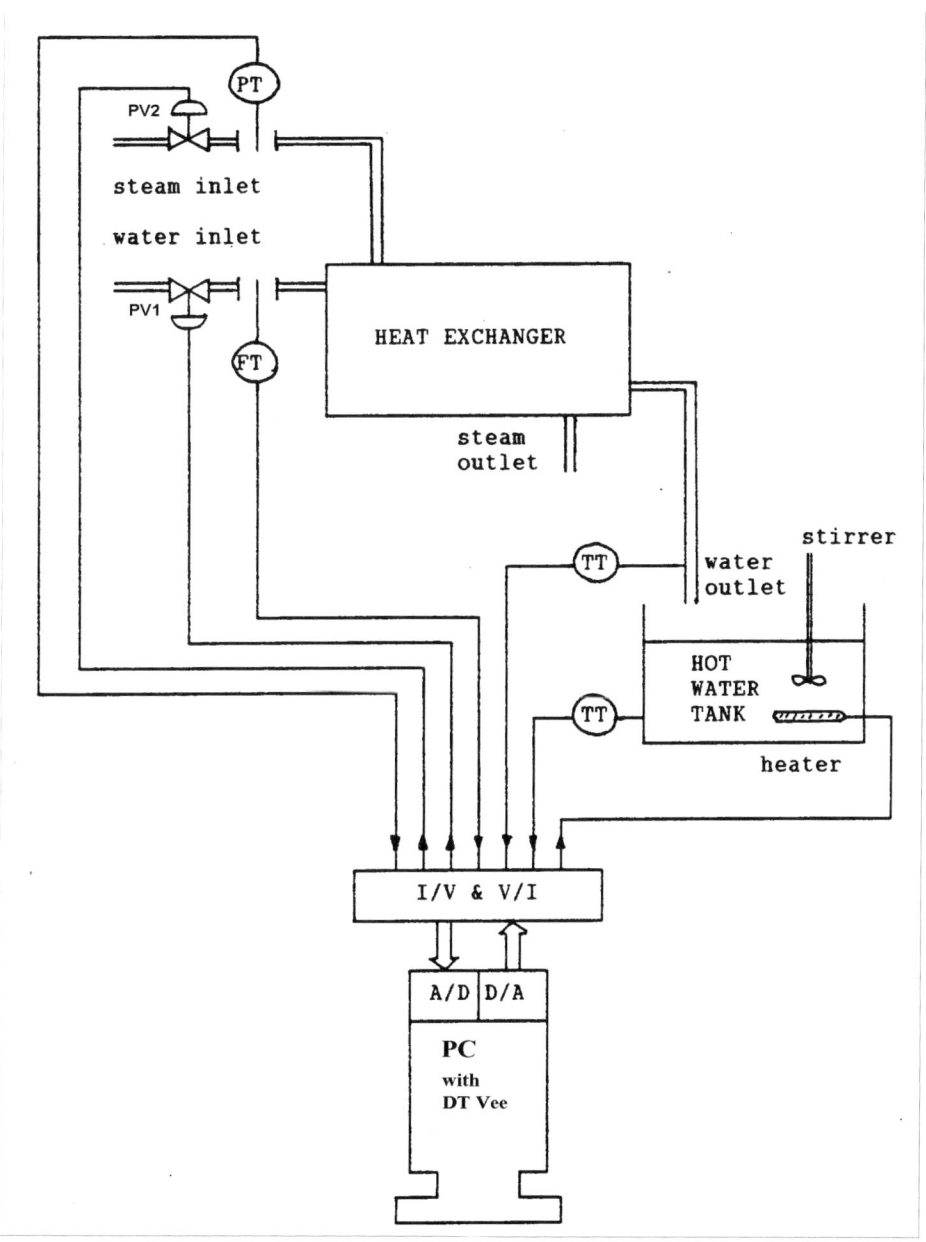

Fig. 6.15. Schematic diagram of the heat exchanger

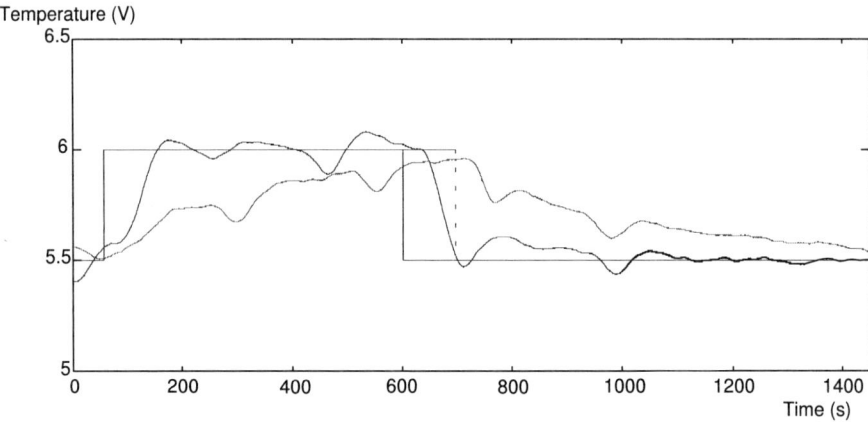

Fig. 6.16. Auto-tuner performance for a heat exchange application

6.5 Extension to the MIMO case

We have so far considered single-variable systems only. But the methods developed are readily extendable to the multivariable case. An $m \times m$ multivariable process may be described in terms of its frequency response matrix as

$$\begin{bmatrix} Y_1(j\omega) \\ \vdots \\ Y_m(j\omega) \end{bmatrix} = \begin{bmatrix} g_{11}(j\omega) & \cdots & g_{1m}(j\omega) \\ \vdots & \ddots & \vdots \\ g_{m1}(j\omega) & \cdots & g_{mm}(j\omega) \end{bmatrix} \begin{bmatrix} U_1(j\omega) \\ \vdots \\ U_m(j\omega) \end{bmatrix}. \qquad (6.41)$$

There are basically three different schemes for conducting multiple relay tests on a multivariable process. See a detailed discussion on this in Chapter 5. The sequential relay test scheme is addressed in Chapter 8. This scheme essentially leads to SISO-like problems, and the extension is easier, compared with the decentralized relay scheme. The sequential relay test is only a partial closed-loop method. A decentralized relay feedback test as shown in Figure 6.17 is a more desirable tuning test since it is a complete closed-loop method for a MIMO system. It causes less perturbation to the process, and tends to have the shorter test duration since all loops are closed simultaneously. The extension of the SISO methods to **decentralized relay feedback** is thus more challenging, and we shall concentrate on this for the rest of this section.

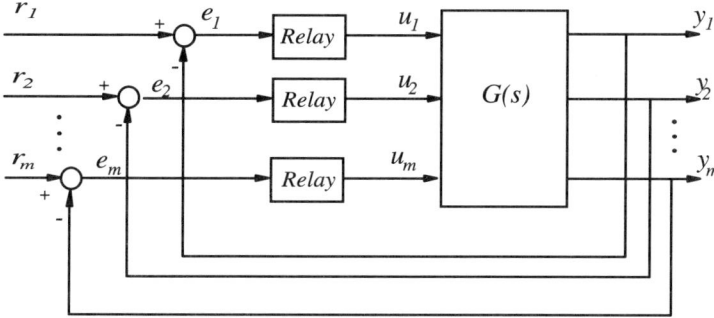

Fig. 6.17. Decentralized relay feedback

For ease of presentation, we first consider a 2 by 2 process:

$$\begin{bmatrix} Y_1(s) \\ Y_2(s) \end{bmatrix} = \begin{bmatrix} g_{11}(s) & g_{12}(s) \\ g_{21}(s) & g_{22}(s) \end{bmatrix} \begin{bmatrix} U_1(s) \\ U_2(s) \end{bmatrix}. \tag{6.42}$$

Assume that the system is at rest at $t = 0$. Decentralized relay feedback as shown in Figure 6.17 is then applied to the process. We wait for the system to reach a stationary oscillation at $t = t_1$, then change the relay amplitude of one loop and continue the relay test until all process outputs reach stationary oscillation again, at $t = t_2$. The process inputs $u_i(t), i = 1, 2$ and outputs $y_i(t), i = 1, 2$, from $t = 0$ to $t = t_2$ are recorded. The data after $t \geq t_2$, can be duplicated due to the periodic nature of the signals if they are needed.

The signals $u_i(t)$ and $y_i(t)$ are multiplied by a decay exponential $e^{-\alpha t}$ to form

$$\tilde{u}_i(t) = u_i(t)e^{-\alpha t} \text{ and } \tilde{y}_i(t) = y_i(t)e^{-\alpha t}, \quad \alpha > 0.$$

Applying the Fourier transform yields

$$\tilde{U}_i(j\omega) = \int_0^\infty \tilde{u}_i(t)e^{-j\omega t}dt = \int_0^\infty u_i(t)e^{-\alpha t}e^{-j\omega t}dt = U_i(j\omega+\alpha), \quad i = 1, 2,$$

and

$$\tilde{Y}_i(j\omega) = \int_0^\infty \tilde{y}_i(t)e^{-j\omega t}dt = \int_0^\infty y_i(t)e^{-\alpha t}e^{-j\omega t}dt = Y_i(j\omega+\alpha), \quad i = 1, 2,$$

which satisfies

$$\begin{bmatrix} \tilde{Y}_1(j\omega) \\ \tilde{Y}_2(j\omega) \end{bmatrix} = \begin{bmatrix} g_{11}(j\omega+\alpha) & g_{12}(j\omega+\alpha) \\ g_{21}(j\omega+\alpha) & g_{22}(j\omega+\alpha) \end{bmatrix} \begin{bmatrix} \tilde{U}_1(j\omega) \\ \tilde{U}_2(j\omega) \end{bmatrix}. \tag{6.43}$$

Like the SISO case in Section 6.3, $\tilde{Y}_i(j\omega)$ and $\tilde{U}_i(j\omega)$ can be computed at discrete frequencies $\omega = \omega_l$, $l = 1, 2, \cdots, M$ using the standard FFT technique from $\tilde{y}_i(t)$ and $\tilde{u}_i(t)$, where M is the user-specified number of frequency response points.

Since (6.43) is a vector equation, it is obvious that it is not sufficient to determine the process shifted frequency response $G(j\omega + \alpha)$. We now look into the signals $u_i(t)$ and $y_i(t)$ again. If the relay amplitude were not changed at $t = t_1$, we can expect that the subsequent process inputs and outputs would continue the stationary oscillation with the previous period. We extract the process inputs and outputs data for $t \leq t_1$ from $u_i(t)$ and $y_i(t)$ and duplicate inputs and outputs data for $t > t_1$ to get a new series of process inputs $u'_i(t)$ and $y'_i(t)$ as if there were no change in the relay amplitude at $t = t_1$. Applying the Fourier transform to them results in

$$\begin{bmatrix} \tilde{Y}'_1(j\omega) \\ \tilde{Y}'_2(j\omega) \end{bmatrix} = \begin{bmatrix} g_{11}(j\omega+\alpha) & g_{12}(j\omega+\alpha) \\ g_{21}(j\omega+\alpha) & g_{22}(j\omega+\alpha) \end{bmatrix} \begin{bmatrix} \tilde{U}'_1(j\omega) \\ \tilde{U}'_2(j\omega) \end{bmatrix}. \tag{6.44}$$

In view of (6.43) and (6.44), the shifted frequency response matrix $G(j\omega + \alpha)$ of the multivariable process is obtained as

$$\begin{bmatrix} g_{11}(j\omega+\alpha) & g_{12}(j\omega+\alpha) \\ g_{21}(j\omega+\alpha) & g_{22}(j\omega+\alpha) \end{bmatrix} = \begin{bmatrix} \tilde{Y}'_1(j\omega) & \tilde{Y}_1(j\omega) \\ \tilde{Y}'_2(j\omega) & \tilde{Y}_2(j\omega) \end{bmatrix} \begin{bmatrix} \tilde{U}'_1(j\omega) & \tilde{U}_1(j\omega) \\ \tilde{U}'_2(j\omega) & \tilde{U}_2(j\omega) \end{bmatrix}^{-1} \tag{6.45}$$

from which, the process frequency response $G(j\omega)$ may be recovered by using (6.32)–(6.34) elementwise.

It should be noted that our method is not confined to the relay feedback case. It can also be applied to many other input excitative signals such as steps and ramps. In the proposed method, no iteration calculation is performed, no prior knowledge of the process is required, and as many points on the process frequency response as desired can be identified.

The proposed identification method can be extended to a general m-input and m-output system with obvious modifications as follows. A decentralized relay test with $m - 1$ relay amplitude changes during the test is performed on the process, and the resultant process inputs and outputs which contain

m stationary oscillation limit cycles are recorded as $u_i^1(t)$ and $y_i^1(t)$. From the $u_i^1(t)$ and $y_i^1(t)$, $m-1$ series data $\{u_i^2(t),\cdots,u_i^m(t)\}$ and $\{y_i^2(t),\cdots,y_i^m(t)\}$ can be constructed as before, which contain $1,\cdots,m-1$ stationary limit cycles respectively. They are processed with the proposed transform to obtain $\{\tilde{U}^1(j\omega),\cdots,\tilde{U}^m(j\omega)\}$ and $\{\tilde{Y}^1(j\omega),\cdots,\tilde{Y}^m(j\omega)\}$. The shifted frequency response matrix of the process is then given by

$$G(j\omega+\alpha) = [\tilde{Y}^1(j\omega) \quad \cdots \quad \tilde{Y}^m(j\omega)][\tilde{U}^1(j\omega) \quad \cdots \quad \tilde{U}^m(j\omega)]^{-1}, \qquad (6.46)$$

from which we then compute the frequency response $G(j\omega)$ if necessary.

Example 6.5.1. Consider the well-known Wood/Berry binary distillation column plant (Wood and Berry, 1973):

$$G(s) = \begin{bmatrix} \frac{12.8e^{-s}}{1+16.7s} & \frac{-18.9e^{-3s}}{1+21s} \\ \frac{6.6e^{-7s}}{1+10.9s} & \frac{-19.4e^{-3s}}{1+14.4s} \end{bmatrix}.$$

For our method, a decentralized relay test with the same unit relay amplitudes in both loop 1 and loop 2 was performed first. The relay amplitude of loop 2 was increased to 3.0 at $t=60$ and the test continued. The resulting $u_i(t)$ and $y_i(t)$ were recorded, and $u_i'(t)$ and $y_i'(t)$ generated. The frequency response identified by the proposed method is shown in Figure 6.18. For this example, the ERR for elements $g_{11}(s)$, $g_{12}(s)$, $g_{21}(s)$ and $g_{22}(s)$ are respectively 2.4%, 0.59%, 0.62% and 0.58%. The accuracy of the proposed method is evident.

The fitting method for SISO PID tuning in the preceding section can also be extended to the MIMO case. Consider a multivariable control system in a conventional feedback configuration with a multivariable process $G(s)$ and a multivariable cross-coupled PID-type regulator $K(s)$ given by

$$K(s) = \{k_{ij}\}, \quad k_{ij} = k_{Pij} + k_{Iij}\frac{1}{s} + k_{Dij}s. \qquad (6.47)$$

One can choose the objective closed-loop transfer function matrix as

$$H(s) = \text{diag}\{h_i\} = \text{diag}\{\frac{\omega_{ni}^2}{s^2+2\zeta\omega_{ni}s+\omega_{ni}^2}e^{-L_is}\}. \qquad (6.48)$$

Then, the objective open-loop transfer function matrix is given by

$$Q(s) = \text{diag}\{q_i\} = \text{diag}\{\frac{h_i}{1-h_i}\}. \qquad (6.49)$$

With the process frequency response matrix G available and objective loop Q given in (6.49), the multivariable regulator K is designed such that GK is fitted

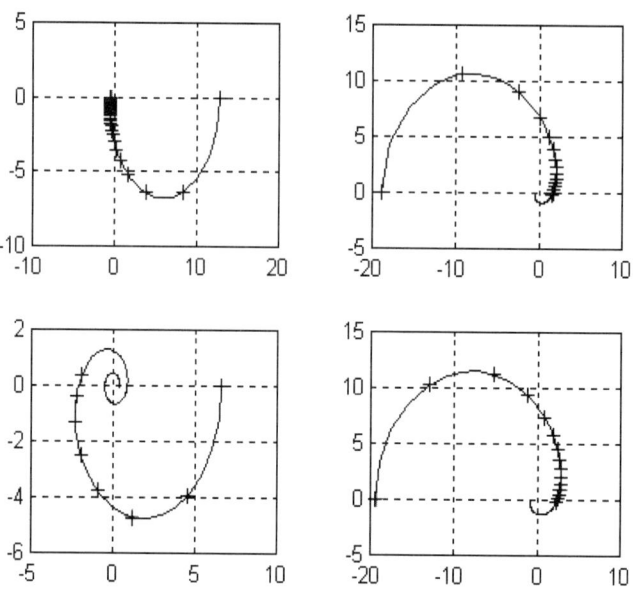

Fig. 6.18. Process frequency responses (+estimated; —actual)

to Q as well as possible. A matrix version of the linear least squares frequency response fitting can be used to determine K. For multivariable processes with high dimensions, the matrix version of the linear least squares tends to be time consuming. To simplify this matrix fitting into a set of scalar fitting problems, the relationship between the diagonal and off-diagonal elements of a multivariable controller which achieves complete decoupling is investigated. For ease of presentation, we partition an m by m system as

$$\begin{bmatrix} Y_1(s) \\ \bar{Y}_2(s) \end{bmatrix} = \begin{bmatrix} G_{11}(s) & \bar{G}_{12}(s) \\ \bar{G}_{21}(s) & \bar{G}_{22}(s) \end{bmatrix} \begin{bmatrix} U_1(s) \\ \bar{U}_2(s) \end{bmatrix} \quad (6.50)$$

where $\bar{Y}_2(s)$, $\bar{U}_2(s)$, $\bar{G}_{12}(s)$, $\bar{G}_{21}(s)$, $\bar{G}_{22}(s)$ are matrices with dimensions $(m-1) \times 1$, $(m-1) \times 1$, $1 \times (m-1)$, $(m-1) \times 1$, $(m-1) \times (m-1)$ respectively, and the multivariable controller matrix as

$$K(s) = \begin{bmatrix} K_{11} & \bar{K}_{12} \\ \bar{K}_{21} & \bar{K}_{22} \end{bmatrix}, \quad (6.51)$$

where $\bar{K}_{12}, \bar{K}_{21}, \bar{K}_{22}$ are matrices with dimensions $1 \times (m-1)$, $(m-1) \times 1$, $(m-1) \times (m-1)$ respectively. For complete decoupling of GK, its off-diagonal elements at column 1 have to satisfy

$$\bar{G}_{21}K_{11} + \bar{G}_{22}\bar{K}_{21} = 0, \tag{6.52}$$

or

$$\bar{K}_{21} = -\bar{G}_{22}^{-1}\bar{G}_{21}K_{11}, \tag{6.53}$$

supposing non-singularity of $\bar{G}_{22}$. It follows that the diagonal element of GK at position (1,1) is

$$G_{11}K_{11} + \bar{G}_{12}\bar{K}_{21} = (G_{11} - \bar{G}_{12}\bar{G}_{22}^{-1}\bar{G}_{21})K_{11} =: \tilde{G}_{11}K_{11}. \tag{6.54}$$

We regard $\tilde{G}_{11}$ as a generalized process. Matching $\tilde{G}_{11}K_{11}$ to q_1 in the frequency domain yields

$$\tilde{G}_{11}(j\omega_l)K_{11}(j\omega_l) = \tilde{G}_{11}(j\omega_l)\begin{bmatrix} 1 & \frac{1}{j\omega_l} & j\omega_l \end{bmatrix}\begin{bmatrix} k_{P11} \\ k_{I11} \\ k_{D11} \end{bmatrix} = q_1(j\omega_l), \tag{6.55}$$

for $l = 1, 2, \cdots, M$. This is a SISO frequency response fitting problem as solved in Section 6.4, and the standard least squares method can be used to determine the controller parameters $(k_{P11}, k_{I11}, k_{D11})$.

With $\bar{G}_{21}$, $\bar{G}_{22}$ and K_{11} available, the same frequency response fitting method is applied to (6.53) to obtain each element of $\bar{K}_{21}$, i.e. one needs to solve the following fitting problem:

$$\bar{K}_{21}(j\omega_l) = -\bar{G}_{22}^{-1}(j\omega_l)\bar{G}_{21}(j\omega_l)K_{11}(j\omega_l), \quad l = 1, 2, \cdots, M, \tag{6.56}$$

for each element of $\bar{K}_{21}$. Similarly, we determine other columns of K.

Example 6.5.1 (cont'd). Based on the process frequency response matrix identified earlier, the generalized process $\tilde{G}_{11}$ is obtained from (6.54). The apparent dead time L_1 is 1.06. Choose gain margin and phase margin $A_{m1} = 5$ and $\varphi_{m1} = \frac{\pi}{3}$, respectively, then we have $\zeta_1 = 0.61$ and $\omega_{n1}L_1 = 0.33$. Solving (6.55) yields

$$k_{11}(s) = 0.0942 + 0.0483\frac{1}{s} - 0.0237s.$$

The parameters of k_{21} are then obtained by solving (6.56) using a least squares method to get

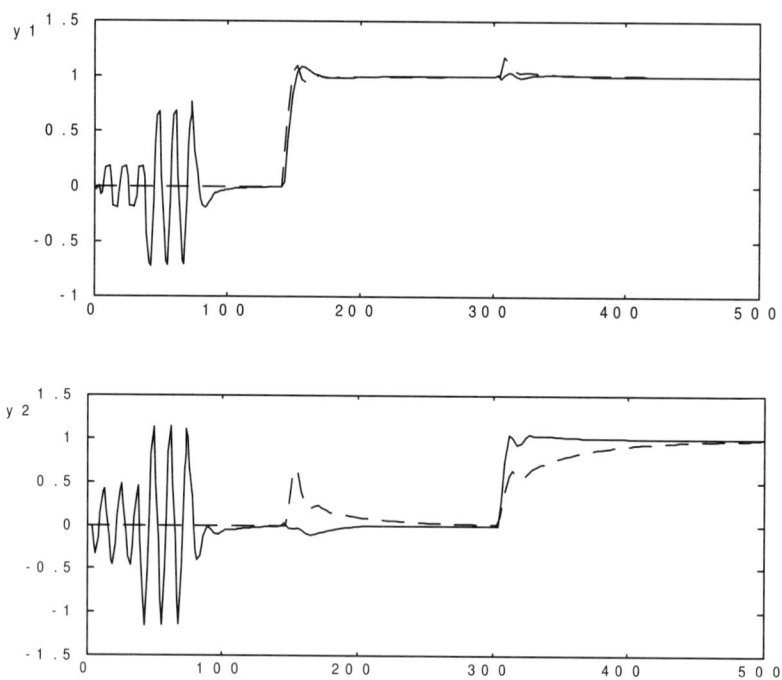

Fig. 6.19. Autotuning process (—proposed method, - - -BLT tuning method)

$$k_{21}(s) = -0.0463 + 0.0221\frac{1}{s} + 0.0347s.$$

Similarly, with $\zeta_2 = 0.707$ and $\omega_{n2}L_2 = 2$ for loop 2, we get

$$k_{22}(s) = -0.0371 - 0.0117\frac{1}{s} + 0.113s,$$

and

$$k_{12}(s) = 0.0855 - 0.0156\frac{1}{s} + 0.140s.$$

The whole tuning process and the resultant performance are shown in Figure 6.19, where the set-point change of loop 1 and loop 2 occur at $t = 141.5$ and $t = 301.5$ respectively. For comparison, the BLT tuning method of Luyben (1986) is also considered. For this multivariable process, the method gives the parameters of multiloop PI controllers as $K_c = (0.375, -0.075)$ and $T_i = (8.29, 23.6)$. Its set-point change responses are also shown in Figure 6.19. One sees that the proposed method gives significant improvement both in the system decoupling and the diagonal loop performances.

Example 6.5.2. Consider the 3 × 3 plant in Loh and Vasnani (1994):

$$G(s) = \begin{bmatrix} \frac{119e^{-5s}}{21.7s+1} & \frac{40e^{-5s}}{337s+1} & \frac{-2.1e^{-5s}}{10s+1} \\ \frac{77e^{-5s}}{50s+1} & \frac{76.7e^{-3s}}{28s+1} & \frac{-5e^{-5s}}{10s+1} \\ \frac{93e^{-5s}}{50s+1} & \frac{-36.7e^{-5s}}{166s+1} & \frac{-103.3e^{-4s}}{23s+1} \end{bmatrix}.$$

With the default values $\zeta_i = 0.707$ and $\omega_n L_i = 2$ for all three loops, the multivariable PID controller is obtained with our MIMO PID tuner as

$K(s) =$

$$\begin{bmatrix} 0.0218 + 0.000995\frac{1}{s} + 0.00967s & -.0959 - 0.000131\frac{1}{s} - 0.000477s & -0.00108 - 0.0000153\frac{1}{s} - 0.00113s \\ -0.00841 - 0.000760\frac{1}{s} + 0.0126s & 0.0720 + 0.00256\frac{1}{s} + 0.0254s & -0.00462 - 0.0000397\frac{1}{s} + 0.00607s \\ 0.00776 + 0.000747\frac{1}{s} - 0.00419s & -0.00355 - 0.000330\frac{1}{s} + 0.00266s & -0.0343 - 0.00142\frac{1}{s} - 0.0176s \end{bmatrix}.$$

The responses for a set-point change in loop 1, 2 and 3 are shown in Figure 6.20. The results show that our closed-loop is nearly decoupled and the diagonal loops have excellent performances.

168   6. Use of Relay Transient Responses

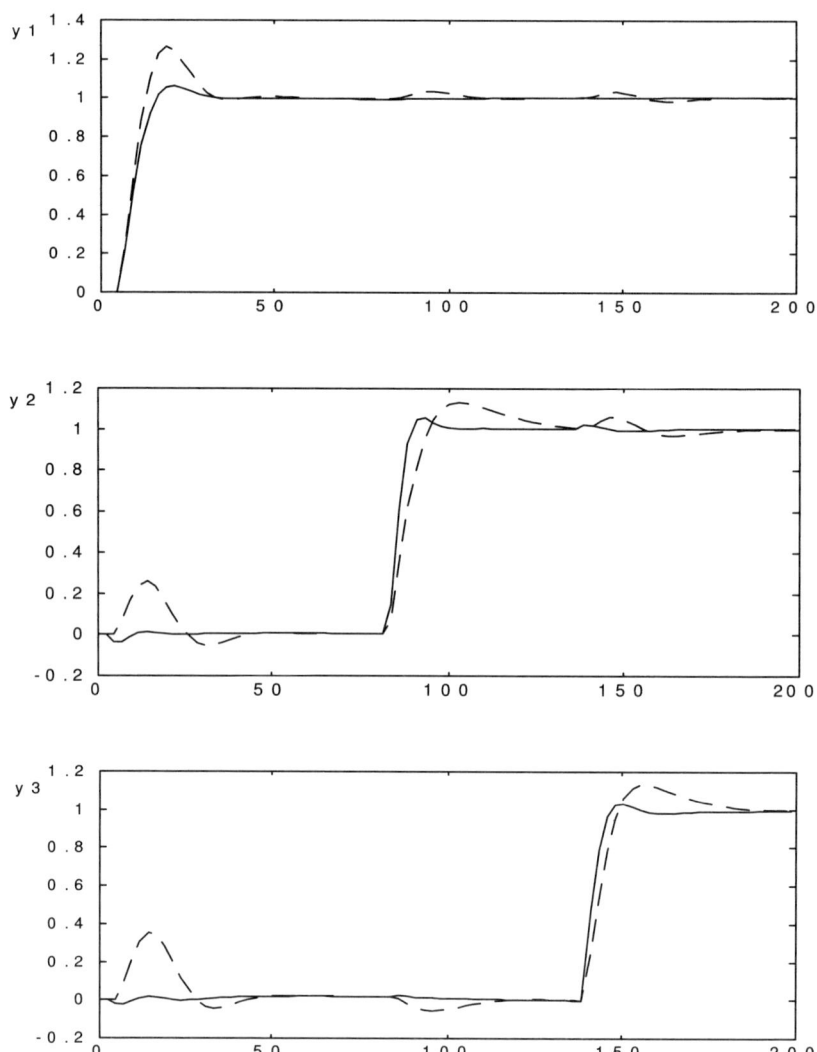

Fig. 6.20. Step responses (—proposed method, - - - Loh's method)

7. Transfer Function Modelling

The preceding two chapters indicate that relay identification usually yields information on process frequency response. However, most modern control designs make use of parametric models. Transfer functions might be the most welcome parametric model. Fitting parametric models to frequency data is thus important (Ninness, 1996). Most existing methods for transfer function identification are for delay-free processes (Ljung, 1985; Sagara and Zhao, 1990; Pintelon et al., 1994; Tugnait and Tontiruttananon, 1998) or assume that the delay is known *a priori*. It has long been recognized that the inclusion of a time-delay term in a transfer function can drastically reduce the model order and facilitate cheap and more efficient implementation of both conventional and model-based controllers (Taiwo, 1999). A frequently used method for dealing with unknown delays has been to use a shift operator model with an expanded numerator polynomial (Kurz and Goedecke, 1981). Another popular approach is based on the approximation of the dead time by a rational transfer function such as the polynomial approximation (Gawthrop and Nihtila, 1985), Padé approximation (Souza et al., 1992) or Laguerre expansion (Malti et al., 1998). Such approaches require estimation of more parameters because of the increased model order, and could result in large unacceptable approximation errors, especially when the system has a large delay. These methods have been developed for discrete systems while continuous systems are more familiar to practising control engineers. Pintelon and Biesen (1990) present a Gaussian frequency domain maximum likelihood estimator of the transfer function of linear continuous time systems with time delay based on a test using a multisine input signal.

Model reduction is closely related to this transfer function modelling problem as the frequency response of a given high-order model can be used as a nonparametric model for the latter problem. Reduced-order models are often required to simplify the design and implementation of control systems. A reduced-order model is usually adequate, provided it has dynamic characteristics close to those of the original high-order system in the frequency range which is most relevant to control system design. For plants with rational transfer functions, balanced truncation and the optimal Hankel norm approximation are popular model reduction methods. However, transfer functions encountered in the process industry usually contain time delay and may not fit into the form of a rational function plus time delay, due to multivariable interactions (see, for example, Chapter 9 of this book).

In this chapter, we present two methods for transfer function modelling from the frequency response. One is to match the given frequency response to a transfer function response using recursive least squares fitting in the frequency domain. The other identifies a transfer function model from the time domain step response constructed from the given frequency response. The models under consideration are in the form of a rational function plus possible time delay. Great attention is paid to the preservation of stability for the resulting transfer function models. These two methods are applicable provided that the transfer function or frequency response is available. The design philosophy which Part III of this book takes is to first find an ideal controller which can achieve the best possible performance in some sense but could be too complicated or even not physically realizable, and then to exploit the methods in this chapter to get a much simpler yet good approximation to the ideal one for actual implementation.

7.1 From Frequency Response

Consider a single-variable system of possibly complicated dynamics with its transfer function $G(s)$ (probably non-rational) or frequency response $G(j\omega)$ available. The problem at hand is to find an nth-order rational function plus dead time model

$$\hat{G}(s) = \hat{g}_0(s)e^{-Ls} = \frac{\beta_n s^n + \cdots + \beta_1 s + \beta_0}{s^n + \alpha_{n-1}s^{n-1} + \cdots + \alpha_1 s + \alpha_0} e^{-Ls}, \qquad (7.1)$$

such that the approximation error e defined by

$$e = \sum_{i=1}^{M} \left| W(j\omega_i)(G(j\omega_i) - \hat{G}(j\omega_i)) \right|^2 \qquad (7.2)$$

is minimized, where the interval $[\omega_1, \omega_M]$ defines the frequency range of interest and $W(j\omega_k)$ serves as the weighting. Note that (7.1) contains the unknown time delay L to be estimated and this makes the problem nonlinear. The solution to our model reduction is obtained by minimizing the approximation error over the possible range of L. This is a one-dimensional search problem and can be easily solved if an estimation of the range of L is given. A reasonable search range for L is $0.5 \sim 2.0$ times L_0 which is the initial estimate of the time delay of $G(s)$, and evaluate $15 \sim 20$ points in that range to find the optimal estimate $\hat{L}$. An alternative is to use a Newton–Raphson type of algorithm to search for the optimal L (Lilja, 1989), which requires no prior range for L.

If the time delay L is known, then the problem becomes to approximate a modified plant $g_0(s) = G(s)e^{Ls}$ with a rational transfer function

$$\hat{g}_0(s) = \frac{\beta_n s^n + \cdots + \beta_1 s + \beta_0}{s^n + \alpha_{n-1} s^{n-1} + \cdots + \alpha_1 s + \alpha_0},$$

such that

$$e_0 \triangleq \sum_{i=1}^{M} |W(j\omega_i)(g_0(j\omega_i) - \hat{g}_0(j\omega_i))|^2 \qquad (7.3)$$

is minimized. Equation (7.3) falls into the framework of transfer function identification in the frequency domain. For this identification, a number of methods are available (Pintelon et al., 1994); the recursive least squares (RLS) algorithm is simple and effective and is briefly described as follows:

$$e_0^{(k)} \triangleq \sum_{i=1}^{M} |\bar{W}_i^{(k)} \{ g_0(j\omega_i)[(j\omega_i)^n + \alpha_{n-1}^{(k)}(j\omega_i)^{n-1} + \ldots + \alpha_1^{(k)}(j\omega_i) + \alpha_0^{(k)}]$$
$$- [\beta_n^{(k)}(j\omega_i)^n + \ldots + \beta_1^{(k)}(j\omega_i) + \beta_0^{(k)}] \}|^2, \qquad (7.4)$$

where

$$\bar{W}_i^{(k)} \triangleq \frac{W(j\omega_i)}{(j\omega_i)^n + \alpha_{n-1}^{(k-1)}(j\omega_i)^{n-1} + \ldots + \alpha_1^{(k-1)}(j\omega_i) + \alpha_0^{(k-1)}} \qquad (7.5)$$

acts as a weighting function in the standard least squares problem in braces. Equation (7.4) is re-arranged to yield

$$e_0^{(k)} \triangleq \sum_{i=1}^{M} |\eta_i^{(k)} - \phi_i^{(k)T} \theta^{(k)}|^2,$$

where

$$\eta_i^{(k)} = -g_0(j\omega_i)(j\omega_i)^n \bar{W}_i^{(k)}, \qquad (7.6)$$

$$\theta^{(k)} = [\alpha_{n-1}^{(k)} \ldots \alpha_0^{(k)} \ \beta_n^{(k)} \ldots \beta_1^{(k)} \ \beta_0^{(k)}]^T, \qquad (7.7)$$

$$\phi_i^{(k)} = [g_0(j\omega_i)(j\omega_i)^{n-1} \ldots g_0(j\omega_i) \ -(j\omega_i)^n \ldots -(j\omega_i) \ -1]^T \bar{W}_i^{(k)}. \qquad (7.8)$$

Then, we have the recursive equation for $\theta^{(i)}$ as

$$\theta^{(k,i)} = \theta^{(k,i-1)} + K^{(k,i)}\varepsilon^{(k,i)}, \quad i = 1, 2, \cdots, M, \tag{7.9}$$

where

$$\varepsilon^{(k,i)} = \eta_i^{(k)} - \phi_i^{(k)T}\theta^{(k,i-1)}, \tag{7.10}$$

$$K^{(k,i)} = P^{(k,i-1)}\phi_i^{(k)}(I + \phi_i^{(k)T}P^{(k,i-1)}\phi_i^{(k)})^{-1}, \tag{7.11}$$

$$P^{(k,i)} = (I - K^{(k,i)}\phi_i^{(k)T})P^{(k,i-1)}. \tag{7.12}$$

Once the above RLS in (7.9)–(7.12) has been completed, the resultant parameter vector $\theta^{(k)} = \theta^{(k,M)}$ is used to update $\bar{W}_i^{(k)}$ to

$$\bar{W}_i^{(k+1)} = \frac{1}{(j\omega_i)^n + \alpha_{n-1}^{(k)}(j\omega_i)^{n-1} + \ldots + \alpha_1^{(k)}(j\omega_i) + \alpha_0^{(k)}}, \tag{7.13}$$

and (7.9)–(7.12) are repeated to calculate $\theta^{(k+1)}$. On convergence, the resultant parameter vector will form one solution to (7.3).

Different weighting functions are employed in the various methods (Pintelon et al., 1994). For simplicity, it is recommended that $\bar{W}_i^{(k)}$ is chosen as

$$\frac{1}{(j\omega_i)^n + \alpha_{n-1}^{(k-1)}(j\omega_i)^{n-1} + \ldots + \alpha_1^{(k-1)}(j\omega_i) + \alpha_0^{(k-1)}}.$$

Simulation shows that the most important frequency range for model reduction is a decade above and below ω_g, where ω_g is the unity gain crossover frequency of the transfer function $G_0(s)$. Therefore, the frequency range $[\omega_1, \omega_M]$ in the optimal fitting problem (7.3) is chosen to span M logarithmically equally spaced points between $0.1\omega_g$ and $10\omega_g$.

Once a reduced-order model $\hat{G}(s)$ is found, the following frequency response maximum relative estimation error can be evaluated:

$$ERR := \max_{\omega \in (0,\, \omega_M)} |\frac{\hat{G}(j\omega) - G(j\omega)}{G(j\omega)}|. \tag{7.14}$$

The following criterion is used to validate the solution:

$$ERR \leq \epsilon, \tag{7.15}$$

where ϵ is the user-specified fitting error threshold. ϵ is specified according to the desired accuracy of the RLR approximation to the original dynamics $G(s)$. Usually ϵ may be set between 1% and 10%. If (7.15) is met, the procedure stops. Otherwise, n is increased by 1 to the smallest integer n that satisfies $ERR \leq \epsilon$.

Algorithm 7.1 *Seek a reduced-order model $\hat{G}(s)$ of order n in (7.1) given $G(s)$ or $G(j\omega)$, approximation threshold ϵ, initial time delay L_0 and parameter vector θ^0.*

Step 1. Choose N between 15 and 20, set $\Delta L = \frac{1.5L_0}{N}$, and obtain $L_k = 0.5L_0 + k\Delta L$, $k = 0, 1, \cdots, N$.

Step 2. Start with $\hat{G}(s)$ as in (7.1) with $n = 1$.

Step 3. For each L_k, find the nth order rational approximation solution $\hat{g}_0(s)$ to the modified process $g_0(s) = G(s)e^{L_k s}$ with the RLS method in (7.9)–(7.12) and evaluate the corresponding approximation error ERR in (7.15) for $\hat{G}(s) = \hat{g}_0(s)e^{-L_k s}$.

Step 4. Take as the solution $\hat{G}(s)$ that yields the minimum error ERR and $ERR \leq \epsilon$. Otherwise, $n+1 \to n$ and go to Step 3.

The preservation of stability is a crucial issue in frequency domain based model reduction. Let $\hat{G}(s) = \hat{g}_0(s)e^{-Ls}$ be the value that yields the smallest approximation error e in (7.2). It should be pointed out that Algorithm 7.1 with zero initial parameter vector might result in unstable $\hat{G}(s)$, especially for high-order models, even though $G(s)$ is stable. One can use stability tests and the projection algorithm (Ljung and Söderström, 1983) to constrain the poles of the model to be in the stable region. However, simulation shows that this method can slow down convergence of the recursive least squares considerably, and may result in large modelling error if the dynamics of the transfer function to be modelled is complicated. Here, we notice that since the problem of minimizing (7.2) is nonlinear, Algorithm 7.1 may give different local optimal solutions if it starts with different initial settings. Among those solutions, only stable $\hat{G}(s)$ models are required, given a stable $G(s)$. If we initiate the algorithm with a stable model, the algorithm is likely to reach a stable approximate upon convergence.

When $n = 1$, we set the initial model of $\hat{G}(s)$,

$$\hat{G}_0 = \frac{\beta_0}{s + \alpha_0} e^{-L_0 s}, \tag{7.16}$$

as follows. Matching $\hat{G}_0(j\omega)$ to $G(j\omega)$ at $\omega = 0$ and $\omega = \omega_c$, where ω_c is the phase crossover frequency of $G(s)$, i.e., $\angle G(j\omega_c) = -\pi$, we obtain

$$\begin{cases} \alpha_0 = \omega_c \sqrt{\frac{|G(j\omega_c)|^2}{G^2(0) - |G(j\omega_c)|^2}}, \\ \beta_0/\alpha_0 = G(0), \\ L_0 = \frac{1}{\omega_c}\{-\arg[G(j\omega_c)] - \tan^{-1}(\frac{\omega_c}{\alpha_0})\}. \end{cases} \tag{7.17}$$

When $n = 2$, one may adopt the following structure:

$$\hat{G}_0 = \frac{\beta_0}{s^2 + \alpha_1 s + \alpha_0} e^{-L_0 s}. \tag{7.18}$$

Similarly, match $\hat{G}_0(j\omega)$ to $G(j\omega)$ at the two points $\omega = \omega_b$ and $\omega = \omega_c$, where $\angle G(j\omega_b) = -\frac{\pi}{2}$ and $\angle G(j\omega_c) = -\pi$. It then follows (Wang et al., 1999d) that the parameters β_0, α_1, α_0 and L_0 can be determined as

$$\begin{cases} \frac{\sin(\omega_c L_0)}{\cos(\omega_b L_0)} = \frac{\omega_c |G(j\omega_c)|}{\omega_b |G(j\omega_b)|}, \\ \beta_0 = (\omega_c^2 - \omega_b^2) \left[\frac{\sin(\omega_b L_0)}{|G(j\omega_b)|} + \frac{\cos(\omega_c L_0)}{|G(j\omega_c)|} \right]^{-1}, \\ \alpha_1/\beta_0 = \frac{\sin(\omega_c L_0)}{\omega_c |G(j\omega_c)|}, \\ \alpha_0/\beta_0 = (\omega_c^2 - \omega_b^2) \left[\frac{\omega_c^2 \sin(\omega_b L_0)}{|G(j\omega_b)|} + \frac{\omega_b^2 \cos(\omega_c L_0)}{|G(j\omega_c)|} \right]^{-1}. \end{cases} \tag{7.19}$$

For the cases $n > 2$, we can set the initial model $\hat{G}_0(s)$ of the respective orders with β_0, α_1, α_0 and L_0 determined as in (7.19), while all the remaining high-degree coefficients are set to 0. Extensive simulations show that this technique works very well.

Example 7.1.1. Consider a high-order plant (Maffezzoni and Rocco, 1997):

$$G(s) = 2.15 \frac{(-2.7s + 1)(158.5s^2 + 6s + 1)}{(17.5s + 1)^4 (20s + 1)} e^{-14s}.$$

With zero initial parameter vector, Algorithm 7.1 gives a reduced-order model

$$\hat{G}(s) = \frac{-0.0275s^2 - 0.0010s - 0.0001}{s^3 - 0.0129s^2 - 0.0029s - 0.0001} e^{-63.20s}, \tag{7.20}$$

which is unstable. When the parameters obtained from (7.19) are adopted for the initial model, we get

$$\hat{G}(s) = \frac{0.4687s^2 + 0.0038s + 0.0024}{s^3 + 1.2160s^2 + 0.0665s + 0.0011} e^{-41.90s}, \tag{7.21}$$

which is stable with $ERR = 2.71\%$. The frequency responses of the actual and estimated models are shown in Figure 7.1.

7.2 From Step Response

In this section, we present a simple yet robust algorithm for the identification of linear continuous time-delay processes from step responses. New linear regression equations are derived from the solution and its various-order integrals of the process differential equation. The regression parameters are then estimated

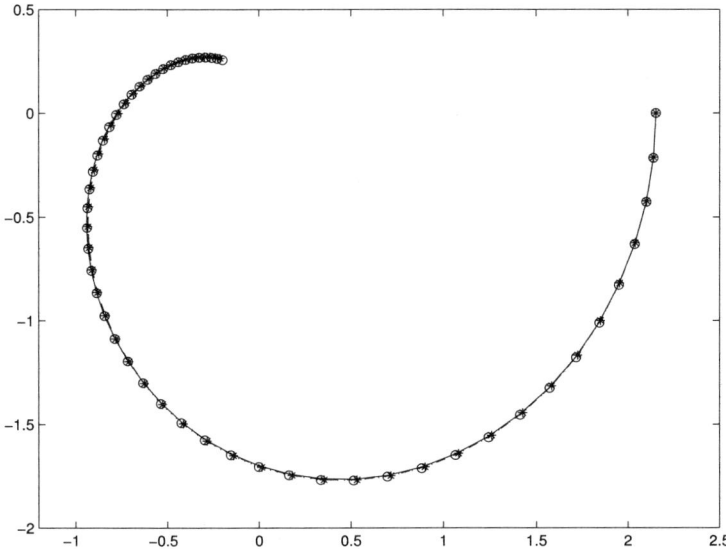

Fig. 7.1. Frequency domain model reduction

(——— actual process; ∗ ∗ ∗ unstable model; − ∘ − ∘ − stable model)

without iteration, and explicit relationships between the regression parameters and those in the process are given. Due to use of the process output integrals in the regression equations, the resulting parameter estimation is very robust in the face of large measurement noise in the output. The method is first detailed in Subsection 7.2.1 for a second-order plus dead time model (SOPDT) with one zero, which can approximate most practical industrial processes, covering monotonic or oscillatory dynamics of minimum-phase or non-minimum-phase processes. Such a model can be obtained without any iteration. A general model is addressed and theoretical supports are provided in Subsection 7.2.2. This is followed by implementation issues in Subsection 7.2.3 and simulation and real-time testing in Subsection 7.2.4.

7.2.1 Second-order Modelling

This subsection focuses on SOPDT modelling. It serves for motivation of the general method to be described in the next subsection and recommended for use in applications since such SOPDT models essentially cover most practical industrial processes (Luyben, 1990).

Assume that a stable process is represented by

176   7. Transfer Function Modelling

$$Y(s) = G(s)U(s) = \frac{b_1 s + b_2}{s^2 + a_1 s + a_2} e^{-Ls} U(s). \tag{7.22}$$

Assume also that the process has distinct (possibly complex) poles and that the input is of step type, i.e., $U(s) = h/s$. Then the process output is

$$Y(s) = K_p h \lambda_1 \lambda_2 \frac{\beta s + 1}{s(s + \lambda_1)(s + \lambda_2)} e^{-Ls}$$
$$= (\frac{\alpha_0}{s} + \frac{\alpha_1}{s + \lambda_1} + \frac{\alpha_2}{s + \lambda_2}) e^{-Ls}, \quad \lambda_1 \neq \lambda_2, \tag{7.23}$$

where $\alpha_0 = K_p h$, $\alpha_1 = -K_p h \lambda_1 \lambda_2 \frac{\beta \lambda_1 - 1}{\lambda_1(\lambda_1 - \lambda_2)}$ and $\alpha_2 = K_p h \lambda_1 \lambda_2 \frac{\beta \lambda_2 - 1}{\lambda_2(\lambda_1 - \lambda_2)}$. Under zero initial conditions, the output time response is

$$y(t) = \begin{cases} 0, & \text{if } t < L; \\ K_p h + \alpha_1 e^{-\lambda_1(t-L)} + \alpha_2 e^{-\lambda_2(t-L)}, & \text{if } t \geq L. \end{cases} \tag{7.24}$$

In real applications, such a step test is conducted on the process. The resulting input and output responses, $\{u(t), y(t), t = t_1, t_2, \cdots\}$, are logged for the identification of model (7.22). The static gain, K_p, is easily found from the steady-state response as

$$K_p = \frac{y(\infty)}{h}.$$

The output transient response, $\Delta y(t) = y(t) - y(\infty) = y(t) - K_p h$, can then be obtained from (7.24) as

$$\Delta y(t) = \begin{cases} -K_p h, & \text{if } t < L; \\ \alpha_1 e^{-\lambda_1(t-L)} + \alpha_2 e^{-\lambda_2(t-L)}, & \text{if } t \geq L. \end{cases} \tag{7.25}$$

For any time $t \geq L$, one integrates $\Delta y(t)$ once and twice to give

$$\int_0^t \Delta y(\tau_1) d\tau_1 = -K_p h L - \frac{\alpha_1}{\lambda_1} e^{-\lambda_1(t-L)} - \frac{\alpha_2}{\lambda_2} e^{-\lambda_2(t-L)} + \frac{\alpha_1}{\lambda_1} + \frac{\alpha_2}{\lambda_2}, \tag{7.26}$$

$$\int_0^t \int_0^{\tau_2} \Delta y(\tau_1) d\tau_1 d\tau_2 = \frac{1}{2} K_p h L^2 - K_p h L t + \frac{\alpha_1}{\lambda_1^2} e^{-\lambda_1(t-L)} + \frac{\alpha_2}{\lambda_2^2} e^{-\lambda_2(t-L)}$$
$$- \frac{\alpha_1}{\lambda_1^2} - \frac{\alpha_2}{\lambda_2^2} + (\frac{\alpha_1}{\lambda_1} + \frac{\alpha_2}{\lambda_2})(t - L). \tag{7.27}$$

Equations (7.25) and (7.26) are combined into the matrix form:

$$\begin{bmatrix} \Delta y(t) \\ \int_0^t \Delta y(\tau_1) d\tau_1 \end{bmatrix} = \begin{bmatrix} 1 & 1 \\ -\frac{1}{\lambda_1} & -\frac{1}{\lambda_2} \end{bmatrix} \begin{bmatrix} \alpha_1 e^{-\lambda_1(t-L)} \\ \alpha_2 e^{-\lambda_2(t-L)} \end{bmatrix} + \begin{bmatrix} 0 \\ -K_p h L + \frac{\alpha_1}{\lambda_1} + \frac{\alpha_2}{\lambda_2} \end{bmatrix}$$
$$:= Ax + B. \tag{7.28}$$

One notes that the matrix A is nonsingular due to the assumed distinct λ_1 and λ_2, and thus (7.28) is solvable for x. Substituting the solution x of (7.28) into (7.27) yields

$$\int_0^t \int_0^{\tau_2} \Delta y(\tau_1) d\tau_1 d\tau_2$$
$$= -\frac{1}{\lambda_1 \lambda_2} \Delta y(t) - (\frac{1}{\lambda_1} + \frac{1}{\lambda_2}) \int_0^t \Delta y(\tau_1) d\tau_1$$
$$- (\frac{1}{\lambda_1 \lambda_2} - \frac{1}{2}L^2 + \beta L) K_p h - (\frac{1}{\lambda_1} + \frac{1}{\lambda_2} + L - \beta) K_p h t. \quad (7.29)$$

Define

$$\gamma(t) = -\int_0^t \int_0^{\tau_2} \Delta y(\tau_1) d\tau_1 d\tau_2,$$
$$\phi(t) = \left[\Delta y(t), \int_0^t \Delta y(\tau_1) d\tau_1, K_p h, K_p h t\right]^T,$$
$$\theta = \left[\frac{1}{\lambda_1 \lambda_2}, \frac{1}{\lambda_1} + \frac{1}{\lambda_2}, \frac{1}{\lambda_1 \lambda_2} - \frac{1}{2}L^2 + \beta L, \frac{1}{\lambda_1} + \frac{1}{\lambda_2} + L - \beta\right]^T. \quad (7.30)$$

Then, (7.29) can be expressed as

$$\gamma(t) = \phi^T(t) \theta(t).$$

Invoke (7.29) for $t = t_i \geqslant L$, $i = 1, 2, \cdots, N$, to form the regression:

$$\Gamma = \Phi \theta, \quad (7.31)$$

where $\Gamma = [\gamma(t_1), \gamma(t_2), \cdots, \gamma(t_N)]^T$, and $\Phi = [\phi(t_1), \phi(t_2), \cdots, \phi(t_N)]^T$.

One can readily see that four columns in Φ are independent of each other and $\Phi^T \Phi$ is nonsingular (the formal proof is given in the next subsection). Thus, the ordinary least squares (LS) can be applied to (7.31) to find its solution:

$$\hat{\theta} = (\Phi^T \Phi)^{-1} \Phi^T \Gamma. \quad (7.32)$$

The first question which arises from (7.32) is whether or not the estimate $\hat{\theta}$ is consistent when there is measurement noise. To address this, suppose that the actual measurement of the process output $\hat{y}(t)$ is corrupted by measurement noise $v(t)$, which is assumed to be a strictly stationary stochastic process with zero mean; then we have

$$y(t) = \hat{y}(t) + v(t),$$

or

$$\Delta y(t) = \alpha_1 e^{-\lambda_1(t-L)} + \alpha_2 e^{-\lambda_2(t-L)} + v(t), \quad \text{for } t \geqslant L.$$

Equation (7.31) will accordingly be changed to

$$\Phi\theta = \Gamma + \Delta, \qquad (7.33)$$

where $\Delta = [\delta(t_1), \delta(t_2), \cdots, \delta(t_N)]^T$ and

$$\delta(t) = \int_0^t \int_0^{\tau_2} v(\tau_1)d\tau_1 d\tau_2 - \left(\frac{1}{\lambda_1} + \frac{1}{\lambda_2}\right)\int_0^t v(\tau_1)d\tau_1 - \frac{1}{\lambda_1\lambda_2}v(t).$$

It follows from Söderström and Stoica (1989) that in such a case, the ordinary LS estimate in (7.32) is not consistent, because $\Delta(t)$ is now correlated with $\Phi(t)$. One solution is to use the instrumental variable (IV) method (Young, 1970; Strejc, 1980).

Proposition 7.2.1. *For the system in (7.33), if the instrumental variable matrix Z is chosen to satisfy the following two limiting properties (Söderström and Stoica, 1989):*

(i) *the inverse of $\lim_{N\to\infty} \frac{1}{N} Z^T \Phi$ exists;*
(ii) $\lim_{N\to\infty} \frac{1}{N} Z^T \Delta = 0$,

then the estimate given by

$$\hat{\theta} = (Z^T\Phi)^{-1} Z^T \Gamma \qquad (7.34)$$

is consistent.

There might be many choices for Z. The condition number,

$$\text{cond}(Z^T\Phi) = \frac{\overline{\sigma}(Z^T\Phi)}{\underline{\sigma}(Z^T\Phi)},$$

may be used as a criterion for choosing a suitable Z, where $\underline{\sigma}(Z^T\Phi)$ and $\overline{\sigma}(Z^T\Phi)$ are the smallest and largest singular values of the square matrix $Z^T\Phi$, respectively. With these requirements in mind, we choose Z as

$$Z = \begin{bmatrix} \frac{1}{t_1^2} & \frac{1}{t_1^3} & t_1^2 & t_1^3 \\ \frac{1}{t_2^2} & \frac{1}{t_2^3} & t_2^2 & t_2^3 \\ \vdots & \vdots & \vdots & \vdots \\ \frac{1}{t_N^2} & \frac{1}{t_N^3} & t_N^2 & t_N^3 \end{bmatrix}. \qquad (7.35)$$

The formal proof that it satisfies conditions (i) and (ii) is given in the next subsection.

Once θ is found from (7.34), one has to recover a_i, b_i and L to obtain the process model (7.22). It follows from (7.30) and simple algebra that

$$\begin{bmatrix} a_1 \\ a_2 \\ \beta \\ L \end{bmatrix} = \begin{bmatrix} \theta_2/\theta_1 \\ 1/\theta_1 \\ \pm\sqrt{2(\theta_3 - \theta_1) + (\theta_4 - \theta_2)^2} \\ \theta_4 - \theta_2 + \beta \end{bmatrix}. \tag{7.36}$$

A negative β corresponds to a non-minimum-phase process and causes an inverse response, i.e., the output step response first moves in an opposite direction to its final value. A positive β corresponds to a minimum-phase process. Thus, by observing the output time response, one is able to determine β from

$$\beta = \begin{cases} -\sqrt{2(\theta_3 - \theta_1) + (\theta_4 - \theta_2)^2}, & \text{if inverse response is detected;} \\ \sqrt{2(\theta_3 - \theta_1) + (\theta_4 - \theta_2)^2}, & \text{otherwise.} \end{cases}$$

This rule always works well if the underlying process is of second order or is second-order dominated.

Remark 7.2.1. It is also noted that the number $2(\theta_3 - \theta_1) + (\theta_4 - \theta_2)^2$ may sometimes be negative if the process behaves much differently from second-order dynamics. This gives rise to complex β and L, which is unacceptable. Usually, a higher-order model should be utilized in such cases and the problem is then overcome. If, on the other hand, a SOPDT model is still preferred for practical reasons (say, fixed hardware and software structure in the given controllers), the problem can formally be formulated as an LS estimation with the constraint

$$\min_\theta J = (\Gamma - \Phi\theta)^T(\Gamma - \Phi\theta)$$
$$\text{s.t.} \quad 2(\theta_3 - \theta_1) + (\theta_4 - \theta_2)^2 \geq 0, \tag{7.37}$$

or

$$\min_\theta J = (\Gamma - \Phi\theta)^T(\Gamma - \Phi\theta)$$
$$\text{s.t.} \quad \theta^T A\theta + B\theta \geq 0,$$

where

$$A = \begin{bmatrix} 0 & 0 & 0 & 0 \\ 0 & 1 & 0 & -1 \\ 0 & 0 & 0 & 0 \\ 0 & -1 & 0 & 1 \end{bmatrix}, \quad B = \begin{bmatrix} -2 & 0 & 2 & 0 \end{bmatrix}.$$

The resulting solution guarantees real values for β and L, but requires more complicated calculations. A practical and natural choice for β is to simply take $\beta = 0$, and this leads to $L = \theta_4 - (\frac{1}{\lambda_1} + \frac{1}{\lambda_2})$, or $L = \sqrt{2(\lambda_1 \lambda_2 - \theta_3)}$. A suitable choice for L could be one that minimizes the total squared error $(\Gamma^T - \Phi\theta)^T(\Gamma^T - \Phi\theta)$.

One can conclude from the development presented above that a SOPDT model can be identified from a process step response by applying a LS or IV algorithm only once without any iteration. The simulation and real-time implementation in Subsection 7.2.4 further show that the proposed identification is also very robust in noisy environments. It is thus appealing to extend this method to the general case of nth-order modelling.

7.2.2 nth-order Modelling

Suppose that a time-invariant stable process with distinct poles (possibly complex) is represented by a NOPDT model:

$$Y(s) = G(s)U(s) = \frac{b_1 s^{n-1} + b_2 s^{n-2} + \cdots + b_{n-1} s + b_n}{s^n + a_1 s^{n-1} + \cdots + a_{n-1} s + a_n} e^{-Ls} U(s)$$

$$= K_p \prod_{i=1}^{n} \lambda_i \frac{\prod_{i=1}^{n-1}(\beta_i s + 1)}{\prod_{i=1}^{n}(s + \lambda_i)} e^{-Ls} U(s), \tag{7.38}$$

where

$$\begin{cases} a_k = \sum_{1 \leqslant i_1 < i_2 < \cdots < i_k \leqslant n} \lambda_{i_1} \lambda_{i_2} \cdots \lambda_{i_k}, & k = 1, 2, \cdots, n; \\ b_k = K_p \prod_{i=1}^{n} \lambda_i \sum_{1 \leqslant i_1 < i_2 < \cdots < i_{n-k} \leqslant n-1} \beta_{i_1} \beta_{i_2} \cdots \beta_{i_{n-k}}, \\ \quad k = 1, 2, \cdots, n-1; \\ b_n = K_p \prod_{i=1}^{n} \lambda_i. \end{cases} \tag{7.38a}$$

Assume that the input is of step type, i.e., $U(s) = h/s$, then the process output becomes

$$Y(s) = (\frac{\alpha_0}{s} + \frac{\alpha_1}{s + \lambda_1} + \frac{\alpha_2}{s + \lambda_2} + \cdots + \frac{\alpha_n}{s + \lambda_n}) e^{-Ls}, \tag{7.39}$$

where

$$\alpha_j = -K_p h \prod_{i=1}^{n} \lambda_i \frac{\prod_{i=1}^{n-1}(1 - \beta_i \lambda_j)}{\lambda_j \prod_{i=1, i \neq j}^{n}(\lambda_i - \lambda_j)}, \quad j = 1, 2, \cdots, n. \tag{7.40}$$

Under zero initial conditions, the output time response is

$$y(t) = \begin{cases} 0, & \text{if } t < L; \\ K_p h + \sum_{i=1}^{n} \alpha_i e^{-\lambda_i(t-L)}, & \text{if } t \geq L. \end{cases} \quad (7.41)$$

Suppose that the static gain, K_p, is obtained first and separately from the output steady state as

$$K_p = \frac{y(\infty)}{h}.$$

Otherwise, refer to Subsection 7.2.3 for our modification for simultaneous estimation of all the parameters. The output transient response, $\Delta y(t) = y(t) - y(\infty) = y(t) - K_p h$, can be written as

$$\Delta y(t) = \begin{cases} -K_p h, & \text{if } t < L; \\ \sum_{i=1}^{n} \alpha_i e^{-\lambda_i(t-L)}, & \text{if } t \geq L. \end{cases} \quad (7.42)$$

For an integer $m \geq 1$, define

$$\int_{[t_1,\, t_2]}^{(m)} \Delta y = \underbrace{\int_{t_1}^{t_2} \int_{t_1}^{\tau_n} \cdots \int_{t_1}^{\tau_2} \Delta y(\tau_1) d\tau_1 d\tau_2 \cdots d\tau_{n-1} d\tau_n}_{m}.$$

Lemma 7.2.1.

$$\int_{[0,\, t]}^{(m)} \Delta y = \begin{cases} -\frac{1}{m!} K_p h t^m, & \text{if } t < L; \\ -\sum_{i=0}^{m-1} \frac{1}{i!(m-i)!} K_p h L^{m-i}(t-L)^i + \sum_{i=1}^{n} \alpha_i [\frac{(-1)^m}{\lambda_i^m} e^{-\lambda_i(t-L)} \\ \quad + \sum_{k=1}^{m} \frac{(-1)^{k-1}}{(m-k)!} \frac{1}{\lambda_i^k} (t-L)^{m-k}], & \text{if } t \geq L. \end{cases}$$
(7.43)

Proof. The proof is by induction. For $m = 1$, the integration of Δy in (7.42) gives

$$\int_0^t \Delta y(\tau_1) d\tau_1 = \begin{cases} -K_p h t, & \text{if } t < L, \\ -K_p h L - \sum_{i=1}^{n} \frac{\alpha_i}{\lambda_i} e^{-\lambda_i(t-L)} + \sum_{i=1}^{n} \frac{\alpha_i}{\lambda_i}, & \text{if } t \geq L, \end{cases}$$

which coincides with (7.43) for $m = 1$. Suppose now that (7.43) holds for m. By definition, we have $\int_{[0,\, t]}^{(m+1)} \Delta y = \int_0^t \int_{[0,\, \tau]}^{m} \Delta y d\tau$. For $t < L$, one sees that

$$\int_{[0,\, t]}^{(m+1)} \Delta y = \int_0^t (-\frac{1}{m!} K_p h \tau^m) d\tau = -\frac{1}{(m+1)!} K_p h t^{m+1}. \quad (7.44)$$

For $t \geq L$, it follows that

$$\int_{[0,\,t]}^{(m+1)} \Delta y = \int_0^L \int_{[0,\,\tau]}^m \Delta y d\tau + \int_L^t \int_{[0,\,\tau]}^m \Delta y d\tau$$

$$= \int_0^L (-\frac{1}{m!} K_p h \tau^m) d\tau + \int_L^t \left[-\sum_{i=0}^{m-1} \frac{1}{i!(m-i)!} K_p h L^{m-i}(\tau-L)^i d\tau \right.$$

$$+ \int_L^t \sum_{i=1}^n \alpha_i [\frac{(-1)^m}{\lambda_i^m} e^{-\lambda_i(\tau-L)} + \sum_{k=1}^m \frac{(-1)^{k-1}}{(m-k)!} \frac{1}{\lambda_i^k} (t-L)^{m-k}] d\tau$$

$$= -\frac{1}{(m+1)!} K_p h L^{m+1} - \sum_{i=1}^m \frac{1}{i!(m+1-i)!} K_p h L^{m+1-i}(t-L)^i$$

$$+ \sum_{i=1}^n \alpha_i [\frac{(-1)^{m+1}}{\lambda_i^{m+1}} e^{-\lambda_i(t-L)}$$

$$+ \frac{(-1)^m}{\lambda_i^{m+1}} + \sum_{k=1}^m \frac{(-1)^{k-1}}{(m+1-k)!} \frac{1}{\lambda_i^k} (t-L)^{m+1-k}]$$

$$= -\sum_{i=0}^m \frac{1}{i!(m+1-i)!} K_p h L^{m+1-i}(t-L)^i + \sum_{i=1}^n \alpha_i [\frac{(-1)^{m+1}}{\lambda_i^{m+1}} e^{-\lambda_i(t-L)}$$

$$+ \sum_{k=1}^{m+1} \frac{(-1)^{k-1}}{(m+1-k)!} \frac{1}{\lambda_i^k} (t-L)^{m+1-k}]. \tag{7.45}$$

Formulas (7.44) and (7.45) verify (7.43) for $m+1$. The proof is thus completed.

For any time $t \geqslant L$, $n+1$ equations in (7.42) and (7.43) for $m = 0, 1, \cdots, n$, are arranged into

$$\begin{bmatrix} f \\ \int_{[0,t]}^{(n)} \Delta y \end{bmatrix} = \begin{bmatrix} A \\ p^T \end{bmatrix} x + \begin{bmatrix} q \\ \mu \end{bmatrix}, \tag{7.46}$$

where

$$f = \left[\Delta y, \int_{[0,t]}^{(1)} \Delta y, \cdots, \int_{[0,t]}^{(n-1)} \Delta y \right]^T,$$

$$A = \begin{bmatrix} 1 & 1 & \cdots & 1 & 1 \\ -\frac{1}{\lambda_1} & -\frac{1}{\lambda_2} & \cdots & -\frac{1}{\lambda_{n-1}} & -\frac{1}{\lambda_n} \\ \vdots & \vdots & \cdots & \vdots & \vdots \\ (-\frac{1}{\lambda_1})^{n-1} & (-\frac{1}{\lambda_2})^{n-1} & \cdots & (-\frac{1}{\lambda_{n-1}})^{n-1} & (-\frac{1}{\lambda_n})^{n-1} \end{bmatrix}, \tag{7.47}$$

$$p = \left[(-\frac{1}{\lambda_1})^n, (-\frac{1}{\lambda_2})^n, \cdots, (-\frac{1}{\lambda_n})^n \right]^T,$$

$$x = \left[\alpha_1 e^{-\lambda_1(t-L)} \quad \alpha_2 e^{-\lambda_2(t-L)} \quad \cdots \quad \alpha_{n-1} e^{-\lambda_{n-1}(t-L)} \quad \alpha_n e^{-\lambda_n(t-L)} \right]^T,$$

$$q = \left[0, \ -K_phL + \sum_{i=1}^{n}\frac{\alpha_i}{\lambda_i}, \ \cdots, \ -\sum_{i=0}^{n-2}\frac{1}{i!(n-1-i)!}K_phL^{n-1-i}(t-L)^i\right.$$

$$\left.+\sum_{i=1}^{n}\alpha_i[\sum_{k=1}^{n-1}\frac{(-1)^{k-1}}{(n-1-k)!}\frac{1}{\lambda_i^k}(t-L)^{n-1-k}]\right]^T,$$

$$\mu = -\sum_{i=0}^{n-1}\frac{1}{i!(n-i)!}K_phL^{n-i}(t-L)^i+\sum_{i=1}^{n}\alpha_i[\sum_{k=1}^{n}\frac{(-1)^{k-1}}{(n-k)!}\frac{1}{\lambda_i^k}(t-L)^{n-k}].$$

Lemma 7.2.2. The matrix A in (7.47) is nonsingular if λ_i, $i=1,2,\cdots,n$, are distinct. Its inverse is

$$A^{-1} = \frac{1}{\det(A)}\begin{bmatrix} A_{11} & A_{21} & \cdots & A_{n1} \\ A_{12} & A_{22} & \cdots & A_{n2} \\ \vdots & \vdots & \ddots & \vdots \\ A_{1n} & A_{2n} & \cdots & A_{nn} \end{bmatrix}, \tag{7.48}$$

where

$$\det(A) = \prod_{1\leqslant j<i\leqslant n}(-\frac{1}{\lambda_i}+\frac{1}{\lambda_j}) \neq 0, \tag{7.48a}$$

$$A_{ij} = (-1)^{n-j}\prod_{\substack{1\leqslant k<l\leqslant n \\ k,l\neq j}}(-\frac{1}{\lambda_l}+\frac{1}{\lambda_k})\sum_{\substack{1\leqslant j_1<\cdots<j_{n-i}\leqslant n \\ j_1,\cdots,j_{n-i}\neq j}}\frac{1}{\lambda_{j_1}}\cdots\frac{1}{\lambda_{j_{n-i}}}. \tag{7.48b}$$

Proof. Note that matrix A is a Vandermonda-like matrix and its determinant (Broyden, 1975) is thus given by (7.48a). It is non-zero in the case of distinct λ_i, $i=1,2,\cdots,n$. A^{-1} is given formally by (7.48), where A_{ij} is the cofactor of the (ij)th element of A.

Introduce an auxiliary matrix:

$$M(\rho) = \begin{bmatrix} 1 & \cdots & 1 & 1 & 1 & \cdots & 1 \\ -\frac{1}{\lambda_1} & \cdots & -\frac{1}{\lambda_{j-1}} & \rho & -\frac{1}{\lambda_{j+1}} & \cdots & -\frac{1}{\lambda_n} \\ \vdots & \ddots & \vdots & \vdots & \vdots & \ddots & \vdots \\ -(\frac{1}{\lambda_1})^{i-2} & \cdots & -(\frac{1}{\lambda_{j-1}})^{i-2} & \rho^{i-2} & -(\frac{1}{\lambda_{j+1}})^{i-2} & \cdots & -(\frac{1}{\lambda_n})^{i-2} \\ -(\frac{1}{\lambda_1})^{i-1} & \cdots & -(\frac{1}{\lambda_{j-1}})^{i-1} & \rho^{i-1} & -(\frac{1}{\lambda_{j+1}})^{i-1} & \cdots & -(\frac{1}{\lambda_n})^{i-1} \\ -(\frac{1}{\lambda_1})^{i} & \cdots & -(\frac{1}{\lambda_{j-1}})^{i} & \rho^{i} & -(\frac{1}{\lambda_{j+1}})^{i} & \cdots & -(\frac{1}{\lambda_n})^{i} \\ \vdots & \ddots & \vdots & \vdots & \vdots & \ddots & \vdots \\ -(\frac{1}{\lambda_1})^{n-1} & \cdots & -(\frac{1}{\lambda_{j-1}})^{n-1} & \rho^{n-1} & -(\frac{1}{\lambda_{j+1}})^{n-1} & \cdots & -(\frac{1}{\lambda_n})^{n-1} \end{bmatrix}.$$

One immediately sees

$$M_{ij} = A_{ij}, \tag{7.49}$$

where M_{ij} is the cofactor of the (ij)th element of $M(\rho)$. Since $M(\rho)$ is again a Vandermonda-like matrix, its determinant is given by

$$\det[M(\rho)] = \{\prod_{\substack{1 \leqslant k < l \leqslant n \\ k,l \neq j}} (-\frac{1}{\lambda_l} + \frac{1}{\lambda_k})\}\{\prod_{\substack{1 \leqslant k \leqslant n \\ k \neq j}} (\rho - \frac{1}{\lambda_k})\}. \tag{7.50}$$

For this polynomial in ρ, the coefficient of ρ^{i-1} is

$$(-1)^{n-j} \prod_{\substack{1 \leqslant k < l \leqslant n \\ k,l \neq j}} (-\frac{1}{\lambda_l} + \frac{1}{\lambda_k}) \sum_{\substack{1 \leqslant j_1 < \cdots < j_{n-i} \leqslant n \\ j_1,\cdots,j_{n-i} \neq j}} \frac{1}{\lambda_{j_1}} \cdots \frac{1}{\lambda_{j_{n-i}}}. \tag{7.51}$$

On the other hand, one may also apply the Laplace expansion theorem to the jth column of $M(\rho)$ to find its determinant:

$$\det[M(\rho)] = M_{1j} + M_{2j}\rho + \cdots + M_{ij}\rho^{i-1} + \cdots + M_{nj}\rho^{n-1}. \tag{7.52}$$

Collecting (7.49) – (7.52) yields (7.48b).

Since A is invertible, x is solved from the first n rows in (7.46) as

$$x = A^{-1}(f - q),$$

and is substituted into the last row in (7.46) to give

$$\int_{[0,t]}^{(n)} \Delta y = p^T A^{-1} f + (\mu - p^T A^{-1} q).$$

After lengthy but straightforward algebra, one reaches

$$\phi^T(t)\theta = \gamma(t), \quad t \geqslant L, \tag{7.53}$$

where $\gamma(t) \triangleq -\int_{[0,t]}^{(n)} \Delta y$, $\theta(t) = [\theta_1, \theta_2, \cdots, \theta_{2n}]^T$,

$$\theta_k \triangleq \begin{cases} \dfrac{\sum_{1 \leqslant i_1 < \cdots < i_{k-1} \leqslant n} \lambda_{i_1} \cdots \lambda_{i_{k-1}}}{\prod_{i=1}^n \lambda_i}, & \text{for } k = 1, 2, \cdots, n; \\ \dfrac{1}{\prod_{i=1}^n \lambda_i} + \sum_{j=0}^{n-1} \dfrac{(-1)^{n+1-j}}{(n-j)!} L^{n-j} \\ \quad \cdot \sum_{1 \leqslant i_1 < i_2 < \cdots < i_j \leqslant n-1} \beta_{i_1} \beta_{i_2} \cdots \beta_{i_j}, & \text{for } k = n+1; \\ \dfrac{\sum_{1 \leqslant i_1 < i_2 < \cdots < i_{k-(n+1)} \leqslant n} \lambda_{i_1} \lambda_{i_2} \cdots \lambda_{i_{k-(n+1)}}}{\prod_{i=1}^n \lambda_i} \\ \quad + \sum_{j=0}^{2n+1-k} \dfrac{(-1)^{2n-k-j}}{(2n+1-k-j)!} L^{2n+1-k-j} \\ \quad \cdot \sum_{1 \leqslant i_1 < i_2 < \cdots < i_j \leqslant n-1} \beta_{i_1} \beta_{i_2} \cdots \beta_{i_j}, & \text{for } k = n+2, \cdots, 2n, \end{cases} \tag{7.54}$$

7.2 From Step Response   185

are dependent of the parameters of the NOPDT model in (7.38), and

$$\phi(t)^T \triangleq \left[\Delta y(t),\ \int_{[0,t]}^{(1)} \Delta y,\ \cdots,\ \int_{[0,t]}^{(n-1)} \Delta y,\ K_p h,\ K_p h t,\ \cdots,\ \frac{1}{(n-1)!} K_p h t^{n-1} \right],$$

can be obtained from the output response. Equation (7.53) for $t = t_i \geq L$, $i = 1, 2, \cdots, N$, $t_1 < t_2 < \cdots < t_N$, is arranged into the matrix form:

$$\Phi \theta = \Gamma, \tag{7.55}$$

where

$$\Gamma = \left[\gamma^T(t_1), \gamma^T(t_2), \cdots, \gamma^T(t_N) \right]^T, \quad \Phi = \left[\phi(t_1), \phi(t_2), \cdots, \phi(t_N) \right]^T.$$

Lemma 7.2.3. $\Phi^T \Phi$ *is nonsingular if the process is of nth or higher order with distinct poles and the sampling number satisfies* $N \geq 2n$.

Proof. For simplicity, we consider only the uniform sampling case, i.e., $t_{i+1} - t_i = T_s$, $i = 1, 2, \cdots, N-1$, where T_s is the sampling period. The proof can be extended to the case of different sampling intervals.

Suppose that there are real $c_i \in \mathbb{R}$, $i = 1, 2, \cdots, 2n$, such that

$$\Phi [c_1, c_2, \cdots, c_{2n}]^T = 0,$$

or

$$c_1 \begin{bmatrix} \Delta y_1 \\ \Delta y_2 \\ \vdots \\ \Delta y_N \end{bmatrix} + \cdots + c_n \begin{bmatrix} \int_{[0,t_1]}^{(n-1)} \Delta y \\ \int_{[0,t_2]}^{(n-1)} \Delta y \\ \vdots \\ \int_{[0,t_N]}^{(n-1)} \Delta y \end{bmatrix} + c_{n+1} \begin{bmatrix} K_p h \\ K_p h \\ \vdots \\ K_p h \end{bmatrix} + \cdots + c_{2n} \begin{bmatrix} \frac{1}{(n-1)!} K_p h t_1^{n-1} \\ \frac{1}{(n-1)!} K_p h t_2^{n-1} \\ \vdots \\ \frac{1}{(n-1)!} K_p h t_N^{n-1} \end{bmatrix}$$
$$= 0, \tag{7.56}$$

where $\Delta y_i = \Delta y(t_i)$, $i = 1, 2, \cdots, N$. If not all c_i, $i = 1, 2, \cdots, n$, are zero, subtracting the $(k-1)$th row in (7.56) from the kth row, $k = N, N-1, \cdots, 3, 2$, yields

$$c_1 \begin{bmatrix} \Delta y_1 \\ \Delta y_2 - \Delta y_1 \\ \vdots \\ \Delta y_N - \Delta y_{N-1} \end{bmatrix} + \cdots + c_n \begin{bmatrix} \int_{[0,t_1]}^{(n-1)} \Delta y \\ \int_{[0,t_2]}^{(n-1)} \Delta y - \int_{[0,t_1]}^{(n-1)} \Delta y \\ \vdots \\ \int_{[0,t_N]}^{(n-1)} \Delta y - \int_{[0,t_{N-1}]}^{(n-1)} \Delta y \end{bmatrix} + c_{n+1} \begin{bmatrix} K_p h \\ 0 \\ \vdots \\ 0 \end{bmatrix}$$

$$+ c_{n+2} \begin{bmatrix} K_p h t_1 \\ K_p h T_s \\ \vdots \\ K_p h T_s \end{bmatrix} + \cdots + c_{2n} \begin{bmatrix} \frac{1}{(n-1)!} K_p h t_1^{n-1} \\ \frac{1}{(n-1)!} K_p h (t_2^{n-1} - t_1^{n-1}) \\ \vdots \\ \frac{1}{(n-1)!} K_p h (t_N^{n-1} - t_{N-1}^{n-1}) \end{bmatrix} = 0. \quad (7.57)$$

Extract the last $N - 1$ rows in (7.57) and subtract the $(k - 1)$th row from the kth row, $k = N, N - 1, \cdots, 4, 3$, to give

$$c_1 \begin{bmatrix} \Delta y_2 - \Delta y_1 \\ \Delta y_3 - 2\Delta y_2 + \Delta y_1 \\ \vdots \\ \Delta y_N - 2\Delta y_{N-1} + \Delta y_{N-2} \end{bmatrix} + \cdots +$$

$$+ c_n \begin{bmatrix} \int_{[0,t_2]}^{(n-1)} \Delta y - \int_{[0,t_1]}^{(n-1)} \Delta y \\ \int_{[0,t_3]}^{(n-1)} \Delta y - 2\int_{[0,t_2]}^{(n-1)} \Delta y + \int_{[0,t_1]}^{(n-1)} \Delta y \\ \vdots \\ \int_{[0,t_N]}^{(n-1)} \Delta y - 2\int_{[0,t_{N-1}]}^{(n-1)} \Delta y + \int_{[0,t_{N-2}]}^{(n-1)} \Delta y \end{bmatrix}$$

$$+ c_{n+2} \begin{bmatrix} K_p h T_s \\ 0 \\ \vdots \\ 0 \end{bmatrix} + \cdots + c_{2n} \begin{bmatrix} \frac{1}{(n-1)!} K_p h (t_2^{n-1} - t_1^{n-1}) \\ \frac{1}{(n-1)!} K_p h (t_3^{n-1} - 2t_2^{n-1} + t_1^{n-1}) \\ \vdots \\ \frac{1}{(n-1)!} K_p h (t_N^{n-1} - 2t_{N-1}^{n-1} + t_{N-2}^{n-1}) \end{bmatrix} = 0.$$

Repeat this operation $n - 2$ more times to produce

$$c_1 \begin{bmatrix} \Delta y_{n+1} - \binom{n}{1} \Delta y_n + \binom{n}{2} \Delta y_{n-1} + \cdots + (-1)^n \binom{n}{n} \Delta y_1 \\ \Delta y_{n+2} - \binom{n}{1} \Delta y_{n+1} + \binom{n}{2} \Delta y_n + \cdots + (-1)^n \binom{n}{n} \Delta y_2 \\ \vdots \\ \Delta y_N - \binom{n}{1} \Delta y_{N-1} + \binom{n}{2} \Delta y_{N-2} + \cdots + (-1)^n \binom{n}{n} \Delta y_{N-n} \end{bmatrix} + \cdots +$$

$$+c_n \begin{bmatrix} \int_{[0,t_{n+1}]}^{n-1} \Delta y - \binom{n}{1}\int_{[0,t_n]}^{n-1} \Delta y + \binom{n}{2}\int_{[0,t_{n-1}]}^{n-1} \Delta y + \cdots + (-1)^n \binom{n}{n}\int_{[0,t_1]}^{n-1} \Delta y \\ \int_{[0,t_{n+2}]}^{n-1} \Delta y - \binom{n}{1}\int_{[0,t_{n+1}]}^{n-1} \Delta y + \binom{n}{2}\int_{[0,t_n]}^{n-1} \Delta y + \cdots + (-1)^n \binom{n}{n}\int_{[0,t_2]}^{n-1} \Delta y \\ \vdots \\ \int_{[0,t_N]}^{n-1} \Delta y - \binom{n}{1}\int_{[0,t_{N-1}]}^{n-1} \Delta y + \binom{n}{2}\int_{[0,t_{N-2}]}^{n-1} \Delta y + \cdots + (-1)^n \binom{n}{n}\int_{[0,t_{N-n}]}^{n-1} \Delta y \end{bmatrix}$$
$$= 0, \tag{7.58}$$

which has $(N-n)$ rows left, where $\binom{n}{i} = \frac{n!}{i!(n-i)!}$.

Suppose that the order of the true process is $\hat{n}$ and note from (7.41) that

$$\Delta y(t) = y - K_p h = \sum_{i=1}^{\hat{n}} \alpha_i e^{-\lambda_i(t-L)}, \qquad \text{for } t \geqslant L.$$

With a tedious but straightforward calculation, one can express (7.58) as

$$c_1 \begin{bmatrix} \sum_{i=1}^{\hat{n}} \zeta_i \\ \sum_{i=1}^{\hat{n}} \zeta_i e^{-\lambda_i T_s} \\ \vdots \\ \sum_{i=1}^{\hat{n}} \zeta_i e^{-(N-n-1)\lambda_i T_s} \end{bmatrix} + \cdots + c_n \begin{bmatrix} (-1)^{n-1} \sum_{i=1}^{\hat{n}} \frac{\zeta_i}{\lambda_i^{n-1}} \\ (-1)^{n-1} \sum_{i=1}^{\hat{n}} \frac{\zeta_i}{\lambda_i^{n-1}} e^{-\lambda_i T_s} \\ \vdots \\ (-1)^{n-1} \sum_{i=1}^{\hat{n}} \frac{\zeta_i}{\lambda_i^{n-1}} e^{-(N-n-1)\lambda_i T_s} \end{bmatrix}$$
$$= 0, \tag{7.59}$$

where

$$\zeta_i = \alpha_i e^{\lambda_i L} e^{-(n+1)\lambda_i T_s}(1 - e^{-\lambda_i T_s})^n \neq 0, \qquad i = 1, 2, \cdots, \hat{n}.$$

Under the assumed $\hat{n} \geqslant n$ and $N \geqslant 2n$, the first n rows in (7.59), after Guassian elimination, become

$$\begin{bmatrix} 1 & -\frac{1}{\lambda_1} & \cdots & (-\frac{1}{\lambda_1})^{n-1} \\ 1 & -\frac{1}{\lambda_2} & \cdots & (-\frac{1}{\lambda_2})^{n-1} \\ \vdots & \vdots & \ddots & \vdots \\ 1 & -\frac{1}{\lambda_n} & \cdots & (-\frac{1}{\lambda_n})^{n-1} \end{bmatrix} \begin{bmatrix} c_1 \\ c_2 \\ \vdots \\ c_n \end{bmatrix} = 0, \tag{7.60}$$

and thus c_i, $i = 1, 2, \cdots, n$, are zero since λ_i, $i = 1, 2, \cdots, n$, have been assumed to be distinct. Therefore, (7.56) reduces to

$$c_{n+1} \begin{bmatrix} K_p h \\ K_p h \\ \vdots \\ K_p h \end{bmatrix} + c_{n+2} \begin{bmatrix} K_p h t_1 \\ K_p h t_2 \\ \vdots \\ K_p h t_N \end{bmatrix} + \cdots + c_{2n} \begin{bmatrix} \frac{1}{(n-1)!} K_p h t_1^{n-1} \\ \frac{1}{(n-1)!} K_p h t_2^{n-1} \\ \vdots \\ \frac{1}{(n-1)!} K_p h t_N^{n-1} \end{bmatrix} = 0,$$

or

$$K_{ph}\begin{bmatrix} 1 & t_1 & \cdots & \frac{1}{(n-1)!}t_1^{n-1} \\ 1 & t_2 & \cdots & \frac{1}{(n-1)!}t_2^{n-1} \\ \vdots & \vdots & \ddots & \vdots \\ 1 & t_N & \cdots & \frac{1}{(n-1)!}t_N^{n-1} \end{bmatrix}\begin{bmatrix} c_{n+1} \\ c_{n+2} \\ \vdots \\ c_{2n} \end{bmatrix} = 0$$

The n columns of the coefficient matrix are linearly independent, which leads to $c_i = 0$, $i = n+1, n+2, \cdots, 2n$. Thus, all c_i, $i = 1, 2, \cdots, 2n$, are zero and the columns of Φ are linearly independent. Since the matrix Φ has $2n$ independent columns, $\Phi^T\Phi$ is nonsingular (Broyden, 1975).

The least squares solution of (7.55) for θ which minimizes the following equation error:

$$\min_\theta (\Gamma - \Phi\theta)^T(\Gamma - \Phi\theta),$$

is given by

$$\hat{\theta} = (\Phi^T\Phi)^{-1}\Phi^T\Gamma. \tag{7.61}$$

In practice, the true process output $\hat{y}$ is corrupted by measurement noise $v(t)$ as

$$y = \hat{y} + v. \tag{7.62}$$

Here v is assumed to be a strictly stationary stochastic process with zero mean. In this case, (7.55) is modified to

$$\Phi\theta = \Gamma + \Delta, \tag{7.63}$$

where $\Delta = [\delta_1, \delta_2, \cdots, \delta_N]^T$, and

$$\delta_i = \left[\int_{[0,t_i]}^{(n)} v - \frac{1}{\prod_{i=1}^n \lambda_i}v(t_i) - \frac{\sum_{i=1}^n \lambda_i}{\prod_{i=1}^n \lambda_i}\int_{[0,t_i]}^{(1)} v - \frac{\sum_{1\leq i_1 < i_2 \leq n} \lambda_1\lambda_2}{\prod_{i=1}^n \lambda_i}\int_{[0,t_i]}^{(2)} v \right.$$
$$\left. - \cdots - \frac{\sum_{1\leq i_1 < i_2 < \cdots < i_{n-1} \leq n} \lambda_1\lambda_2\cdots\lambda_{i_{n-1}}}{\prod_{i=1}^n \lambda_i}\int_{[0,t_i]}^{(n-1)} v \right].$$

For our case, the instrumental variable matrix Z is chosen as

$$Z = \begin{bmatrix} \frac{1}{t_1^n} & \frac{1}{t_1^{n+1}} & \cdots & \frac{1}{t_1^{2n-1}} & t_1^2 & t_1^3 & \cdots & t_1^{n+1} \\ \frac{1}{t_2^n} & \frac{1}{t_2^{n+1}} & \cdots & \frac{1}{t_2^{2n-1}} & t_2^2 & t_2^3 & \cdots & t_2^{n+1} \\ \vdots & \ddots & \vdots & \vdots & \vdots & \ddots & \vdots \\ \frac{1}{t_N^n} & \frac{1}{t_N^{n+1}} & \cdots & \frac{1}{t_N^{2n-1}} & t_N^2 & t_N^3 & \cdots & t_N^{n+1} \end{bmatrix}. \tag{7.64}$$

Lemma 7.2.4. *For the system (7.63), the matrix Z in (7.64) is an instrumental variable matrix which satisfies the conditions in Proposition 7.2.1.*

Proof. Partition Z and Φ as

$$Z = [z_1, z_2, \cdots, z_N]^T, \quad \Phi = [\phi_1, \phi_2, \cdots, \phi_N]^T.$$

Denote

$$\psi_{k,i} = \begin{cases} t_i^{-(n+k-1)} \phi_i^T, & \text{if } k \leq n, \\ t_i^{k-n+1} \phi_i^T, & \text{if } k > n, \end{cases}$$

where $i = 1, 2, \cdots, N$ and $k = 1, 2, \cdots, 2n$, we have

$$Z^T \Phi = \sum_{i=1}^{N} z_i \phi_i^T = \sum_{i=1}^{N} \left[\psi_{1,i}, \psi_{2,i}, \cdots, \psi_{2n,i} \right]^T$$

$$= \left[\sum_{i=1}^{N} \psi_{1,i}, \sum_{i=1}^{N} \psi_{2,i}, \cdots, \sum_{i=1}^{N} \psi_{2n,i} \right]^T.$$

Suppose that there are $c_i \in \mathbb{R}$, $i = 1, 2, \cdots, 2n$, such that

$$c_1 \sum_{i=1}^{N} \psi_{1,i} + \cdots + c_n \sum_{i=1}^{N} \psi_{n,i} + c_{n+1} \sum_{i=1}^{N} \psi_{n+1,i} + \cdots + c_{2n} \sum_{i=1}^{N} \psi_{2n,i} = 0, \quad (7.65)$$

or

$$\sum_{i=1}^{N} \left[c_1 t_i^{-n} + c_2 t_i^{-(n+1)} + \cdots + c_n t_i^{-(2n-1)} + c_{n+1} t_i^2 + \cdots + c_{2n} t_i^{n+1} \right] \phi_i^T = 0.$$

Since Δy is corrupted by measurement noise v, $\phi_i^T, i = 1, 2, \cdots, N$, are linearly independent with probability one, (7.65) leads to

$$c_1 t_i^{-n} + \cdots + c_n t_i^{-(2n-1)} + c_{n+1} t_i^2 + \cdots + c_{2n} t_i^{n+1} = 0, \quad i = 1, 2 \cdots, N,$$

or in matrix form:

$$\begin{bmatrix} \frac{1}{t_1^n} & \frac{1}{t_1^{n+1}} & \cdots & \frac{1}{t_1^{2n-1}} & t_1^2 & t_1^3 & \cdots & t_1^{n+1} \\ \frac{1}{t_2^n} & \frac{1}{t_2^{n+1}} & \cdots & \frac{1}{t_2^{2n-1}} & t_2^2 & t_2^3 & \cdots & t_2^{n+1} \\ \vdots & \ddots & \vdots & \vdots & \vdots & \ddots & \vdots \\ \frac{1}{t_N^n} & \frac{1}{t_N^{n+1}} & \cdots & \frac{1}{t_N^{2n-1}} & t_N^2 & t_N^3 & \cdots & t_N^{n+1} \end{bmatrix} \begin{bmatrix} c_1 \\ c_2 \\ \vdots \\ c_{2n} \end{bmatrix} = 0.$$

It is clear that the columns of the coefficient matrix are linearly independent. This implies $c_i = 0$, $i = 1, 2, \cdots, 2n$, and therefore all columns of $Z^T \Phi$ are linearly independent.

190   7. Transfer Function Modelling

Let $g(N)$ be a function of N. Denote $O(g(N)) \sim N^k$ (k is an integer) if $\lim_{N\to\infty} \frac{g(N)}{N^k} \to c$, a nonzero constant. Noting that $\Delta \hat{y}(t)$ meets (7.43) and is corrupted by zero-mean stochastic process $v(t)$, it follows that

$$\lim_{N\to\infty} \frac{1}{N} Z^T \Phi$$

$$= \lim_{N\to\infty} \frac{1}{N} \begin{bmatrix} \sum_{i=1}^N \frac{\Delta y_i}{t_i^n} & \cdots & \sum_{i=1}^N \frac{\int_{[0,t_i]}^{(n-1)} \Delta y_i}{t_i^n} & \sum_{i=1}^N \frac{K_p h}{t_i^n} & \cdots & \frac{K_p h}{(n-1)!} \sum_{i=1}^N \frac{1}{t_i} \\ \vdots & \ddots & \vdots & \vdots & \ddots & \vdots \\ \sum_{i=1}^N \frac{\Delta y_i}{t_i^{2n-1}} & \cdots & \sum_{i=1}^N \frac{\int_{[0,t_N]}^{(n-1)} \Delta y_i}{t_i^n} & \sum_{i=1}^N \frac{K_p h}{t_i^{2n-1}} & \cdots & \frac{K_p h}{(n-1)!} \sum_{i=1}^N \frac{1}{t_i^n} \\ \sum_{i=1}^N t_i^2 \Delta y_i & \cdots & \sum_{i=1}^N t_i^2 \int_{[0,t_N]}^{(n-1)} \Delta y_i & \sum_{i=1}^N K_p h t_i^2 & \cdots & \frac{K_p h}{(n-1)!} \sum_{i=1}^N K_p h t_i^{n+1} \\ \vdots & \ddots & \vdots & \vdots & \ddots & \vdots \\ \sum_{i=1}^N t_i^{n+1} \Delta y_i & \cdots & \sum_{i=1}^N t_i^{n+1} \int_{[0,t_N]}^{(n-1)} \Delta y_i & \sum_{i=1}^N K_p h t_i^{n+1} & \cdots & \frac{K_p h}{(n-1)!} \sum_{i=1}^N K_p h t_i^{2n} \end{bmatrix}$$

$$= \begin{bmatrix} O(\frac{1}{N^{n+1}}) & O(\frac{1}{N^n}) & \cdots & O(\frac{1}{N^2}) & O(\frac{1}{N^n}) & O(\frac{1}{N^{n-1}}) & \cdots & O(\frac{1}{N}) \\ O(\frac{1}{N^{n+2}}) & O(\frac{1}{N^{n+1}}) & \cdots & O(\frac{1}{N^3}) & O(\frac{1}{N^{n+1}}) & O(\frac{1}{N^n}) & \cdots & O(\frac{1}{N^2}) \\ \vdots & \vdots & \ddots & \vdots & \vdots & \vdots & \ddots & \vdots \\ O(\frac{1}{N^{2n}}) & O(\frac{1}{N^{2n-1}}) & \cdots & O(\frac{1}{N^{n+1}}) & O(\frac{1}{N^{2n-1}}) & O(\frac{1}{N^{2n-2}}) & \cdots & O(\frac{1}{N^{n-2}}) \\ O(N) & O(N^2) & \cdots & O(N^n) & O(N^2) & O(N^3) & \cdots & O(N^{n+1}) \\ O(N^2) & O(N^3) & \cdots & O(N^{n+1}) & O(N^3) & O(N^4) & \cdots & O(N^{n+2}) \\ \vdots & \vdots & \ddots & \vdots & \vdots & \vdots & \ddots & \vdots \\ O(N^n) & O(N^{n+1}) & \cdots & O(N^{2n-1}) & O(N^{n+1}) & O(N^{n+2}) & \cdots & O(N^{2n}) \end{bmatrix}.$$

For the above matrix, it can be seen that the product of any n elements, one from each row (or column), is of the same order as N^0, i.e., $O(1)$. The determinant of this $n \times n$ matrix involves the sum of all such $n!$ products. Therefore,

$$\det(\lim_{N\to\infty} \frac{1}{N} Z(N)^T \Phi(N)) = O(1),$$

and the inverse of $\lim_{N\to\infty} \frac{1}{N} Z(N)^T \Phi(N)$ exists.

By the assumption, v is a strictly stationary random process, and its statistics are unchanged by a time shift in the time origin. Equivalently, its distribution of all orders are independent of the time origin (Barkat, 1991). This indicates that $\delta(t)$ and the elements in Z are independent. Therefore, the instrumental variables are uncorrelated with Z, or

$$\lim_{N\to\infty} \frac{1}{N} Z^T \Delta = 0.$$

In view of the above development, we can establish the following theorem.

Theorem 7.2.1. *For a linear time-invariant process of nth or higher order with distinct poles, if the noise in the output measurement is a zero-mean strictly stationary stochastic process and the number of output samples N meets $N \geqslant 2n$, then the estimate given by*

$$\hat{\theta} = (Z^T \Phi)^{-1} Z^T \Gamma \tag{7.66}$$

is consistent, where Z is given in (7.64). Under noise-free circumstances, Z may be replaced by Φ.

Once θ is estimated from (7.66), one has to recover the process model (7.38). It follows from (7.38a) and (7.39) that

$$\begin{cases} a_k = \frac{\theta_{k+1}}{\theta_k}, & k = 1, 2, \cdots, n-1; \\ a_n = \frac{1}{\theta_1}, & \end{cases}$$

and

$$b_n = \frac{K_p}{\theta_1},$$

$$\theta_{n+1} = \theta_1 + \frac{(-1)^{n+1}}{n!} L^n + \frac{1}{b_n} [\sum_{i=1}^{n-1} \frac{(-1)^{n+1-i}}{(n-i)!} L^{n-i} b_{n-i}], \tag{7.67}$$

$$\theta_{n+k} = \theta_k + \frac{(-1)^{n-k}}{(n-k+1)!} L^{n-k+1} + \frac{1}{b_n} [\sum_{i=1}^{n-k+1} \frac{(-1)^{n-k-i}}{(n-k-i+1)!} L^{n-k-i+1} b_{n-i}],$$
$$k = 2, 3, \cdots, n. \tag{7.68}$$

For $k = n$, (7.68) gives

$$\theta_{2n} = \theta_n + L - \frac{\theta_1}{K_p} b_{n-1},$$

from which b_{n-1} can be expressed as a linear function of L as

$$b_{n-1} = \frac{K_p}{\theta_1} (\theta_n + L - \theta_{2n}).$$

Substituting this expression into (7.68) for $k = n-1$, b_{n-2} can be expressed as a linear function of L and L^2. Repeat this operation for all other k in (7.68); each b_i can then be expressed as a linear function of $L, \cdots, L^i$, $i = 1, \cdots, n-1$. Further substituting all these expressions into (7.67), an nth degree polynomial equation in L is derived, from which n roots of L can be easily found. In determining a suitable solution, a rule of thumb is to choose the one that leads to the following minimal estimation error standard deviation between the actual step response recorded from the actual process, $y(kT_s)$, and the step response of the estimated model, $\tilde{y}(kT_s)$, where the estimation error standard deviation is defined by

$$err = \frac{1}{N} \sum_{k=1}^{N} [y(kT_s) - \tilde{y}(kT_s)]^2. \tag{7.69}$$

Remark 7.2.2. In the proposed algorithm, instead of using noise-accentuating derivative operations on noisy signals, numerical integration is used to make this method robust to noise. Using the IV method, the proposed method also gives consistent estimates when the output is corrupted by measurement noise.

Remark 7.2.3. The efficiency of an estimator can be evaluated by its closeness to the Cramer–Rao Lower Bound (CRLB). However, for our algorithm, we are unable to do so. To our best knowledge, the methods available for calculating CRLB assume additive noise, where the noise is supposed to be white Gaussian noise or colored noise (Kerr, 1989; Kay, 1993; Ghogho and Swami, 1999). The noise part Δ in our case is, however, a nonlinear transformation of v, where v is assumed to be a strictly stationary stochastic process with zero mean. Therefore, the existing methods cannot be applied directly to calculate CRLB. Further, the statistical properties hold for the parameter θ, but not for the original transfer function parameters, since the transformation from θ to model parameters is nonlinear. The efficiency of the estimator for θ does not necessarily imply efficiency for the transfer function parameters and it is hard to address statistical properties for the latter. In view of these difficulties, CRLB is not calculated or checked here. Instead, the effectiveness of the proposed method is demonstrated by simulation examples as well as implementation tests.

Remark 7.2.4. It should be pointed out that numerical integration techniques introduce errors, especially close to the Nyquist frequency. Numerical integration has been extensively studied in the literature. In general, there is no fixed rule for deciding which numerical integration method is the best, and it depends mainly on the integrands and the individual problem itself (Dai and Sinha, 1991). To match the ideal integrator characteristics as closely as possible within the frequency band of interest, digital filter design methods (Pintelon and Schoukens, 1990) can be applied.

Remark 7.2.5. In practice, most processes are inevitable of high order with inherent nonlinearity, and it is rare to have duplicated or even close poles in the reduced-order model. However, complex symmetric mechanical structures may have multiple complex poles (multiple resonance frequencies), which are very close to each other. This will cause bad numerical conditioning and is a topic for future work.

7.2.3 Implementation Issues

In this subsection, several practical issues concerning implementation of the proposed algorithm are discussed.

Choice of t_1   It is noted from the above development that the first sample $y(t_1)$ should not be taken into the algorithm until $t_1 \geqslant L$, when the output deviates from the previous steady state. In practice, the selection of the lagged $y(t)$ after $t_i \geqslant L$ can be made as follows. Before the step test starts, the process output is monitored for a period called the 'listening period', during which time the noise band B_n can be found. After a step change in the process input is applied, the time t at which $y(t)$ satisfies

$$\text{abs}(\text{mean}(y(t-\tau \text{ to } t))) > 2B_n$$

is considered to meet $t > L$, where τ is a user-specified time interval and is used for averaging.

Choice of t_N   The initial part of the step response contains more extensive frequency information than the later part. In fact, the part of the response after steady state contains only zero frequency information, i.e., at $\omega = 0$. Therefore, it is meaningless to use more data in the algorithm after the steady state, and t_N should satisfy

$$L < t_N < T_{set},$$

where T_{set} is the settling time, defined as the time required for the process to settle within $\pm 2\%$ of its steady state.

For most processes, the frequency response usually shows higher gain at the low frequencies than at the high frequencies. Also, a step test excites slow dynamics more effectively. Thus, the relative error of frequency response identification is likely to be dominated by high frequency errors. It can be seen that the transient part of the time response truncated at smaller t_N contains more information at high frequencies, and identification based on best fitting will naturally lead to smaller error in frequency response estimation at high frequencies and thus over the whole frequency range. In such a case, however, time response beyond this time is not taken into account and thus the time domain error in identification measured over the whole time response period will become larger due to possibly poor matching of data outside the span. In view of this analysis, there exists a trade-off between the estimation accuracy in frequency and time domains. After extensive simulations, it is recommended that t_N be set at $0.8T_{set}$.

Choice of N   The computational effort becomes heavy if too many samples are taken into consideration, leading to large Φ. Moreover, $\Phi^T \Phi$ or $Z^T \Phi$ may tend to become ill-conditioned for very large N and this may cause computational difficulty in estimating $\hat{\theta}$. Therefore, N has to be limited. For the case with a large number of recorded data points, the default value of N is recommended to be 200, and t_i may be set as

$$t_i = t_1 + \frac{i-1}{N}(t_N - t_1), \qquad i = 1, 2, \cdots, N.$$

Computation of Solution The solution formulas (7.32), (7.34), (7.61) and (7.66) are used for theoretical purposes only. In real applications, to compute the solution in a numerically stable way, QR decomposition or single-value decomposition methods are applied to get the solution of the linear equation system:

$$\Phi\theta = \Gamma$$

for the LS problem, or

$$Z^T \Phi \theta = Z^T \Gamma$$

for the IV problem.

Recursive Solution The LS solution in (7.61) and IV solution in (7.66) are non-recursive and require matrix inversion. Alternatively, one can easily cast them into a recursive version that requires no matrix inversion (Young, 1970; Strejc, 1980; Söderström and Stoica, 1983). This is a standard practice and thus requires no further discussion.

Model Order In general, the order of the process is unknown. For most identification schemes, a model of known order is assumed (Strejc, 1981). Our method may also follow such a strategy. Alternatively, the model order may be pre-set to two since a SODPT model with a zero introduced, as discussed in Subsection 7.2.1, can describe most dynamics well. If the mean square error in (7.69) is too large, the model order is increased by one and a higher-order model is estimated following the general identification method in Subsection 7.2.2.

Estimation of K_p It is common practice in process control to read off K_p directly using the steady-state output as $K_p = y(\infty)/h$. However, there do exist cases in which the step test stops before the steady state is reached, or the user wishes to estimate all parameters together in a consistent way. Then, with obvious modifications to the derivation in Subsection 7.2.2, the following regression equation can be derived:

$$\Phi\theta = \Gamma,$$

where $\theta(t) = [\theta_1 \; \theta_2 \; \cdots \; \theta_{2n+1}]^T$ and

$$\theta_k = \begin{cases} \frac{\sum_{1\leqslant i_1<\cdots<i_{k-1}\leqslant n}\lambda_{i_1}\cdots\lambda_{i_{k-1}}}{\prod_{i=1}^{n}\lambda_i}, & \text{for } k=1,2,\cdots,n, \\ K_p\sum_{j=0}^{n-1}\frac{(-1)^{n+1-j}}{(n-j)!}L^{n-j}\sum_{1\leqslant i_1<i_2<\cdots<i_j\leqslant n-1}\beta_{i_1}\beta_{i_2}\cdots\beta_{i_j}, \\ & \text{for } k=n+1, \\ K_p\sum_{j=0}^{2n+1-k}\frac{(-1)^{2n-k-j}}{(2n+1-k-j)!}L^{2n+1-k-j}\sum_{1\leqslant i_1<i_2<\cdots<i_j\leqslant n-1}\beta_{i_1}\beta_{i_2}\cdots\beta_{i_j}, \\ & \text{for } k=n+2,\cdots,2n, \\ K_p, & \text{for } k=2n+1, \end{cases}$$

$$\Gamma = \left[-\int_{[0,t_1]}^{(n)} y, -\int_{[0,t_2]}^{(n)} y, \cdots, -\int_{[0,t_N]}^{(n)} y\right]^T,$$

and

$$\Phi = \begin{bmatrix} y(t_1) & \int_{[0,t_1]}^{(n-1)} y & \cdots & \int_{[0,t_1]}^{(n-1)} y\,h & ht_1 & \cdots & \frac{1}{(n-1)!}ht_1^{n-1} \\ y(t_2) & \int_{[0,t_2]}^{(n-1)} y & \cdots & \int_{[0,t_2]}^{(n-1)} y\,h & ht_2 & \cdots & \frac{1}{(n-1)!}ht_2^{n-1} \\ \vdots & \vdots & \ddots & \vdots & \vdots & \ddots & \vdots \\ y(t_N) & \int_{[0,t_N]}^{(n-1)} y & \cdots & \int_{[0,t_N]}^{(n-1)} y\,h & ht_N & \cdots & \frac{1}{(n-1)!}ht_N^{n-1} \end{bmatrix}.$$

Thus, a LS-like algorithm can be applied and K_p can be estimated simultaneously with other parameters.

Data Screening In the course of a step test, it is possible that the process is perturbed by a sudden rapid disturbance. This corrupts the process response and the corresponding samples are obviously unreliable and should not be used for identification. For the proposed identification method, a data screening technique, i.e., discarding those bad samples and retaining only the reliable ones, is applicable since essentially t_1 and t_N can be any time between L and T_{set}. To see this, the proposed algorithm before calculating $\hat{\theta}$ is applied to each batch of good data to form respective Φ_i, Γ_i, $i=1,2$. They are combined to form

$$\begin{bmatrix} \Gamma_1 \\ \Gamma_2 \end{bmatrix} = \begin{bmatrix} \Phi_1 \\ \Phi_2 \end{bmatrix}\theta,$$

which is already in the normal regression form and used to find $\hat{\theta}$. This feature of data screening is quite attractive and time saving, especially for slow processes frequently encountered in process control.

Impact of Offset Errors Low frequency noise and offset errors could cause estimation errors to the proposed method. This is a common problem to any identification method using step tests. As stated in Schoukens and Pintelon (1991), the test signal should enable us to inject as much energy as possible, or a high signal-to-noise ratio is required. This implies that offsets should be small. The current control engineering practice is that any control test has to be carefully planned and prepared to ensure that the process is in a good steady-state condition with all major inputs well controlled or monitored before a test is actually applied. In other words, practising engineers try to avoid any significant disturbances coming into the process during the test, and will be alerted when a significant disturbance is detected (this usually requires the test to be redone). If there exist inherent offsets, they may cause significant estimation error in K_p, which further leads to estimation errors in other parameters. In such a case, the problem can be solved if K_p is known or obtained *a priori*. Sometimes, the steady-state gain may be found from physical/chemical principles. If K_p is not available, techniques such as the relay test (Hang et al., 1993a) may be employed to get an accurate K_p. With a known K_p, the effect of the offset on the step response can be eliminated from the recorded data, and the proposed method is then applied to the modified step response.

7.2.4 Simulation and Real-time Test

The proposed step identification method is now applied to several typical processes. Without loss of generality, a unit step is employed in all the simulations below. For a better assessment of its accuracy, identification errors in both the time domain and the frequency domain are considered. This is because some step identification methods are found to be able to fit the process response well in the time domain, but the frequency response of the model sometimes deviates too far from the real process frequency response. To achieve better control performance, the estimation error should be small in both time and frequency domains. Comparison is made with the area method and graphical methods, to show the performance enhancement.

The time domain identification error is measured over the transient period by the standard deviation, *err* as defined in (7.69). Once the identification is carried out, the model $\hat{G}(s)$ is available and the frequency domain identification error is measured by the *worst-case error*, ERR as defined in (7.14), where, the frequency range $[0, \omega_c]$ is considered with $\angle G(j\omega_c) = -\pi$, since this range is the most significant for controller design. To test the robustness of the proposed method, noise is introduced into the process output. In the context of system identification, noise-to-signal ratio defined (Haykin, 1989) by

$$NSR = \frac{\text{mean(abs(noise))}}{\text{mean(abs(signal))}}, \qquad (7.70)$$

is used to represent noise level.

Example 7.2.1. Consider a high-order monotonic process:

$$G(s) = \frac{1}{(s+1)^5}.$$

A unit step test is performed, and the process input and output step response are recorded. In the noise-free case, the model estimated with the proposed method is

$$\hat{G}(s) = \frac{e^{-1.4491s}}{4.3988s^2 + 3.4499s + 1}$$

for the default $t_N = 0.8T_{set} \approx 8.47\text{s}$, where the errors are $err = 1.1034 \times 10^{-4}$ and $ERR = 4.5728\%$. To see the effect of t_N on identification, smaller and larger values of t_N are tested as follows. If t_N is set to $0.7T_{set} \approx 7.41\text{s}$, the errors become $err = 1.8294 \times 10^{-4}$ and $ERR = 4.5063\%$. If t_N is set to be $T_{set} \approx 10.59\text{s}$, the identification errors become $err = 7.1736 \times 10^{-5}$ and $ERR = 9.5715\%$. As stated in the previous subsection, a smaller t_N leads to a larger err and a smaller ERR, while a larger t_N gives a smaller err and a larger ERR. The recommended t_N here is a good trade-off.

For comparison, the area identification method and the graphical methods are also applied. The area method gives

$$\hat{G}(s) = \frac{e^{-2.6400s}}{2.3772s + 1},$$

with $err = 1.9276 \times 10^{-3}$ and $ERR = 50.1586\%$, and the graphical method gives

$$\hat{G}(s) = \frac{e^{-1.3526s}}{4.7143s^2 + 3.5774s + 1},$$

with $err = 1.1215 \times 10^{-4}$ and $ERR = 7.6723\%$.

To improve the accuracy, the proposed method yields a third-order model:

$$\hat{G}(s) = \frac{0.2711s^2 + 0.1953s + 1}{2.6222s^3 + 5.2668s^2 + 3.8870s + 1} e^{-1.3264s}$$

with $err = 1.0858 \times 10^{-6}$ and $ERR = 1.5359\%$, and a fourth-order model:

$$\hat{G}(s) = \frac{0.0076s^3 + 0.1722s^2 + 0.2948s + 1}{1.7732s^4 + 5.9048s^3 + 7.6344s^2 + 4.4782s + 1} e^{-0.8365s}$$

with $err = 6.0577 \times 10^{-7}$ and $ERR = 1.3001\%$. The results are quite accurate and the effectiveness of the proposed method is obvious.

The identification results for a variety of different dynamics are listed in Table 7.1, and the effectiveness of the proposed method is evident.

Table 7.1. Identification results

		$G(s)$	$\frac{1}{(s+1)^5}$	$\frac{1.08e^{-10s}}{(s+1)^2(2s+1)^3}$	$\frac{(-4s+1)e^{-s}}{9s^2+2.4s+1}$	$\frac{(s+1)(s-1)(s+10)}{(s+2)^3(s+3)(s+4)}e^{-0.5s}$	$\frac{2.15(-2.7s+1)(158.5s^2+6s+1)}{(17.5s+1)^4(20s+1)}e^{-14s}$
The proposed method	$\hat{G}(s)$		$\frac{e^{-1.45s}}{4.40s^2+3.45s+1}$	$\frac{1.08e^{-12.06s}}{1.90s^2+5.87s+1}$	$\frac{(-4.00s+1)e^{-1.00s}}{9.00s^2+2.40s+1}$	$\frac{0.10(1.00s-1)}{0.33s^2+0.82s+1}e^{-0.68s}$	$\frac{2.15e^{-28.58s}}{1939.41s^2+69.63s+1}$
	err		1.10×10^{-4}	5.48×10^{-5}	2.78×10^{-13}	5.04×10^{-6}	8.84×10^{-4}
	ERR		4.57%	4.59%	2.23×10^{-4}%	1.97%	5.45%
The area method	$\hat{G}(s)$		$\frac{e^{-2.63s}}{2.38s+1}$	$\frac{1.08e^{-14.04s}}{3.96s+1}$	—	—	$\frac{2.15e^{-53.90s}}{46.69s+1}$
	err		1.93×10^{-3}	1.26×10^{-3}	—	—	1.15×10^{-2}
	ERR		50.16%	4.69%	—	—	60.87%
The graphical method	$\hat{G}(s)$		$\frac{e^{-1.35s}}{4.71s^2+3.58s+1}$	$\frac{1.08e^{-12.17s}}{1.76s^2+5.64s+1}$	$\frac{e^{-4.58s}}{6.79s^2+1.57s+1}$	$\frac{0.10e^{-1.63s}}{0.11s^2+0.63s+1}$	$\frac{2.15e^{-25.28s}}{2232.67s^2+72.02s+1}$
	err		1.12×10^{-4}	9.81×10^{-5}	6.79×10^{-13}	8.11×10^{-3}	2.48×10^{-4}
	ERR		7.67%	4.59%	32.16%	39.77%	14.06%

Example 7.2.2. To demonstrate the robustness of the proposed method, step tests with measurement noise $NSR = 3\%, 5\%, 10\%, 15\%, 20\%, 30\%, 40\%$ or 50%, were performed for the process:

$$G(s) = \frac{-4s+1}{9s^2 + 2.4s + 1}e^{-s}.$$

The results are listed in Table 7.2. The robustness of the proposed method with the IV solution is obvious. To compare the LS and IV methods, consider $NSR = 10\%$ for instance; the LS solution is

$$\hat{G}(s) = \frac{-3.8119s + 1.0248}{7.4998s^2 + 2.2273s + 1}e^{-1.3252s},$$

with $err = 5.47 \times 10^{-2}$ and $ERR = 16.63\%$, while the IV method produces a better solution:

$$\hat{G}(s) = \frac{-3.8610s + 1.0248}{9.0035s^2 + 2.4207s + 1}e^{-1.0457s},$$

with $err = 3.11 \times 10^{-2}$ and $ERR = 4.04\%$.

Example 7.2.3. To see the applicability of the proposed method to nonlinear systems, consider the system (Slotine and Li, 1991):

$$\begin{cases} \dot{x}_1 = \sin x_2, \\ \dot{x}_2 = x_1^4 \cos x_2 + u, \\ y(t) = x_1(t-2). \end{cases}$$

With $NSR = 10\%$, the SOPDT model is estimated as

$$\hat{G}(s) = \frac{5.4676}{4.7184s^2 + 3.1780s + 1}e^{-2.0784s}.$$

The step responses of the process and estimated model are shown in Figure 7.2. The effectiveness of the proposed method can be observed.

Example 7.2.4. The proposed identification method for the SOPDT model was applied to a pilot plant, a coupled-tank control apparatus, made by KentRidge Instrument Ptd. Ltd, Singapore. The equipment consists of two small tower-type tanks mounted above a reservoir which functions as storage for the water (Figure 7.3). In this implementation, the baffle between the two tanks is raised to allow water to flow between the tanks. The objective is to control the water level in the second tank by manipulating the voltage to the pump in the

Table 7.2. Identification results for $G(s) = \frac{-4s+1}{9s^2+2.4s+1}e^{-s}$ under different noise levels

NSR	The proposed method			The graphical method		
	$\hat{G}(s)$	err	ERR	$\hat{G}(s)$	err	ERR
0	$\frac{-4s+1}{9s^2+2.4s+1}e^{-s}$	2.78×10^{-13}	$2.23 \times 10^{-4}\%$	$\frac{e^{-4.5830s}}{6.7908s^2+1.5686s+1}$	6.79×10^{-3}	32.16%
3%	$\frac{-3.9591s+1.0074}{9.0011s^2+2.4062s+1}e^{-1.0133s}$	1.22×10^{-3}	0.80%	$\frac{1.0074e^{-4.7111}}{6.6981s^2+1.2781s+1}$	9.40×10^{-3}	33.64%
5%	$\frac{-3.9314s+1.0124}{9.0011s^2+2.4104s+1}e^{-1.0224s}$	3.40×10^{-3}	1.34%	$\frac{1.0124e^{-5.1052s}}{5.7129s^2+1.0258s+1}$	1.46×10^{-2}	39.20%
10%	$\frac{-3.8610s+1.0248}{9.0035s^2+2.4207s+1}e^{-1.0457s}$	1.36×10^{-2}	2.68%	$\frac{1.0248e^{-4.5815s}}{8.0858s^2+0.7943s+1}$	3.47×10^{-2}	88.70%
15%	$\frac{-3.7876s+1.0374}{9.0052s^2+2.4311s+1}e^{-1.0702s}$	3.11×10^{-2}	4.04%	$\frac{1.0374e^{-4.2788s}}{8.8604s^2+0.3878s+1}$	1.07×10^{-1}	284.94%
20%	$\frac{-3.7118s+1.0501}{9.0068s^2+2.4415s+1}e^{-1.0958s}$	5.59×10^{-2}	5.41%	$\frac{1.0501e^{-11.0898s}}{2.5997s^2+18.3952s+1}$	2.11×10^{-1}	109.32%
30%	$\frac{-3.5505s+1.0762}{9.0098s^2+2.4626s+1}e^{-1.1515s}$	1.29×10^{-1}	8.21%	$\frac{1.0762e^{-5.6244s}}{3.4034s^2+2.4626s+1}$	1.63×10^{-1}	82.39%
40%	$\frac{-3.3708s+1.1036}{9.0127s^2+2.4845s+1}e^{-1.2155s}$	2.39×10^{-1}	1.15%	$\frac{1.1036e^{-9.3729s}}{0.7107s^2+0.2881s+1}$	2.55×10^{-1}	85.32%
50%	$\frac{-3.1575s+1.1341}{9.0155s^2+2.5084s+1}e^{-1.2941s}$	4.00×10^{-1}	14.42%	$\frac{1.1341e^{-6.3534s}}{2.4955s^2+0.7447s+1}$	4.05×10^{-1}	43.20%

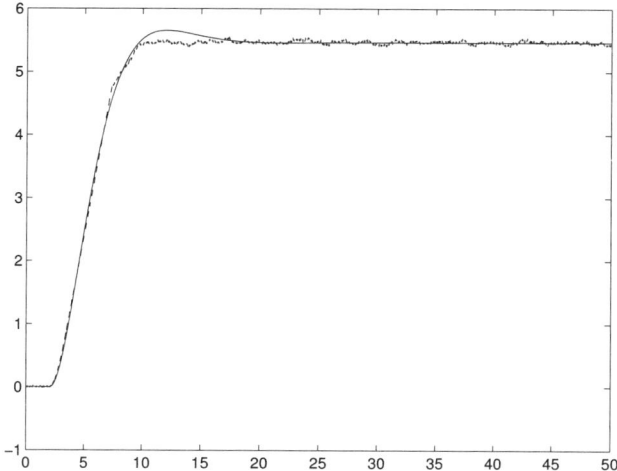

Fig. 7.2. Step responses for Example 7.2.3
($- - -$: actual process; ———: estimated model from the proposed method)

Fig. 7.3. Coupled-tank level control system

first tank. Using the proposed identification method, the process was modelled as

$$\hat{G}(s) = \frac{1068.5638s + 62.2012}{626.5379s^2 + 66.8088s + 1} e^{-1.4292s},$$

and the response for this $\hat{G}(s)$ under the same step input is shown with the solid line in Figure 7.4, where the dashed line is from the actual process. The effectiveness of the proposed method is clear.

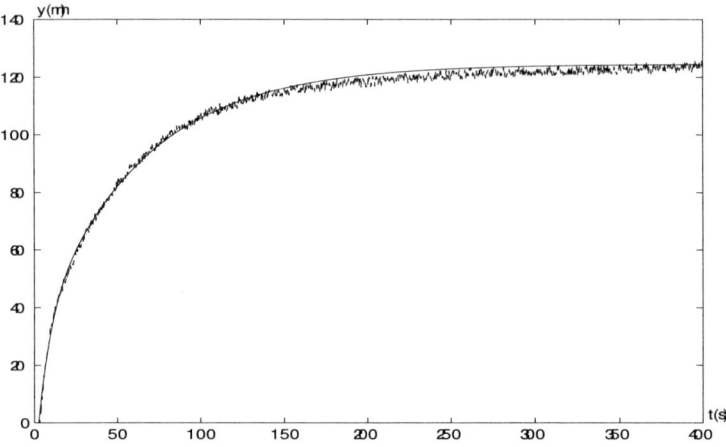

Fig. 7.4. Step response of coupled-tank
(− − −: from process; ———: from model)

7.3 A Hybrid Approach

The method described in Section 7.1 cannot guarantee the preservation of stability of transfer function models. This is a typical problem associated with frequency domain methods. On the other hand, with the time domain approach it is easier to deal with stability preservation. One sees that the step response of a stable $G(s)$ will remain finite while that of an unstable $\hat{G}(s)$ will tend to infinity. The squared error between them will diverge, and an unstable $\hat{G}(s)$ must be excluded from the solutions to the problem of minimizing such an error, in other words, the solution must be stable. Based on this idea, we present a time domain transfer function modelling algorithm in this section. This algorithm not only preserves stability but also enables non-iterative estimation of all model parameters including the time delay. It constructs the step response from the frequency response and then uses the identification algorithm described in the preceding section to obtain the transfer function model.

Construction of Step Response from $G(s)$ or $G(j\omega)$ Suppose that the plant is stable with $G(s)$ or $G(j\omega)$ given. Let the plant input $u(t)$ be of step type with size h. Then the plant output step response in the Laplace domain is

$$Y(s) = G(s)\frac{h}{s}. \tag{7.71}$$

It seems very easy to obtain the corresponding time response $y(t)$ by simply applying the inverse Fourier transform ($\mathcal{F}^{-1}$). However, since the steady-state part of $y(t)$ is not absolutely integrable, such a calculation is inapplicable and meaningless (Wang et al., 1997c). To solve this problem, $y(t)$ is decomposed into

$$y(t) = y(\infty) + \Delta y(t) = G(0)h + \Delta y(t).$$

Applying the Laplace transform to both sides gives

$$Y(s) = \frac{G(0)}{s} + \mathcal{L}\{\Delta y(t)\}. \tag{7.72}$$

Bringing (7.71) in, we have

$$\mathcal{L}\{\Delta y(t)\} = h\frac{G(s) - G(0)}{s}.$$

Applying the inverse Laplace transform yields

$$\Delta y(t) = h\mathcal{L}^{-1}\{\frac{G(s) - G(0)}{s}\}.$$

Thus, the plant step response is constructed as

$$y(t) = h[G(0) + \mathcal{F}^{-1}\{\frac{G(j\omega) - G(0)}{j\omega}\}], \tag{7.73}$$

where $\mathcal{F}^{-1}$ may easily be implemented by the inverse fast Fourier transform (IFFT).

Identification from Step Response Apply the algorithm in Section 7.2 to the step response $y(t)$ in (7.73) to get

$$\hat{G}(s) = \frac{b_1 s^{n-1} + b_2 s^{n-2} + \cdots + b_{n-1} s + b_n}{s^n + a_1 s^{n-1} + \cdots + a_{n-1} s + a_n} e^{-Ls}. \tag{7.74}$$

Example 7.3.1.   Reconsider the high-order plant in Example 7.1.1. The output step response is first constructed using the IFFT. The area method (Rake, 1980) gives the model:

$$\hat{G}(s) = \frac{2.15}{46.69s + 1} e^{-53.90s},$$

with $err = 1.15 \times 10^{-2}$ and $ERR = 60.87\%$. The FOPDT model estimated by our hybrid algorithm in this section is

$$\hat{G}(s) = \frac{0.0396}{s + 0.0184} e^{-49.9839s},$$

with $err = 8.6 \times 10^{-3}$ and $ERR = 48.12\%$. Our SOPDT model is

$$\hat{G}(s) = \frac{0.0011}{s^2 + 0.0343s + 0.0005} e^{-28.8861s},$$

with $err = 4.0631 \times 10^{-4}$ and $ERR = 5.81\%$. For a third-order plus dead time model (TOPDT), we have

$$\hat{G}(s) = \frac{0.2059s^2 + 0.0005s + 0.0001}{s^3 + 0.9969s^2 + 0.0382s + 0.0001} e^{-31.9265s},$$

with $err = 2.0009 \times 10^{-4}$ and $ERR = 1.27\%$. It can be seen that the results from our algorithm are consistently better than that of the area method. The estimation errors decrease with model order n. Especially from FOPDT to SOPDT, errors decrease dramatically, but the error decrease slows down from SOPDT to TOPDT. In this case, an SOPDT model is good enough. Step and frequency responses of the actual and estimated models are shown in Figure 7.5.

The proposed method is also tested for many other examples. Typical results are given in Table 7.3. They show that consistent enhancement of the reduced-order model accuracy is achieved with the proposed method over the recursive least squares method of Section 7.1 if the latter starts with zero initial parameters. The former does always preserve stability, as expected, while the latter is not capable of that. However, the recursive least squares method is simpler in implementation, and with those stability measures introduced in Section 7.1, the chance of getting stable models is quite high for low-order modelling, though not guaranteed.

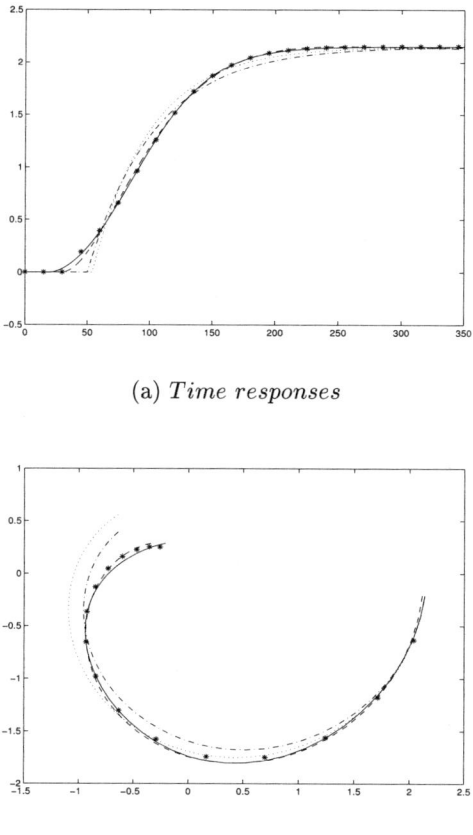

(a) *Time responses*

(b) *Frequency responses*

Fig. 7.5. Model reduction via hybrid algorithm

(—— actual process; ··· area method; — · — · —· proposed FOPDT;— — — proposed SOPDT; * * * proposed TOPDT)

Table 7.3. Summary of simulation results

Plant	Scheme	Model order	Results	err	ERR
$2.15\dfrac{(-2.7s+1)(158.5s^2+6s+1)}{(17.5s+1)^4(20s+1)}e^{-14s}$	RLS	1	$\dfrac{0.0046}{s+0.0019}e^{-29.3s}$	0.291	138.72%
		2	$\dfrac{0.0010}{s^2+0.0318s+0.0005}e^{-33.60s}$	0.044	52.86%
		3	$\dfrac{(42.3333s^2+2.5081s+1.1548)e^{-30.50s}}{10^4s^3+995.3655s^2+38.8783s+0.5371}$	0.058	1.21%
	Proposed	1	$\dfrac{0.0396}{s+0.0184}e^{-49.9839s}$	0.200	48.12%
		2	$\dfrac{0.0011}{s^2+0.03443s+0.0005}e^{-28.8861s}$	0.050	5.81%
		3	$\dfrac{(205.94s^2+5.25s+1.12)e^{-31.9265s}}{10^4s^3+996.85s^2+38.24s+0.52}$	0.039	1.27%
$\dfrac{(s+1)(s-1)(s+10)}{(s+2)^3(s+3)(s+4)}e^{-0.5s}$	RLS	2	$\dfrac{-0.1288s-0.1479}{s^2-1.4828s+1.4114}e^{-4.41s}$	80.858	7.90%
	Proposed	2	$\dfrac{0.2173s-0.2176}{s^2+1.81s+2.124}e^{-0.59s}$	2.232×10^{-4}	5.58%
$\dfrac{(s+1)^{10}}{(s+0.5)^4(s+2)^6}e^{-3s}$	RLS	2	$\dfrac{1.4605s^2+1.3761s+0.8004}{s^2+8.6101s+3.2016}e^{-3.03s}$	8.240×10^{-4}	9.14%
	Proposed	2	$\dfrac{0.6796s^2+0.6897s+0.3868}{s^2+4.458s+1.549}e^{-2.9s}$	5.575×10^{-4}	10.82%

8. A General Identification Approach

Chapters 5 through 7 have covered in detail process frequency response estimation using relay feedback and conversion of frequency responses to transfer function models. The methods developed there are very useful in applications. There are, however, a few other issues which deserve attention. First, many industrial control systems already run in closed-loop before an identification test and/or controller retuning is carried out. In some cases, performing identification experiments under closed-loop conditions may be necessary for safety or economic reasons, or if the system contains inherent feedback mechanisms (Forssell and Ljung, 1999). It is thus desirable not to disconnect the existing controller during the test. This gives rise to the closed-loop identification problem, requiring that a test be conducted on a closed-loop system, but not on the process itself.

Second, the types of test signals could be other than relay, say, pulse, pseudo-random binary sequence, step, ramp and sinusoidal functions (Unbehauen and Rao, 1987). Of all these tests, the step test is probably the simplest. The step test needs little equipment, can be performed manually and dominates in process control applications. Chapter 7 presented an algorithm for transfer function identification from open-loop step responses. Its extension to closed-loop testing and multivariable systems is appealing.

Third, most industrial processes are multivariable in nature. The control of such multivariable systems has always been a challenge due to its complex interactive nature. Control techniques such as model predictive control (Richalet *et al.*, 1978) and internal model control (Garcia and Morari, 1982) explicitly require a process model. SISO identification techniques based on relay or step tests need to be modified to cope with multivaribale nature.

There is thus a great demand for a general identification scheme for multivariable systems that can cover many different experimental tests in a unified framework. In this chapter, a novel robust multivariable process identification method is presented. From the recorded process inputs and outputs, the frequency response matrix is calculated using the fast Fourier transform (FFT),

208   8. A General Identification Approach

the step response is further constructed using the inverse fast Fourier transform (IFFT) and the transfer function is determined using an algorithm from Chapter 7, in an elementwise way. This method is applicable to any kind of test and pre-tuned controller regardless of the transient type, provided that the system is stable and the responses are not chaotic. We present the SISO version in Section 8.1 to motivate the general MIMO version in Section 8.2. Section 8.3 deals solely with unstable processes, which necessitate a different technique.

8.1 SISO Systems

Consider the conventional feedback control system shown in Figure 8.1. The SISO process $G(s)$ is under the control of the SISO controller $K(s)$.

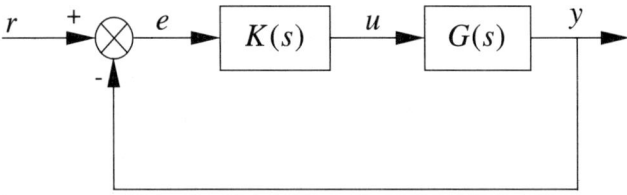

Fig. 8.1. Feedback control system

8.1.1 The Method

The idea behind our method is to always estimate the process frequency response first, regardless of test type, and then to convert it to the step response from which a transfer function model can be obtained using an algorithm from Chapter 7.

Test Suppose that a given process $G(s)$ is stable and already runs in closed-loop with a pre-tuned controller $K(s)$. Suppose that the system has been brought to a constant steady state. A test is conducted either in open-loop or closed-loop (user defined) so that a change in the manipulated variable $u(t)$ for the open-loop test, or a change in the reference $r(t)$, is activated. It should be pointed out that such a signal can be of any type. Suppose that the closed-loop system is stable and produces a steady state at the end of the applied test. The steady state can be a constant or some form of stationary oscillation. During the test, the manipulated variable $u(t)$ and the process output $y(t)$ are recorded until the steady state occurs.

Process Frequency Response In our method, except for an open-loop step test, for which this and the next steps are skipped, the frequency response of the process is first calculated. The open-loop relay test has been thoroughly discussed in Chapter 6. The analysis and techniques developed there are also applicable to other tests. The point to emphasize here is that the frequency responses $Y(j\omega)$ and $U(j\omega)$ of $y(t)$ and $u(t)$ cannot be obtained directly by taking the Fourier transform of these recorded $y(t)$ and $u(t)$. Here, we adopt a technique from Chapter 6, the decomposition method, and generalize it to cover all other tests, provided that the test yields a final finite steady state.

Let $y(t)$ and $u(t)$, respectively, be decomposed into transient parts $\Delta y(t)$, $\Delta u(t)$, and the steady state part or the periodic stationary cycle parts $y_s(t)$, $u_s(t)$ as

$$y(t) = \Delta y(t) + y_s(t),$$

$$u(t) = \Delta u(t) + u_s(t).$$

The definition of the transfer function gives

$$G(s) = \frac{Y(s)}{U(s)} = \frac{Y_s(s) + \Delta Y(s)}{U_s(s) + \Delta U(s)},$$

and the process frequency response is formally

$$G(j\omega) = \frac{Y_s(j\omega) + \Delta Y(j\omega)}{U_s(j\omega) + \Delta U(j\omega)}.$$

Note that $\Delta Y(j\omega)$ and $\Delta U(j\omega)$ can be obtained using the Fourier transform since the transient parts $\Delta u(t)$ and $\Delta y(t)$ decay to zero exponentially. In practice, they are calculated using the FFT. Thus, it follows that

$$G(j\omega) \approx \frac{Y_s(j\omega) + \mathcal{FFT}(\Delta y(t))}{U_s(j\omega) + \mathcal{FFT}(\Delta u(t))}. \tag{8.1}$$

If the test (say, a step test) gives a constant steady state output $y(\infty)$ we have

$$Y_s(j\omega) = \frac{y(\infty)}{j\omega}. \tag{8.2}$$

On the other hand, if the test (e.g. relay) causes stationary oscillations, it follows (Chapter 6) that

$$Y_s(j\omega) = \frac{1}{1 - e^{-j\omega T_c}} \int_0^{T_c} y_s(t) e^{-j\omega t} dt, \tag{8.3}$$

where T_c is the period of the stationary oscillations in the process output y_s. Similarly, the following expression can be obtained:

$$U_s(j\omega) = \begin{cases} \frac{u(\infty)}{j\omega}, & \text{step test;} \\ \frac{1}{1-e^{-j\omega T_c}} \int_0^{T_c} u_s(t)e^{-j\omega t}dt, & \text{relay test.} \end{cases} \quad (8.4)$$

With (8.2) or (8.3) and (8.4), the process frequency response can be calculated from (8.1).

It should be pointed that the formula (8.1) assumes zero initial conditions. If the system has a nonzero operating point of $y(0)$ and $u(0)$, then $y(t)$, $y(\infty)$, $u(t)$ and $u(\infty)$ should be replaced by $y(t) - y(0)$, $y(\infty) - y(0)$, $u(t) - u(0)$ and $u(\infty) - u(0)$, respectively.

Construction of Process Step Response Imagine that the process input $u(t)$ is a unit step function, then

$$Y(s) = G(s)\frac{1}{s}. \quad (8.5)$$

It seems very easy to obtain the corresponding $y(t)$ by simply applying the inverse Fourier transform to $Y(j\omega)$. However, since the the output response, $y(t)$, is not absolutely integrable, such a calculation is meaningless as discussed in Chapter 7. To solve this problem, $y(t)$ is decomposed into

$$y(t) = y(\infty) + \Delta y(t) = G(0) + \Delta y(t).$$

Applying the Laplace transform to both sides gives

$$Y(s) = \frac{G(0)}{s} + \mathcal{L}\{\Delta y(t)\}. \quad (8.6)$$

Bring (8.5) in, we have

$$\mathcal{L}\{\Delta y(t)\} = \frac{G(s) - G(0)}{s}.$$

Applying the inverse Laplace transform yields

$$\Delta y(t) = \mathcal{L}^{-1}\{\frac{G(s) - G(0)}{s}\}.$$

Thus, using the inverse fast Fourier transform (IFFT), the process step response is constructed as

$$y(t) \approx G(0) + \mathcal{IFFT}(\frac{G(j\omega) - G(0)}{j\omega}). \quad (8.7)$$

Process Transfer Function The stable step response constructed above is fitted into the following model:

$$Y(s) = \hat{G}(s)U(s) = \frac{b_1 s^{n-1} + b_2 s^{n-2} + \cdots + b_{n-1} s + b_n}{s^n + a_1 s^{n-1} + \cdots + a_{n-1} s + a_n} e^{-Ls} U(s). \quad (8.8)$$

The algorithm in Section 2 of Chapter 7 is adopted here to obtain a stable $\hat{G}(s)$ from such a step response. No further discussion is needed.

Remark 8.1.1. Compared with existing closed-loop identification methods, the proposed one is more robust because it makes use of all the recorded data. It can also be applied to a wide range of tests, and the controller is not limited to PID types only. Moreover, it can produce a general transfer function model, but is not restricted to first or second order. These advantages make it more appealing than its counterparts in many aspects.

Remark 8.1.2. The combined use of the FFT and IFFT is crucial for enhancement of the identification performance. It looks possible and reasonable to estimate the process frequency response with the FFT and then obtain a transfer function using the response directly, say the recursive least squares method of Chapter 7. However, simple tests such as step or relay cannot excite the process effectively and uniformly over the whole working frequency range and the error in frequency response estimation varies with frequencies. This may in turn cause poor identification of the transfer function model. We find that the error inherent in the process frequency response with the FFT is neutralized or compensated for by the later IFFT, so that the process step response can be recovered accurately, which leads to a better transfer function model, as will be demonstrated in our simulation below.

8.1.2 Simulation

The proposed identification method is now applied to several typical processes. For better assessment of its accuracy, identification errors in both the time domain and the frequency domain are considered. This is because some closed-loop identification methods are found to be good in the time domain, but the frequency response of the model sometimes deviates too far from the real process frequency response. To achieve better control performance, the estimation error should be small in both time and frequency domains.

The time domain identification error is measured over the transient period by standard deviation:

$$err = \frac{1}{N} \sum_{k=1}^{N} [y(kT) - \hat{y}_c(kT)]^2, \quad (8.9)$$

where $y(kT)$ is the recorded output response of the actual process to a given test, while $\hat{y}(kT)$ is the output response of the estimated model under the recorded same control signal. The frequency domain identification error is measured by the *worst-case error*:

$$ERR = \max_i \left\{ \left| \frac{\hat{G}(j\omega_i) - G(j\omega_i)}{G(j\omega_i)} \right| \times 100\%, \quad i = 1, 2, \cdots, M \right\}, \qquad (8.10)$$

where $G(j\omega_i)$ and $\hat{G}(j\omega_i)$ are the actual and estimated process frequency responses respectively. ERR is evaluated over the frequency range with phase from 0 to $-\pi$, since this range is most important for control design.

In practice, measurement noise is generally present. To examine the robustness of our algorithm against noise, noise is introduced into the output for these simulation studies. The noise level is measured by the following noise-to-signal ratio (Haykin, 1989):

$$NSR = \frac{\text{mean(abs(noise))}}{\text{mean(abs(signal))}}.$$

Example 8.1.1. Consider a process

$$G(s) = \frac{e^{-s}}{12s^2 + 8s + 1}.$$

Comparison is made with the method in Suganda *et al.* (1998) since it is the latest method available of its kind. In Suganda's method, different sets of PI parameters are tested for these examples, which give different dynamics, corresponding to under-damping, median dampling and over-damping, respectively. For the different kinds of dynamics, different formulas are applied to give a transfer function for the closed-loop system. Since the proposed method is general and consistent for all these different scenarios, it is not necessary to test all these cases. Here, one sets the controller so as to give an under-damped response for the process; the other cases are shown for two extra processes in Table 8.1.

G in this example is a second-order plus dead time (SOPDT) process, and should be estimated by a SOPDT model without model error in the noise-free case. A unit step is performed in closed-loop on the system with a PI controller ($K_c = 0.4$ and $T_i = 2$). Using the proposed method, the estimated SOPDT model is

$$\hat{G}(s) = \frac{1.0001 e^{-0.9963s}}{12.0643 s^2 + 8.0108 s + 1},$$

Table 8.1. Identification results

Process $G(s)$	NSR	The proposed method $\hat{G}(s)$	err	ERR	Suganda's method $\hat{G}(s)$	err	ERR
$\dfrac{e^{-s}}{12s^2+8s+1}$	0	$\dfrac{1.00e^{-1.00s}}{12.06s^2+8.01s+1}$	4.91×10^{-4}	0.35%	$\dfrac{1.00e^{-0.99s}}{12.04s^2+7.98s+1}$	2.85×10^{-3}	0.45%
	5%	$\dfrac{1.00e^{-1.06s}}{1.89s^2+7.92s+1}$	12.87	4.75%	$\dfrac{1.00e^{-2.94s}}{13.70s^2+8.37s+1}$	42.27	134.79%
$\dfrac{e^{-3s}}{(s+1)^2(2s+1)}$	0	$\dfrac{1.00e^{-3.49s}}{3.26s^2+3.52s+1}$	4.80×10^{-3}	0.48%	$\dfrac{1.00e^{-3.30s}}{3.72s^2+3.59s+1}$	9.08×10^{-2}	2.30%
	5%	$\dfrac{1.00e^{-3.44s}}{3.45s^2+3.49s+1}$	26.0987%	$1.00e^{-0.88s}$ 0.49%	$\dfrac{1.00e^{-2.93s}}{13.84s^2+6.83s+1}$	44.12	65.89%
$\dfrac{0.5s+1}{(s+1)^2(2s+1)}e^{-3s}$	0	$\dfrac{1.00e^{-3.11s}}{2.90s^2+3.40s+1}$	4.85×10^{-3}	0.49%	$\dfrac{1.00e^{-2.93s}}{2.86s^2+3.35s+1}$	2.23×10^{-1}	5.87%
	5%	$\dfrac{1.00e^{-3.43s}}{2.86s^2+3.41s+1}$	23.76	5.49%	$\dfrac{1.00e^{-2.09s}}{1.75s^2+3.87s+1}$	37.92	63.83%

214   8. A General Identification Approach

which is almost identical to the actual process. The identification errors are $err = 4.91 \times 10^{-4}$ and $ERR = 0.35\%$. Under the same test condition, Suganda's method gives the model:

$$\hat{G}(s) = \frac{1.0000e^{-0.9945s}}{12.0409s^2 + 7.9810s + 1},$$

where the identification errors are $err = 2.85 \times 10^{-3}$ and $ERR = 0.45\%$. The improved accuracy is obvious, and this accuracy holds for other scenarios (see Table 8.1).

It can be seen that Suganda's method works well for those cases where the controller has been well tuned. It is possible that the controller is badly tuned. To test the effectiveness of the proposed method in such a situation, the following examples are presented. For these examples, there are no solutions using Suganda's method. In fact, using their graphical method, complex coefficients result when trying to find a SOPDT transfer function for the closed-loop system.

Example 8.1.2. Consider a high-order oscillatory process with large dead time:

$$G(s) = \frac{1.08e^{-10s}}{(s+1)^2(2s+1)^3}.$$

The result from the proposed method is shown in Table 8.2, and the estimated model is very close to the actual process. Other examples are also shown in this table.

Table 8.2. Identification results

$G(s)$	$\hat{G}(s)$	err	ERR
$\frac{1.08e^{-10s}}{(s+1)^2(2s+1)^3}$	$\frac{1.08e^{-12.81s}}{7.54s^2+5.26s+1}$	0.69	2.14%
$\frac{-4s+1}{9s^2+2.4s+1}e^{-s}$	$\frac{-3.98s+1.00}{8.86s^2+2.39s+1}e^{-1.07s}$	0.11	0.82%
$\frac{1}{(s+1)^5}$	$\frac{1.00e^{-1.32s}}{4.75s^2+3.28s+1}$	0.97	9.28%

Since the proposed method makes use of many points rather than one or two points on the process response and adopts a least squares or instrumental variable method, it is expected to be robust to noise. The identification results of the process with different noise levels are listed in Table 8.3, indicating that the proposed method works well even when there is large measurement noise present.

Table 8.3. Identification results with different NSR

NSR	$\hat{G}(s)$	err	ERR
0	$\dfrac{1.0801}{7.5403s^2+5.2612s+1}e^{-12.8110s}$	0.69	2.14%
3%	$\dfrac{1.0828}{7.2238s^2+5.1499s+1}e^{-12.6847s}$	11.23	3.44%
5%	$\dfrac{1.0823}{7.4893s^2+5.1953s+1}e^{-12.5564s}$	21.44	4.29%
10%	$\dfrac{1.0808}{7.6543s^2+5.1754s+1}e^{-12.4387s}$	42.08	5.97%
15%	$\dfrac{1.0794}{8.4881s^2+5.1712s+1}e^{-12.1350s}$	94.30	9.67%
20%	$\dfrac{1.0780}{9.0539s^2+5.2606s+1}e^{-11.9564s}$	131.98	10.59%
30%	$\dfrac{1.0908}{15.5858s^2+5.5866s+1}e^{-10.3358s}$	325.63	24.35%
40%	$\dfrac{1.0932}{16.4593s^2+4.9406s+1}e^{-10.6667s}$	572.71	32.22%

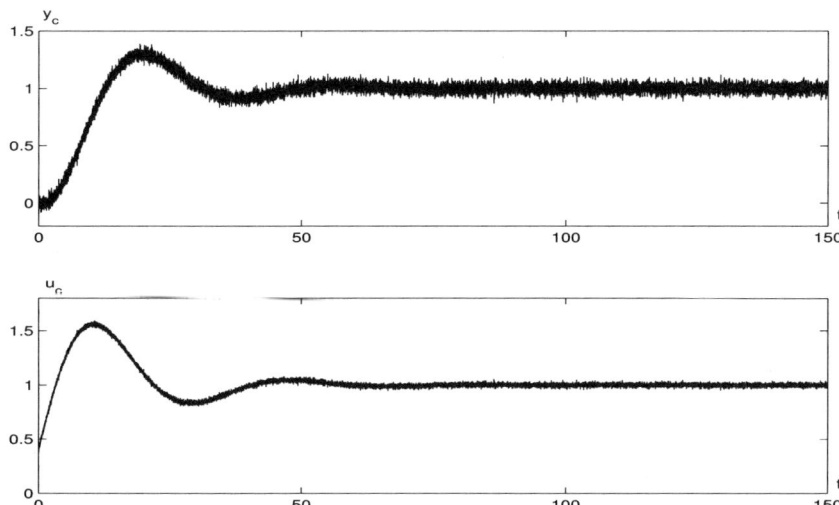

Fig. 8.2. Closed-loop response for $G(s) = \dfrac{e^{-s}}{12s^2+8s+1}$ with $NSR = 3\%$

Example 8.1.3. To illustrate Remark 8.1.2, a step test is done for

$$G = \frac{e^{-s}}{12s^2 + 8s + 1}$$

with $NSR = 3\%$. The response for the closed-loop system is shown in Figure 8.2. From the corrupted output and control signal, the proposed method gives

$$\hat{G}(s) = \frac{1.0100}{11.3097s^2 + 7.9830s + 1}e^{-1.0144s}.$$

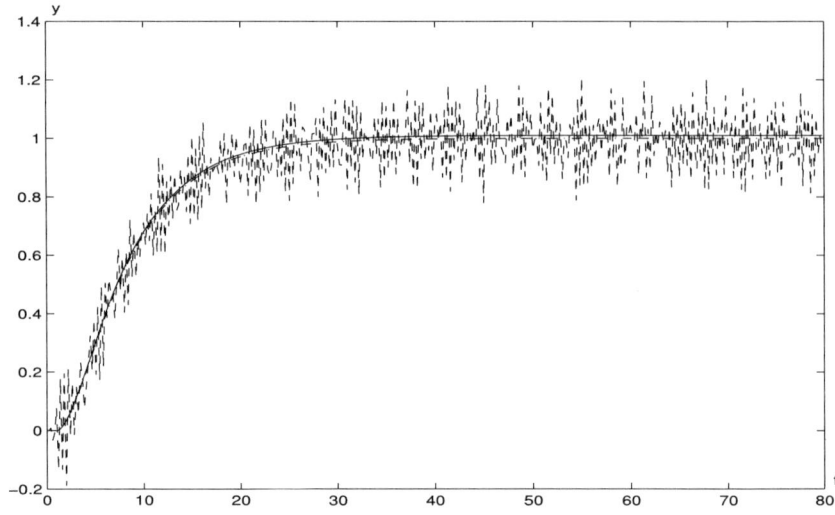

Fig. 8.3. Process step responses
($-\cdot-\cdot-\cdot$: actual process; - - -: constructed using FFT and IFFT; ———: estimated model)

The process step responses are shown in Figure 8.3, where the dash-dot line is for the actual process, and the dashed line is constructed from the FFT and IFFT method, while the solid line is from the estimated model with the proposed method. The effectiveness of the proposed method is clearly observed. It is also noticed that the constructed process step response using the FFT and IFFT techniques is very close to the actual one. Figure 8.4 shows Bode plots of the actual process (dash-dot line), the FFT-based estimation using (8.1) (dashed line) and the final model (solid line). The estimation error between the FFT-based approximation and the actual process is 77.77%, while that between the final model and the actual process is 3.08%. For this SOPDT process, the FFT gives a large approximation error in the high frequency part. However, accurate estimation is recovered after using the IFFT.

8.2 MIMO Systems

We want to extend the SISO method of the preceding section to the multivariable case. Only square systems will be addressed.

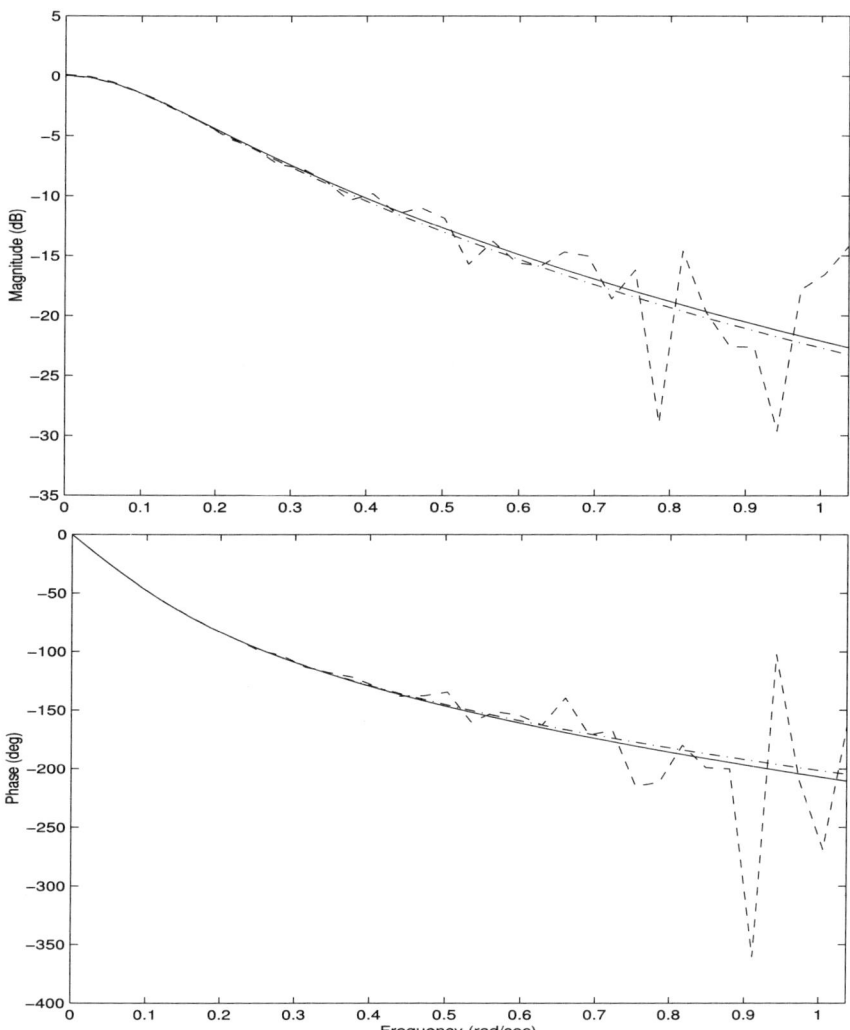

Fig. 8.4. Comparison of frequency responses
($-\cdot-\cdot-$: actual process; - - -: FFT-based estimation; ———: final model)

8.2.1 The Method

Let an $m \times m$ linear process be described by

$$Y(s) = G(s)U(s), \tag{8.11}$$

with output vector

$$Y(s) = \begin{bmatrix} Y_1(s) & Y_2(s) & \cdots & Y_m(s) \end{bmatrix}^T,$$

input vector

$$U(s) = \begin{bmatrix} U_1(s) & U_2(s) & \cdots & U_m(s) \end{bmatrix}^T,$$

and process transfer function matrix $G = \{G_{ij}(s)\}$, $i, j = 1, 2, \cdots, m$. The process may be in the feedback system depicted in Figure 8.1 with an $m \times m$ controller, $K(s)$. For later use, partition the $m \times m$ identity matrix I as

$$I = \begin{bmatrix} e_1 & e_2 & \cdots & e_m \end{bmatrix}. \tag{8.12}$$

Similar to the SISO case in the preceding section, the idea of our MIMO method is to calculate the process frequency response $G(j\omega)$ using the FFT, construct for each entry of $G(s)$ its step response from $G_{ij}(j\omega)$ using the IFFT, and estimate a transfer function from such a response. But before these descriptions, we need to discuss the different tests to which our identification method is applicable.

Test Some test has to be conducted on a given process to enable identification. Various aspects of the test should be considered. The first is the type of test signal. The most popular are step and relay. The second consideration is the operating mode during the experimental test. It could be in open-loop mode without a controller or in closed-loop mode with a controller operating. The third consideration is how to configure the test for a multivariable process, as there could be three possible schemes:

(i) *Independent test*: A test is applied to only one loop at a time and released once a steady state, i.e., constant steady state or stationary oscillations (Chen, 1993), is reached, and is repeated until all m tests on each loop have been performed.

(ii) *Sequential test*: A test is applied to the first loop while all other loops are kept unattached. Then a test is made on the second loop while the previous change in the first loop is still in place. The procedure is repeated until the mth loop has been tested.

(iii) *Decentralized test*: All m test signals are applied to m loops simultaneously.

Independent testing is quite time consuming since each test can be activated only after the process has been brought back to its original steady state following the previous test. Decentralized testing may generate complicated responses (Loh, 1994), which are difficult to analyze and predict. This has been addressed in Chapters 5 and 6. The sequential test is thus considered in this chapter. With such a scheme there are four typical combinations of test:

Case 1: Open-loop (sequential) step test;
Case 2: Closed-loop (sequential) step test;
Case 3: Open-loop (sequential) relay test; (8.13)
Case 4: Closed-loop (sequential) relay test.

Diagrams of open-loop and closed-loop sequential relay feedback tests are shown in Figure 8.5 and Figure 8.6 for illustration, respectively. It should be stressed that the identification method to be presented below is not restricted to these four cases only, but also applicable to other cases provided that each test leads to a steady state.

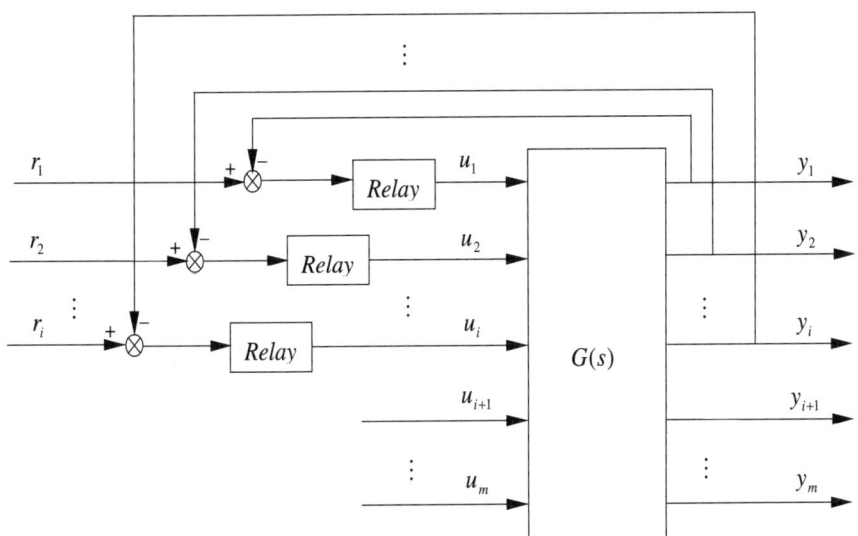

Fig. 8.5. Open-loop sequential relay feedback Test

To unify the presentation for all cases, we introduce a vector $v(t)$:

$$v(t) = \begin{cases} u(t), & \text{open-loop test,} \\ r(t), & \text{closed-loop test,} \end{cases}$$

where $u(t)$ is the input to the plant and $r(t)$ is the set-point for the closed-loop of Figure 8.1. Let $\mathbf{1}(t)$ and $\rho(t)$ be a unit step function and relay function

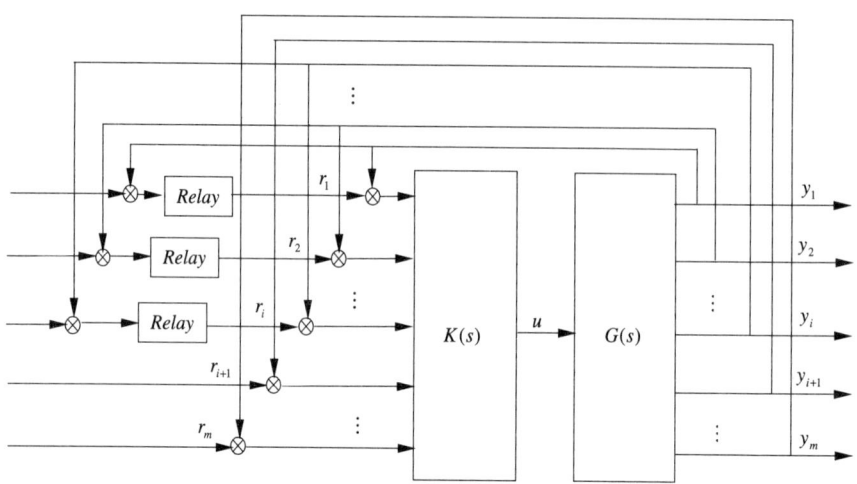

Fig. 8.6. Closed-loop sequential relay feedback test

with unity magnitude, respectively. The types of test signal are reflected by the elements of $v(t)$:

$$v_i(t) = \begin{cases} \alpha_i \mathbf{1}(t), & \text{step}, \\ \beta_i \rho(t) \mathbf{1}(t), & \text{relay}, \end{cases} \tag{8.14}$$

where α_i and β_i are the size and amplitude of the step and relay, respectively. Then the m tests v^i in a sequential experiment over the time span $[t_0, t_m]$ with respective test periods $[t_{i-1}, t_i]$, $i = 1, 2, \cdots, m$, can be expressed as

$$v(t) = \begin{cases} v^1(t) = v_1(t)e_1, & t_0 = 0 \leqslant t < t_1, \\ v^i(t) = v_i(t)e_i + v^{i-1}(t), & t_{i-1} \leqslant t < t_i, i = 2, 3, \cdots, m, \end{cases} \tag{8.15}$$

where e_i is the ith column of the $m \times m$ identity matrix defined in (8.12), and $t_0 = 0$ is the start time for the experiment while t_i is the end time for the ith test when the system has entered the steady state.

The sequential experiment on an $m \times m$ process is actually executed as follows:

Initialization: Bring the process to constant steady states $y(0)$ and $u(0)$ at $t_0 = 0$. Specify the step sizes α_i or the relay amplitudes β_i, $i = 1, 2, \cdots, m$.
Test 1: Apply a test of the given type and magnitude to the first input v_1, while keeping all other inputs v_i unchanged, that is, $v = v^1$. Wait for

the system to settle down at $t = t_1$, that is, the output y has reached a constant steady state during the step test or stationary oscillations during the relay test.

Test i: Let $v = v^i$ as shown in (8.15) and apply it to the system at $t = t_{i-1}$, $i = 2, 3, \cdots, m$, until the system settles down again at $t = t_i$.

Regardless of the different test cases (8.13), we always collect the process output and input responses from all m tests, and denote them by $\{u(t), y(t), t \in [t_0, t_m]\}$.

Process Frequency Response This and the following steps of our method are to find the step response for every entry of the process transfer matrix. They are not required for Case 1 where such step responses are already available. It can be seen from (8.11) that if one has m independent output and input frequency response vectors $\tilde{Y}^i(j\omega)$ and $\tilde{U}^i(j\omega)$, the process frequency response matrix $G(j\omega)$ can then be obtained from the equation:

$$\left[\tilde{Y}^1(j\omega)\ \tilde{Y}^2(j\omega)\ \cdots\ \tilde{Y}^m(j\omega)\right] = G(j\omega)\left[\tilde{U}^1(j\omega)\ \tilde{U}^2(j\omega)\ \cdots\ \tilde{U}^m(j\omega)\right].$$

Our task now is to construct m independent time response vectors $\tilde{y}^i(t)$ and $\tilde{u}^i(t)$ from the recorded process output and input responses $y(t)$ and $u(t)$ such that for each i, $\tilde{y}^i(t)$ and $\tilde{u}^i(t)$ have zero initial conditions at the start and steady states at the end, so that their frequency responses $\tilde{Y}^i(j\omega)$ and $\tilde{U}^i(j\omega)$ meet $\tilde{Y}^i(j\omega) = G(j\omega)\tilde{U}^i(j\omega)$. For Case 2 of a step test, the ith test starts usually with nonzero initial conditions $y(t_{i-1})$ and $u(t_{i-1})$, and ends with new constant steady states $y(t_i)$ and $u(t_i)$. If the output and input have their initial conditions subtracted, the resulting signals will have zero initial conditions at $t = t_{i-1}$ and constant steady states at $t = t_i$. Thus we form from y and u the following modified signals:

$$\tilde{y}^i(t) = y(t) - y(t_{i-1}), \quad t_{i-1} \leqslant t < t_i, \quad i = 1, 2, \cdots, m, \quad (8.16a)$$

$$\tilde{u}^i(t) = u(t) - u(t_{i-1}), \quad t_{i-1} \leqslant t < t_i, \quad i = 1, 2, \cdots, m. \quad (8.16b)$$

For experiments which lead to stationary oscillations (e.g., Case 3 and Case 4), making $\tilde{y}(t_{i-1}) = 0$ only is not enough since its derivations could still be nonzero. Instead, let us imagine that for each i, the ith test lasts beyond $t = t_i$ until $t = t_m$ without activating the subsequent tests, then the last cycle of stationary oscillations $y_s^i(t)$ and $u_s^i(t)$ sustained before $t = t_i$ will be repeated until $t = t_m$. One can thus define for each i

$$\tilde{y}^i(t) = \begin{cases} y(t) - y(0), & 0 \leqslant t < t_i, \\ y_s^i(t) - y(0), & t_i \leqslant t \leqslant t_m, \end{cases} \quad (8.17a)$$

and

$$\tilde{u}^i(t) = \begin{cases} u(t) - u(0), & 0 \leqslant t < t_i, \\ u_s^i(t) - u(0), & t_i \leqslant t \leqslant t_m. \end{cases} \tag{8.17b}$$

Note that in the step case $\tilde{y}^i$ and $\tilde{u}^i$ have different time spaces $[t_{i-1}, t_i]$ while in the relay case they have the same space $[0, t_m]$.

Like the SISO case in the preceding section, decompose the responses into transient parts and steady state parts as

$$\tilde{y}^i(t) = \Delta \tilde{y}^i(t) + \tilde{y}_s^i(t), \quad i = 1, 2, \cdots, m, \tag{8.18a}$$

$$\tilde{u}^i(t) = \Delta \tilde{u}^i(t) + \tilde{u}_s^i(t), \quad i = 1, 2, \cdots, m. \tag{8.18b}$$

For each i, $\tilde{y}^i(t)$ and $\tilde{u}^i(t)$ have almost entered the steady states at $t = t_i$, and both $\Delta \tilde{y}^i(t)$ and $\Delta \tilde{u}^i(t)$ are approximately zero afterwards. The Fourier transform of $\Delta \tilde{y}^i(t)$ then gives its frequency response:

$$\Delta \tilde{Y}^i(j\omega) = \int_0^\infty \Delta \tilde{y}^i(t) e^{-j\omega t} dt. \tag{8.19}$$

In practice, (8.19) is approximated by

$$\Delta \tilde{Y}^i(j\omega) \approx \int_{t_S^i}^{t_E^i} \Delta \tilde{y}^i(t) e^{-j\omega t} dt, \tag{8.20}$$

where the start time t_S^i and the end time t_E^i are defined as

$$t_S^i = \begin{cases} t_{i-1}, & \text{Case 2}, \\ 0, & \text{Case 3 or 4}, \end{cases} \tag{8.21a}$$

and

$$t_E^i = \begin{cases} t_i, & \text{Case 2}, \\ t_m, & \text{Case 3 or 4}. \end{cases} \tag{8.21b}$$

(8.20) can be computed at discrete frequencies using the standard FFT technique.

For $\tilde{y}_s^i(t)$ and $\tilde{u}_s^i(t)$, if the test (say, a step test) gives a constant steady state output, $\tilde{y}_s^i(t_i)$, we have

$$\tilde{Y}_s^i(j\omega) = \frac{\tilde{y}_s^i(t_i)}{j\omega}, \quad i = 1, 2, \cdots, m. \tag{8.22}$$

8.2 MIMO Systems

On the other hand, if the test (e.g. relay) causes stationary oscillations, it follows that

$$\tilde{Y}_s^i(j\omega) = \frac{1}{1 - e^{-j\omega T_i}} \int_{t_i - T_i}^{t_i} \tilde{y}_s^i(t) e^{-j\omega(t + T_i - t_i)} dt, \qquad (8.23)$$

where T_i is the period of the stationary oscillations of $\tilde{y}_s^i$. Similarly, the following expression can be obtained:

$$\tilde{U}_s^i(j\omega) = \begin{cases} \frac{\bar{u}_s^i(t_i)}{j\omega}, & \text{step test,} \\ \frac{1}{1 - e^{-j\omega T_i}} \int_{t_i - T_i}^{t_i} \tilde{u}_s^i(t) e^{-j\omega(t + T_i - t_i)} dt, & \text{relay test.} \end{cases} \qquad (8.24)$$

For each test i, it follows from (8.11) that the following frequency response relation holds:

$$\tilde{Y}^i(j\omega) = G(j\omega)\tilde{U}^i(j\omega), \qquad i = 1, 2, \cdots, m, \qquad (8.25)$$

where

$$\tilde{Y}^i(j\omega) = \Delta\tilde{Y}^i(j\omega) + \tilde{Y}_s^i(j\omega), \qquad (8.26a)$$

$$\tilde{U}^i(j\omega) = \Delta\tilde{U}^i(j\omega) + \tilde{U}_s^i(j\omega) \qquad (8.26b)$$

are obtained from (8.19)–(8.24). Collecting (8.25) for all i gives

$$\bar{Y}(j\omega) = G(j\omega)\bar{U}(j\omega), \qquad (8.27)$$

where

$$\bar{Y}(j\omega) = \begin{bmatrix} \tilde{Y}^1(j\omega) & \tilde{Y}^2(j\omega) & \cdots & \tilde{Y}^m(j\omega) \end{bmatrix},$$
$$\bar{U}(j\omega) = \begin{bmatrix} \tilde{U}^1(j\omega) & \tilde{U}^2(j\omega) & \cdots & \tilde{U}^m(j\omega) \end{bmatrix}.$$

If the inversion of $\bar{U}(j\omega)$ exists, the process frequency response matrix $G(j\omega)$ can then be found from (8.27). The task now is to determine if $\bar{U}(j\omega)$ is invertible or not.

For the closed-loop tests $v = r$ (Case 2 or Case 4), it follows from Figure 8.1 that

$$U(s) = K(s)[I + K(s)G(s)]^{-1} R(s) \triangleq Q(s)R(s). \qquad (8.28)$$

with nonsingular $Q(s)$, or in the frequency domain,

$$\bar{U}(j\omega) = Q(j\omega)\bar{R}(j\omega). \qquad (8.29)$$

For Case 2, $\bar{R}(j\omega)$ is the frequency response of

$$\tilde{r} \triangleq \begin{bmatrix} \tilde{r}^1 & \tilde{r}^2 & \cdots & \tilde{r}^m \end{bmatrix}$$

$$\triangleq \begin{bmatrix} r^1, & r^2 - r^1, & \cdots, & r^m - r^1 - r^2 - \cdots - r^{m-1} \end{bmatrix}$$

$$= \text{diag}\{\alpha_i \mathbf{1}(t)\}$$

and is given by

$$\bar{R}(j\omega) = \begin{bmatrix} \alpha_1/j\omega & 0 & \cdots & 0 \\ 0 & \alpha_2/j\omega & \cdots & 0 \\ \vdots & \vdots & \ddots & \vdots \\ 0 & 0 & \cdots & \alpha_m/j\omega \end{bmatrix},$$

which is nonsingular at any $\omega \in (0, \infty)$. $\bar{U}(j\omega)$ is thus invertible for $\omega \in (0, \infty)$.

For Case 4, the nature of the sequential experiment (8.15) indicates that

$$r^1(t) = \begin{bmatrix} \beta_1 \rho(t)\mathbf{1}(t) \\ 0 \\ \vdots \\ 0 \end{bmatrix}, \quad r^2(t) = \begin{bmatrix} \beta_1 \rho(t)\mathbf{1}(t) \\ \beta_2 \rho(t)\mathbf{1}(t-t_1) \\ \vdots \\ 0 \end{bmatrix},$$

$$\cdots, \quad r^m(t) = \begin{bmatrix} \beta_1 \rho(t)\mathbf{1}(t) \\ \beta_2 \rho(t)\mathbf{1}(t-t_1) \\ \vdots \\ \beta_m \rho(t)\mathbf{1}(t-t_{m-1}) \end{bmatrix},$$

thus $\bar{r} \triangleq \begin{bmatrix} r^1 & r^2 & \cdots & r^m \end{bmatrix}$ has an upper triangular form and so has its frequency response $\bar{R}(j\omega)$. It then follows from (8.29) that $\bar{U}(j\omega)$ is nonsingular for $\omega \in [0, \infty)$.

Case 3 is an open-loop test with $v = u$. Like Case 4, its sequential nature gives an upper triangular $\bar{U}(j\omega)$, which is nonsingular for $\omega \in [0, \infty)$.

In view of the above analysis, all three cases (2,3,4) give an invertible $\bar{U}$, so that the process frequency response can be calculated using

$$G(j\omega) = \begin{bmatrix} \Delta \tilde{Y}^1(j\omega) + \tilde{Y}_s^1(j\omega) & \cdots & \Delta \tilde{Y}^m(j\omega) + \tilde{Y}_s^m(j\omega) \end{bmatrix}$$
$$\cdot \begin{bmatrix} \Delta \tilde{U}^1(j\omega) + \tilde{U}_s^1(j\omega) & \cdots & \Delta \tilde{U}^m(j\omega) + \tilde{U}_s^m(j\omega) \end{bmatrix}^{-1}, \qquad (8.30)$$

which is valid at all $\omega \in [0, \infty)$ for Cases 3 and 4, and at $\omega \in (0, \infty)$ for Case 2. For $\omega = 0$ in Case 2, we can determine the static gain of the process from output and input steady states by

$$G(0) = \begin{bmatrix} \tilde{y}_s^1(t_1) & \cdots & \tilde{y}_s^m(t_m) \end{bmatrix} \begin{bmatrix} \tilde{u}_s^1(t_1) & \cdots & \tilde{u}_s^m(t_m) \end{bmatrix}^{-1}. \tag{8.31}$$

Step Response Construction and Transfer Function Modelling
Once the frequency response matrix $G(j\omega)$ of the process is obtained, the step response of each entry can be constructed separately and can be used to obtain a transfer function which matches the step response best. These two steps are exactly the same as for the SISO case described in Section 8.1 and thus are not detailed here. Just do them in an elementwise way.

Identification Algorithm We can summarize the above development into the following algorithm:

Initialization: Bring the process to the constant steady states. Specify the test type (relay or step), their sizes and the test mode (open-loop or closed-loop).

Step 1: Perform the sequential experiment and record all the process input and process output responses.

Step 2: For Case 1, go to Step 4 directly. Otherwise, calculate the frequency response $G(j\omega)$ from (8.30).

Step 3: Construct the step response from (8.7), one for each entry $G(j\omega)$.

Step 4: Determine a transfer function model from the step response using the algorithm in Section 7.2 of Chapter 7, one for each entry $G(s)$.

8.2.2 Simulation

In this subsection, the proposed identification method is applied to four typical industrial processes to show the effectiveness. Both time domain error, err as defined in (8.9), and frequency domain error, ERR as defined in (8.10), are evaluated for each entry of the model to assess estimation performance.

Example 8.2.1. Consider the Vinante and Luyben plant (Luyben, 1986):

$$G(s) = \begin{bmatrix} \frac{-2.2e^{-s}}{7s+1} & \frac{1.3e^{-0.3s}}{7s+1} \\ \frac{-2.8e^{-1.8s}}{9.5s+1} & \frac{4.3e^{-0.35s}}{9.2s+1} \end{bmatrix}.$$

An open-loop sequential step test (Case 1) is applied. The model estimated with the proposed method is

$$\hat{G}(s) = \begin{bmatrix} \frac{-2.2000e^{-1.0009s}}{7.0050s+1} & \frac{1.3000e^{-0.2987s}}{7.0050s+1} \\ \frac{-2.8000e^{-1.8009s}}{9.5050s+1} & \frac{4.3000e^{-0.3487s}}{9.2050s+1} \end{bmatrix},$$

where the errors are

$$err = \begin{bmatrix} 3.7445 \times 10^{-7}\% & 0.9661\% \\ 2.8127 \times 10^{-7}\% & 0.9690\% \end{bmatrix},$$

and

$$ERR = \begin{bmatrix} 0.1709\% & 0.6973\% \\ 0.1037\% & 0.5959\% \end{bmatrix},$$

respectively, they are both very small.

Example 8.2.2. Consider the 3 × 3 plant in Orgunnaike and Ray (1979):

$$G(s) = \begin{bmatrix} \frac{0.66e^{-2.6s}}{6.7s+1} & \frac{-0.61e^{-3.5s}}{8.64s+1} & \frac{-0.0049e^{-s}}{9.06s+1} \\ \frac{1.11e^{-6.5s}}{3.25s+1} & \frac{-2.36e^{-3s}}{5s+1} & \frac{-0.01e^{-1.2s}}{7.09s+1} \\ \frac{-34.68e^{-9.2s}}{8.15s+1} & \frac{46.2e^{-9.4s}}{10.9s+1} & \frac{0.87(11.61s+1)e^{-s}}{(3.89s+1)(18.8s+1)} \end{bmatrix}.$$

An open-loop sequential relay feedback test (Case 3) is applied. According to (8.17a) and (8.17b), $\tilde{y}^i(t)$ and $\tilde{u}^i(t)$ are constructed by duplicating the final cycles of their respective stationary oscillations before $t = t_i$ over $t \in [t_i, t_m]$. Figure 8.7 shows the first output $\tilde{y}_1^1$, $\tilde{y}_1^2$ and $\tilde{y}_1^3$. The proposed method is then applied to produce

$$\hat{G}(s) = \begin{bmatrix} \frac{0.6600e^{-2.5999s}}{6.7074s+1} & \frac{-0.6100e^{-3.4994s}}{8.6487s+1} & \frac{-0.0049e^{-0.9977s}}{9.0730s+1} \\ \frac{1.1078e^{-6.5287s}}{3.1883s+1} & \frac{-2.3569e^{-3.0194s}}{4.9557s+1} & \frac{-0.0099e^{-1.2056s}}{7.0302s+1} \\ \frac{-34.6121e^{-9.2548s}}{8.0135s+1} & \frac{46.1028e^{-9.4232s}}{10.8171s+1} & \frac{(9.8104s+0.8684)e^{-1.0062s}}{70.7462s^2+22.2668s+1} \end{bmatrix},$$

with errors:

$$err = \begin{bmatrix} 1.0631 \times 10^{-4}\% & 1.6949\% & 1.3163\% \\ 0.0045\% & 1.0160\% & 0.8386\% \\ 0.1585\% & 2.2145\% & 2.0277\% \end{bmatrix},$$

and

$$ERR = \begin{bmatrix} 0.1031\% & 0.0898\% & 0.3450\% \\ 0.7785\% & 1.0260\% & 0.6729\% \\ 1.0572\% & 0.4151\% & 0.8779\% \end{bmatrix},$$

respectively. The resultant model is almost identical to the actual process.

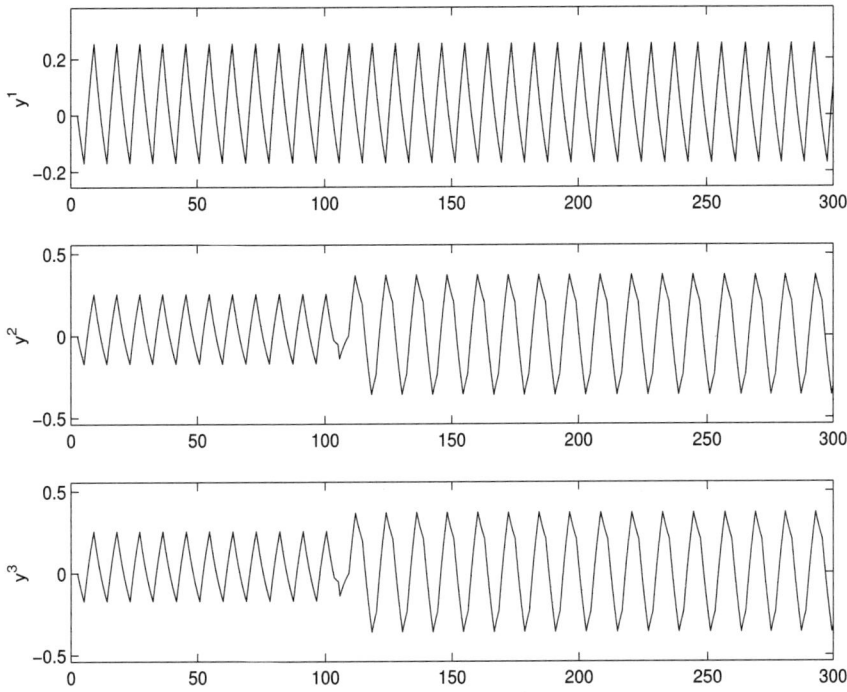

Fig. 8.7. $\tilde{y}_1^i(t)$ constructed from the sequential relay test

Example 8.2.3. Consider the Doukas and Luyben plant (Luyben, 1986):

$$G(s) = \begin{bmatrix} \frac{-9.811e^{-1.59s}}{11.36s+1} & \frac{0.374e^{-7.75s}}{22.22s+1} & \frac{-2.368e^{-27.33s}}{33.3s+1} & \frac{-11.3e^{-3.79s}}{(21.74s+1)^2} \\ \frac{5.984e^{-2.24s}}{14.29s+1} & \frac{-1.986e^{-0.71s}}{66.67s+1} & \frac{0.422e^{-8.72s}}{(250s+1)^2} & \frac{5.24e^{-60s}}{400s+1} \\ \frac{2.38e^{-0.42s}}{(1.43s+1)^2} & \frac{0.0204e^{-0.59s}}{(7.14s+1)^2} & \frac{0.513e^{-s}}{s+1} & \frac{-0.33e^{-0.68s}}{(2.38s+1)^2} \\ \frac{-11.3e^{-3.79s}}{(21.74s+1)^2} & \frac{-0.176e^{-0.48s}}{(6.9s+1)^2} & \frac{15.54e^{-s}}{s+1} & \frac{4.48e^{-0.52s}}{11.11s+1} \end{bmatrix},$$

under the control of the decentralized PI controller (Luyben, 1986):

$$K(s) = \begin{bmatrix} -0.084(1+\frac{s}{33}) & 0 & 0 & 0 \\ 0 & -5.16(1+\frac{s}{15.5}) & 0 & 0 \\ 0 & 0 & 0.305(1+\frac{s}{17.0}) & 0 \\ 0 & 0 & 0 & 0.529(1+\frac{s}{11.2}) \end{bmatrix}.$$

A closed-loop sequential step test (Case 2) is activated. Applying the proposed method yields

228   8. A General Identification Approach

$$\hat{G}(s) = \begin{bmatrix} \frac{-9.8111e^{-1.5867s}}{11.4106s+1} & \frac{0.3740e^{-7.7555s}}{22.2658s+1} \\ \frac{5.9975e^{-2.2277s}}{14.4527s+1} & \frac{-1.9849e^{-0.7108s}}{66.6687s+1} \\ \frac{2.3800e^{-0.4164s}}{2.0578s^2+2.8695s+1} & \frac{0.0204e^{-0.5882s}}{51.0132s^2+14.2885s+1} \\ \frac{-11.3000e^{-3.7347s}}{474.9052s^2+43.5817s+1} & \frac{(-0.0051s-0.1760)e^{-0.5032s}}{47.7076s^2+13.8112s+1} \\ \frac{-2.3678e^{-27.3198s}}{33.3231s+1} & \frac{-11.2999e^{-3.7318s}}{475.0586s^2+43.5833s+1} \\ \frac{0.3993e^{-8.6505s}}{59221s^2+471s+1} & \frac{5.2268e^{-60.4408s}}{397.4322s+1} \\ \frac{0.5130e^{-1.0012s}}{1.0045s+1} & \frac{-0.3300e^{-0.6853s}}{5.6436s^2+4.7611s+1} \\ \frac{15.5400e^{-1.0010s}}{1.0049s+1} & \frac{4.4800e^{-0.5211s}}{11.1146s+1} \end{bmatrix},$$

with the errors

$$err = \begin{bmatrix} 4.7829 \times 10^{-4}\% & 0.9211\% & 0.8915\% & 0.8239\% \\ 0.1343\% & 1.1127\% & 1.8566\% & 2.8123\% \\ 8.2181 \times 10^{-3}\% & 0.1204\% & 1.9589 \times 10^{-3}\% & 0.0738\% \\ 0.0013\% & 2.4990\% & 0.3676\% & 1.4204\% \end{bmatrix}$$

and

$$ERR = \begin{bmatrix} 0.5085\% & 0.2513\% & 0.0635\% & 0.7372\% \\ 1.1258\% & 0.1598\% & 5.3791\% & 0.4579\% \\ 0.6353\% & 0.1221\% & 0.5281\% & 0.3321\% \\ 0.7047\% & 0.3921\% & 0.5328\% & 0.2944\% \end{bmatrix},$$

respectively. The effectiveness of the proposed method is clearly shown.

Example 8.2.4. Consider the well-known Wood/Berry (WB) binary distillation column plant (Wood and Berry, 1973):

$$G(s) = \begin{bmatrix} \frac{12.8e^{-s}}{16.7s+1} & \frac{-18.9e^{-3s}}{21s+1} \\ \frac{6.6e^{-7s}}{10.9s+1} & \frac{-19.4e^{-3s}}{14.4s+1} \end{bmatrix}.$$

Suppose that the process is under the control of a decentralized PI controller (Luyben, 1986):

$$K(s) = \begin{bmatrix} 0.38 + \frac{0.045}{s} & 0 \\ 0 & -0.075 - \frac{0.0032}{s} \end{bmatrix}.$$

With the controller in action, a closed-loop sequential relay feedback test (Case 4) is applied and the proposed method is used to give the estimated model:

$$\hat{G}(s) = \begin{bmatrix} \frac{12.8000e^{-1.0009s}}{16.7050s+1} & \frac{-18.9000e^{-3.0009s}}{21.0050s+1} \\ \frac{6.6000e^{-7.0009s}}{10.9050s+1} & \frac{-19.4000e^{-3.0009s}}{14.4050s+1} \end{bmatrix},$$

where the errors are

$$err = \begin{bmatrix} 0.000279\% & 0.1199\% \\ 0.000287\% & 0.03411\% \end{bmatrix},$$

and

$$ERR = \begin{bmatrix} 0.1458\% & 0.0562\% \\ 0.0563\% & 0.0639\% \end{bmatrix},$$

respectively.

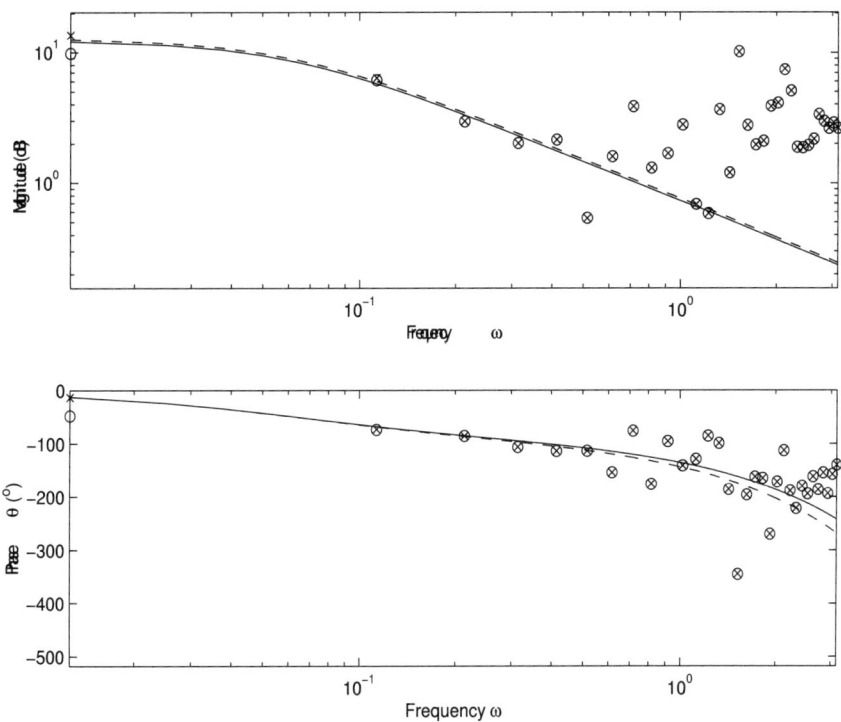

Fig. 8.8. Bode plots for G_{11} with NSR=30%
(- - - plant; ——— model; 'x' FFT only; 'o' Melo)

To demonstrate robustness, the process is again tested under different noise conditions. For comparison, Melo's method (Melo and Friedly, 1992) for the

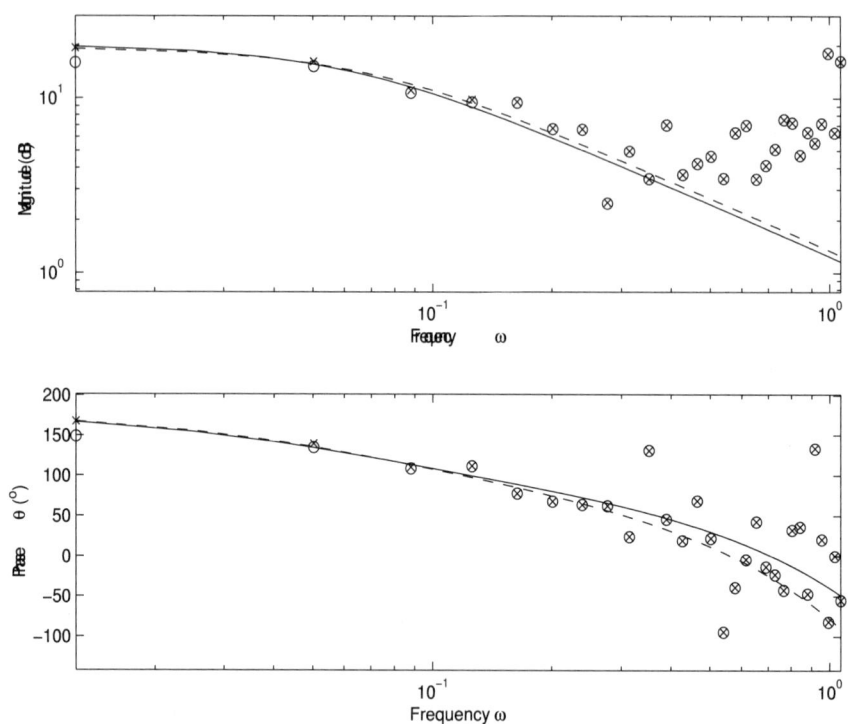

Fig. 8.9. Bode plots for G_{22} with NSR=30%
(- - - plant; ——— model; xxx FFT only; ooo Melo)

estimation of the process frequency response is also tested under the same conditions. The results are listed in Table 8.4. The robustness of the proposed method is evident. Figure 8.8 and Figure 8.9 show the Bode plots for G_{11} and G_{22} of this plant with $NSR = 30\%$, respectively, where the dashed line is for the actual plant, solid line for the model from our method, 'x' for the FFT algorithm without further IFFT and transfer function modelling, and 'o' for Melo's method. As stated in Remark 8.1.2, the FFT alone gives a large approximation error in the high frequency part. However, accurate model is restored after using the IFFT and transfer function modelling.

8.3 Unstable Processes

The method presented so far uses step responses as an intermediate step. It will fail if the step response goes to infinity or the underlying process is unstable.

8.3 Unstable Processes   231

Table 8.4. Identification results with different NSR

NSR	$\hat{G}(s)$	Proposed err	ERR	ERR (Melo's)
0	$\begin{bmatrix} \frac{12.80e^{-1.00s}}{16.71s+1} & \frac{-18.90e^{-3.00s}}{12.01s+1} \\ \frac{6.60e^{-7.00s}}{10.91s+1} & \frac{-19.40e^{-3.00s}}{14.41s+1} \end{bmatrix}$	$\begin{bmatrix} 0.0003\% & 0.1199\% \\ 0.0003\% & 0.0341\% \end{bmatrix}$	$\begin{bmatrix} 0.15\% & 0.06\% \\ 0.06\% & 0.06\% \end{bmatrix}$	$\begin{bmatrix} 0.72\% & 0.27\% \\ 0.29\% & 0.33\% \end{bmatrix}$
5%	$\begin{bmatrix} \frac{12.74e^{-0.98s}}{16.72s+1} & \frac{-19.07e^{-2.96s}}{12.16s+1} \\ \frac{6.50e^{-7.00s}}{10.89s+1} & \frac{-19.51e^{-2.92s}}{14.62s+1} \end{bmatrix}$	$\begin{bmatrix} 0.46\% & 6.81\% \\ 9.83\% & 0.50\% \end{bmatrix}$	$\begin{bmatrix} 3.52\% & 2.25\% \\ 1.52\% & 4.08\% \end{bmatrix}$	$\begin{bmatrix} 88.08\% & 16.06\% \\ 5.07\% & 23.89\% \end{bmatrix}$
10%	$\begin{bmatrix} \frac{12.67e^{-0.95s}}{16.73s+1} & \frac{-19.25e^{-2.91s}}{12.31s+1} \\ \frac{6.39e^{-7.04s}}{10.87s+1} & \frac{-19.62e^{-2.81s}}{14.89s+1} \end{bmatrix}$	$\begin{bmatrix} 1.70\% & 1.32\% \\ 32.23\% & 1.34\% \end{bmatrix}$	$\begin{bmatrix} 7.10\% & 4.61\% \\ 3.14\% & 10.25\% \end{bmatrix}$	$\begin{bmatrix} 173.65\% & 32.56\% \\ 9.49\% & 50.54\% \end{bmatrix}$
20%	$\begin{bmatrix} \frac{12.52e^{-0.93s}}{16.75s+1} & \frac{-19.60e^{-2.87s}}{12.48s+1} \\ \frac{6.16e^{-7.00s}}{10.86s+1} & \frac{-19.85e^{-2.70s}}{15.18s+1} \end{bmatrix}$	$\begin{bmatrix} 5.77\% & 30.70\% \\ 66.99\% & 4.83\% \end{bmatrix}$	$\begin{bmatrix} 1.12\% & 7.22\% \\ 6.73\% & 16.11\% \end{bmatrix}$	$\begin{bmatrix} 334.08\% & 102.7\% \\ 17.71\% & 11.58\% \end{bmatrix}$
30%	$\begin{bmatrix} \frac{12.34e^{-0.86s}}{16.80s+1} & \frac{-19.95e^{-2.71s}}{22.01s+1} \\ \frac{5.88e^{-7.01s}}{10.81s+1} & \frac{-20.07e^{-2.33s}}{16.12s+1} \end{bmatrix}$	$\begin{bmatrix} 14.77\% & 62.37\% \\ 83.15\% & 10.51\% \end{bmatrix}$	$\begin{bmatrix} 22.36\% & 15.12\% \\ 10.90\% & 35.02\% \end{bmatrix}$	$\begin{bmatrix} 41.14\% & 80.04\% \\ 26.70\% & 184.80\% \end{bmatrix}$

For unstable processes, we thus need a different technique for its identification. Consider an SISO unstable process. Suppose that some test like those mentioned in Section 8.1 is performed on it. The resulting input and output responses are recorded. They are employed to find a transfer function model for the process.

Let an unstable process be represented by

$$G(s) = \frac{b_{n-1}s^{n-1} + b_{n-2}s^{n-2} + \ldots + b_1 s + b_0}{s^n + a_{n-1}s^{n-1} + \ldots + a_1 s + a_0} e^{-Ls}, \quad (8.32)$$

or equivalently by

$$y^{(n)}(t) + a_{n-1}y^{(n-1)}(t) + \ldots + a_1 y^{(1)}(t) + a_0 y(t) =$$
$$b_{n-1}u^{(n-1)}(t-L) + b_{n-2}u^{(n-2)}(t-L) + \ldots + b_1 u^{(1)}(t-L) + b_0 u(t-L), \quad (8.33)$$

where $L > 0$.

For an integer $m \geq 1$, define

$$\int_{[0,t]}^{(m)} h = \int_0^t \int_0^{T_m} \cdots \int_0^{T_2} h(\tau_1) d\tau_1 \cdots d\tau_m.$$

Under zero initial conditions, if the time delay L is known, integrating (8.33) n times gives

$$y(t) = -a_{n-1} \int_{[0,t]}^{(1)} y - a_{n-2} \int_{[0,t]}^{(2)} y \cdots - a_1 \int_{[0,t]}^{(n-1)} y - a_0 \int_{[0,t]}^{(n)} y$$
$$+ b_{n-1} \int_{[0,t-L]}^{(1)} u + b_{n-2} \int_{[0,t-L]}^{(2)} u \cdots + b_1 \int_{[0,t-L]}^{(n-1)} u + b_0 \int_{[0,t-L]}^{(n)} u. \quad (8.34)$$

Define

$$\begin{cases} \gamma(t) = y(t), \\ \phi^T(t) = \left[-\int_{[0,t]}^{(1)} y, \cdots, -\int_{[0,t]}^{(n)} y, \int_{[0,t-L]}^{(1)} u, \cdots, \int_{[0,t-L]}^{(n)} u \right], \\ \theta^T = \left[a_{n-1}, \cdots, a_0, b_{n-1}, \cdots, b_0 \right]. \end{cases} \quad (8.35)$$

Then (8.34) can be expressed as

$$\gamma(t) = \phi^T(t)\theta, \quad \text{if } t \geq L.$$

Choose $t = t_i$ and $L \leq t_1 < t_2 < \cdots < t_N$. Then

$$\Gamma = \Phi\theta, \quad (8.36)$$

where $\Gamma = \begin{bmatrix} \gamma(t_1), \gamma(t_2), ..., \gamma(t_N) \end{bmatrix}^T$, and $\Phi = \begin{bmatrix} \phi(t_1), \phi(t_2), ..., \phi(t_N) \end{bmatrix}^T$.
In (8.36), the θ which minimizes the following equation error,

$$\min_{\theta}(\Gamma - \Phi\theta)^T(\Gamma - \Phi\theta),$$

is given by the least squares solution:

$$\hat{\theta} = (\Phi^T \Phi)^{-1} \Phi^T \Gamma. \tag{8.37}$$

Note that the above algorithm requires the unknown time delay L to be estimated. One way of solving this problem is to find the optimal solution of the parameters in (8.32) over the possible range of L. This is a one-dimensional search problem and can easily be solved if an estimate of the range of L is given. A reasonable search range L is $0.5 \sim 2.0$ times L_0, which is the estimated L in the FOPDT model, and we evaluate $15 \sim 20$ points in that range to find the optimal estimate $\hat{L}$.

Simulation has been carried out on a number of processes to illustrate the identification algorithm. The identification results of different processes under different noise levels are given in Table 8.5, indicating good accuracy and robustness for the proposed method.

Table 8.5. Transfer function modeling

Process	NSR	Model	err(%)
$\frac{4e^{-2s}}{-4s+1}$	0%	$\frac{3.9902}{-3.9848s+1.0000}e^{-2.0s}$	1.03
	15%	$\frac{4.0160}{-4.0230s+1.0000}e^{-2.0s}$	5.06
	30%	$\frac{4.0480}{-4.0567s+1.0000}e^{-2.0s}$	9.32
$\frac{e^{-0.5s}}{(-2s+1)(0.5s+1)}$	0%	$\frac{1.0042}{(-2.0285s+1)(0.5119s+1)}e^{-0.50s}$	3.6
	15%	$\frac{0.9947}{(-2.1186s+1.0000)(0.4782s+1)}e^{-0.49s}$	6.7
	30%	$\frac{0.9903}{(-2.2157s+1.0000)(0.4719s+1)}e^{-0.49s}$	19.7
$\frac{e^{-0.5s}}{(-5s+1)(2s+1)(0.5s+1)}$	0%	$\frac{0.9985}{(-5.0097s+1)(1.9853s+1)(0.5062s+1)}e^{-0.50s}$	4.5
	15%	$\frac{0.9942}{(-5.1006s+1)(1.7913s+1)(0.5239s+1)}e^{-0.49s}$	9.7
	30%	$\frac{0.9942}{(-5.1507s+1)(1.6653s+1)(0.5754s+1)}e^{-0.49s}$	25.25

Part III

Controller Design

Introduction to Part III

A process model can be used for system analysis, prediction, design, control, optimization, management and so on. In the context of model-based control system design, there is a large domain of techniques such as pole placement, robust, optimal, and adaptive control approaches. They are usually state-space based. Their specifications are generally quite different from those used in process control. Time delay is frequently encountered in process control, while the state-space based approach usually cannot handle time delay easily. The decoupling requirement, which is very important for process control, is usually not considered explicitly in state-space based approaches. This explains the discrepancy between the many well-developed state-space based multivariable designs (H_∞, LQG, ...) and the rare use of them in process control.

High performance is always the design target in industrial control applications. During the last two decades, model predictive control (MPC) has emerged as a powerful practical control technique in industry due to its ability to compensate for process dead times and include input and output constraints in the formulation of the control law through on-line optimization. Among the MPC techniques is the so-called internal model control (IMC) (Morari and Zafiriou, 1989). The idea inherent in IMC has been around in one form or another for several decades. As early as 1957, Newton *et al.* (1957) converted the closed-loop design problem into an open-loop problem and thus circumvented the closed-loop stability issue, just as is done in the IMC. The Smith predictor (Smith, 1957) contains the reference model idea of the IMC. The IMC structure was formally introduced by Garcia and Morari (1982). It is a powerful control design strategy for linear systems (Garcia and Morari, 1982; Rivera *et al.*, 1986; Morari and Zafiriou, 1989). It uses the process model as the internal model to predict the process output. When the model is perfect, the IMC system becomes an open-loop system and controller design and stability analysis issues become trivial. When a model mismatch exists, by appropriately modifying the difference, robustness can be obtained. The IMC enables the transient response and robustness to be addressed independently. Single-loop control and most of the existing advanced controllers such as the linear quadratic optimal controller and the Smith predictor can equivalently be put into the general IMC form (Garcia and Morari, 1982; Fisher, 1991; Zhu et al., 1995). The advantages of IMC are exploited in many industrial applications (Garcia and Morari, 1982).

It is, however, noted that by far the most widely used controllers in the process industries are PID controllers, so it is worth exploring the relationship between IMC and PID in order to gain insight into the tuning of this simpler controller, its performance and limitations. Note that for a process whose Nyquist curve exhibits a strange shape, especially around the crossover frequency, a PID controller could be difficult to shape satisfactorily. What one can do is de-tune the PID controller by trading off some performance to a sufficient extent to generate a stable closed loop if the original process is stable. But, this necessarily results in sluggish response and poor performance, which may not be desirable. A general, probably high-order single-loop controller is then necessary for high performance and it would be nice to design it from IMC and know when PID is not adequate. Another issue to notice is that there are many multivariable processes in reality. Though individual channels of such processes could be quite simple, such as first-order plus dead time dynamics, multivariable interactions might cause equivalent single loops to be very difficult to compensate for, especially when the number of inputs/outputs is large. Therefore, more effective design is needed.

In our recent developments in control design (Wang et al., 1997a; Wang et al., 2000b; Wang et al., 2001a; Wang et al., 2001c; Wang et al., 2002) to be presented in this part, a unified framework is established to design either IMC or PID or general single-loop controllers for SISO and MIMO systems with possible time delays. Time delay and non-minimum-phase zeros which must be present in and impose performance limitations on closed loops are first characterized, and the IMC control system with best achievable performance under given uncertainties and manipulated variable limits is then designed. Its single-loop controller approximation is determined by an algebraic operation and model reduction and is usually not in PID types. If PID is preferred by users, its parameters are tuned (may be de-tuned automatically if necessary) also using model reduction with possible sacrifice of the closed-loop performance.

9. Single-variable Systems

The internal model control (IMC) is a powerful framework for control system design and implementation (Morari and Zafiriou, 1989), and it has sound theoretical foundation. Its stability analysis is extremely easy to carry out and the design trade-off between performance and robustness is clearly understood. It has attracted the attention of industrial users because there is only one user-defined tuning parameter, which is directly related to the closed-loop time constant or equivalently, the closed-loop bandwidth. On the other hand, the vast majority of controllers being used in industry are of the PID type due to its simplicity and popularity (Åström and Hägglund, 1995). Recently, great efforts have been made to develop PID tuning strategies for more general processes (Barnes *et al.*, 1993; Sung and Lee, 1996; Sung *et al.*, 1996; Datta *et al.*, 2000). Each method was derived for its particular optimization objectives and plant model assumptions, and therefore performs well only for its own class. It is common that practising control engineers may not be certain which tuning method should be chosen to provide good control in a given process. It would hence be desirable to develop a design method that works universally with high performance for general stable linear processes, and is capable of producing a high-order controller when the PID controller is no longer adequate.

This chapter presents a unified framework for control system design. The IMC controller is always designed first. If the IMC scheme cannot be implemented, the equivalent controller in a conventional unity output feedback configuration is derived from the IMC controller and simplified by model reduction to a realizable controller, whose structure can be specified by users as a PID type or general rational function type to suit real situations best. In this chapter, we exclusively consider stable processes except the last section where the method is extended to unstable processes.

9.1 Design Methodology

The schematic of the IMC system is depicted in Figure 9.1, where $G(s)$ is the given stable process to be controlled, $\hat{G}(s)$ a model of the process and $C(s)$ the IMC primary controller. The design procedure for IMC systems is

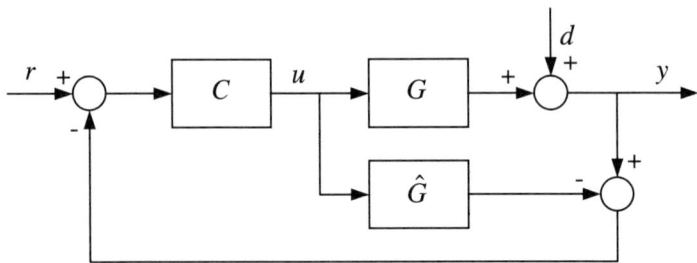

Fig. 9.1. IMC control system

well documented (Morari and Zafiriou, 1989), and highlighted as follows. The model is factorized as

$$\hat{G}(s) = \hat{G}_+(s)\hat{G}_-(s), \tag{9.1}$$

such that $\hat{G}_+(s)$ contains all the dead time and right half-plane zeros of $\hat{G}(s)$:

$$\hat{G}_+(s) = e^{-Ls}\left(\prod_i \frac{1-\beta_i s}{1+\beta_i s}\right), \quad \mathrm{Re}\,(\beta_i) > 0, \tag{9.2}$$

while $\hat{G}_-(s)$ is stable and of minimum phase with no predictors. The primary controller takes the form:

$$C = \hat{G}_-^{-1} f, \tag{9.3}$$

where f is a user-specified low-pass filter and usually chosen as

$$f(s,\tau) = \frac{1}{(\tau s + 1)^m}, \tag{9.4}$$

where m is sufficiently large to guarantee that the IMC controller C is proper. τ is the only tuning parameter to be selected by the user to achieve the appropriate compromise between performance and robustness and to keep the action of the manipulated variable within bounds. A smaller τ provides faster closed-loop response but the manipulated variable is moved more vigorously, while a larger τ provides a slower but smoother response. A larger τ is also less sensitive to model mismatches. In process control practice, the closed-loop bandwidth ω_{cb} can rarely exceed ten times the open-loop process bandwidth ω_{pb} (Morari and Zafiriou, 1989), i.e., $\omega_{cb} \leq 10\omega_{pb}$. Usually, the desired closed-loop bandwidth is chosen as $\omega_{cb} = \gamma\omega_{pb}$, $\gamma \in [0.5, 10]$. Using (9.4), it can be readily seen that

$$\tau = \frac{\sqrt{\sqrt[m]{2}-1}}{\gamma\omega_{pb}}, \quad \gamma \in [0.5,\ 10]. \tag{9.5}$$

In the case of model uncertainty, τ should be increased just enough to meet the condition for which the system is robustly stable (Morari and Zafiriou, 1989).

In order to keep the action of the manipulated variable within bounds, we use a frequency-by-frequency analysis (Skogestad and Postlethwaite, 1996). Assume that at each frequency $|U(j\omega)| \leq \bar{U}$ and $|R(j\omega)| \leq \bar{R}$. The manipulated variable meets

$$U(s) = C(s)R(s) = \frac{1}{(\tau s+1)^m}\hat{G}_-^{-1}(s)R(s).$$

One requires

$$\frac{1}{(\tau j\omega + 1)^m}\hat{G}_-^{-1}(j\omega)R(j\omega) \leq \bar{U}. \tag{9.6}$$

Consider the worst case of $|R(j\omega)| = \bar{R}$, and we require

$$\left|\frac{1}{(\tau j\omega + 1)^m}\right| \leq \left|\hat{G}_-(j\omega)\right|\frac{\bar{U}}{\bar{R}}. \tag{9.7}$$

To derive an inequality on τ imposed by input constraints, let $\omega = \omega_{ob}$, where ω_{ob} is the open-loop bandwidth, and notice that $\left|\hat{G}_-(j\omega_{ob})\right| = \frac{1}{\sqrt{2}}$, we have

$$\left|\frac{1}{(\tau j\omega_{ob}+1)^m}\right| \leq \frac{1}{\sqrt{2}}\frac{\bar{U}}{\bar{R}}, \tag{9.8}$$

i.e.,

$$\tau \geq \sqrt{\sqrt[m]{\frac{2\bar{R}^2}{\bar{U}^2}} - 1} \Big/ \omega_{ob}. \tag{9.9}$$

We choose τ to meet (9.5), (9.9) and any possible robustness specification. Then, the IMC control system has been designed and can be implemented according to Figure 9.1 with the controller in (9.3). To see performance for the case of no plant–model mismatch, the nominal closed-loop transfer function of the IMC system between the set point r and output y is

$$H = \hat{G}_+ f = (\prod_i \frac{1-\beta_i s}{1+\beta_i s}) \frac{1}{(\tau s+1)^m} e^{-Ls}. \tag{9.10}$$

If a user prefers a conventional (or single-loop) feedback control configuration, instead of the IMC scheme, for whatever reason, we can derive a controller for such a configuration from the IMC controller. This involves two issues. One is to convert the IMC system to an equivalent single-loop system. The other is to de-tune the IMC controller parameter to reflect the difference between the two control schemes and thus achievable performance limitations so that the performance from the properly de-tuned IMC system can be achieved by its single-loop (SL) equivalent.

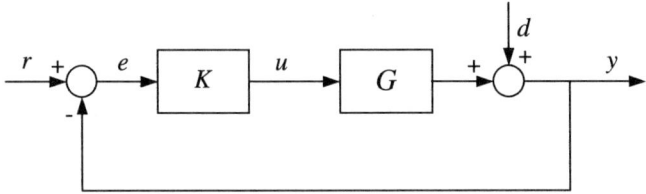

Fig. 9.2. Single-loop control system

The IMC system in Figure 9.1 can be formally redrawn into the equivalent single-loop (SL) feedback system in Figure 9.2, if the SL controller K is related to the IMC controller C via

$$K(s,\tau) = \frac{C(s,\tau)}{1-\hat{G}(s)C(s,\tau)}. \tag{9.11}$$

In Chien (1988), K is chosen as the PID type, and the value of τ is set according to (9.5) as if the single-loop PID controller could achieve the same performance as that of the more complex IMC controller. The dead time is approximated by either a first-order Padé or first-order Taylor series, and the PID controller parameters are obtained by matching the first few *Markov* coefficients of (9.11) for the selected specific process models. The results are listed in Table 9.1. However, it is noted that the use of the Padé approximation or a first-order Taylor expansion introduces extra modelling errors. Furthermore, Chien's rules are applicable only to first-order plus dead time (FOPDT) and second-order plus dead time (SOPDT) processes. This inevitably restricts the general applicability of the method and the performance of the resulting controller.

Our IMC-based design methodology described here can yield the best single-loop controller approximation to the IMC controller regardless of process order

Table 9.1. Chien's IMC–PID rules

Process model	$k_c k_p$	T_I	T_D
$\dfrac{k_p e^{-Ls}}{1+\tau_1 s}$	$\dfrac{\tau_1}{\tau+L}$	τ_1	—
$\dfrac{k_p(\tau_3 s+1)e^{-Ls}}{(\tau_1 s+1)(\tau_2 s+1)}$	$\dfrac{\tau_1+\tau_2-\tau_3}{\tau+L}$	$\tau_1+\tau_2-\tau_3$	$\dfrac{\tau_1\tau_2-(\tau_1+\tau_2-\tau_3)\tau_3}{\tau_1+\tau_2-\tau_3}$
$\dfrac{k_p(-\tau_3 s+1)e^{-Ls}}{(\tau_1 s+1)(\tau_2 s+1)}$	$\dfrac{\tau_1+\tau_2+\frac{\tau_3 L}{\tau+\tau_3+L}}{\tau+\tau_3+L}$	$\tau_1+\tau_2+\dfrac{\tau_3 L}{\tau+\tau_3+L}$	$\dfrac{\tau_3 L}{\tau+\tau_3+L}+\dfrac{\tau_1\tau_2}{\tau_1+\tau_2+\frac{\tau_3 L}{\tau+\tau_3+L}}$
$\dfrac{k_p e^{-Ls}}{s}$	$\dfrac{2\tau+L}{(\tau+L)^2}$	$2\tau+L$	—

and characteristics. The resulting single-loop performance can be better guaranteed and well predicted from the IMC counterpart. Our design idea is very simple: given the equivalent single-loop controller K in (9.11), which may be unnecessarily complicated to implement, apply a suitable model reduction to obtain the best approximation $\hat{K}$ to K. If the user specifies the type of $\hat{K}$ (say, PID), then the model reduction algorithm will generate its parameters. If the approximation accuracy is satisfactory, the design is completed; otherwise, the algorithm will adjust the IMC controller performance until its single-loop approximation is satisfactory. On the other hand, if the user has no preferred controller structure, our algorithm starts with a PID type, and gradually increases the controller complexity such that the simplest approximation $\hat{K}$ is attained with the guaranteed accuracy to K. This allows a unified treatment of all cases and facilitates auto-tuning applications.

A crucial issue in IMC-SL controller design is to get a suitable value for τ which leads to a good single-loop controller approximation to the corresponding IMC one. Note the inherent difference between IMC and SL systems in their configurations (Figures 9.1 and 9.2) where the former has output prediction while the latter does not. In fact, not all IMC systems can be approximated reasonably by single-loop systems (see the remark at the end of Section 9.3). The τ given by (9.5) is suitable for IMC systems, but it does not consider the performance limitations of single-loop feedback systems due to non-minimum-phase zero and dead time. Such limitations are usually expressed by integral relationships (Freudenberg and Looze, 1987). Recently, Åström (2000) proposed the following simple non-integral inequality for the gain crossover frequency ω_{og}

of the open-loop transfer function $\hat{G}K$, where

$$|\hat{G}(j\omega_{og})K(j\omega_{og},\tau)| = 1, \qquad (9.12)$$

to meet

$$\arg \hat{G}_+(j\omega_{og}) \geq -180° + \phi_m - \arg \hat{G}_-(j\omega_{og})K(j\omega_{og},\tau), \qquad (9.13)$$

where ϕ_m is the desired phase margin. The selection of ϕ_m reflects the control system robustness to process uncertainty (Åström, 2000): large ϕ_m is required for large uncertainty. With a lack of information on uncertainty size, a typical range for ϕ_m would be 30°–80°. Our design objective is to achieve a non-oscillatory response as specified by (9.10) and yet have the response as fast as possible. This translates to a damping ratio of approximately $\xi = 0.7$, and the empirical formula $\phi_m = 100\xi$ (Franklin et al., 1990) yields an estimate of $\phi_m = 70°$ for $\xi = 0.7$. Our studies suggest that $\phi_m = 65°$ is usually a good choice and we use this ϕ_m throughout this chapter. With ϕ_m specified, we then find the smallest τ^* which satisfies (9.12) and (9.13).

In short, for single-loop controller design the tuning parameter τ in the filter (9.4) should be, in general, chosen to meet (9.5), (9.9), (9.12) and (9.13) simultaneously. If the process is of minimum phase, (9.12), (9.13) vanish, while (9.5) and (9.9) are in action. On the other hand, if the process has any non-minimum element, our study shows that the τ derived from (9.9), (9.12) and (9.13) always appears in the range given in (9.5) so that (9.9), (9.12) and (9.13) would be enough to determine τ in this case. In the subsequent two sections, PID and general controllers are considered in detail.

9.2 PID Controller

Owing to its simple structure, the PID controller is the most widely used controller in the process industry, even though many advanced control algorithms have been introduced. Consider a PID controller in the form:

$$K_{PID} = k_p + \frac{k_i}{s} + k_d s, \qquad (9.14)$$

where k_p is the proportional gain, k_i the integral gain (units of time), and k_d the derivative gain (units of time). Our task is to find the three PID parameters, so as to match $\hat{K} = K_{PID}$ to $K = \frac{C}{1-C\hat{G}}$ as well as possible. This objective can be realized by minimizing the loss function,

$$\min_{K_{PID}} J \triangleq \min_{K_{PID}} \sum_{i=1}^{M} |K_{PID}(j\omega_i) - K(j\omega_i)|^2, \quad k_p, k_i, k_d > 0, \tag{9.15}$$

whose solution is obtained by standard non-negative least squares to give the optimal PID parameters as $\begin{bmatrix} k_p^* & k_i^* & k_d^* \end{bmatrix}^T = \theta^*$. Our studies suggest that the frequency range $[\omega_1, \omega_M]$ in the optimal fitting (9.15) be chosen as $(0.1\omega_{cb}, \omega_{cb})$ with steps of $(\frac{1}{100} \sim \frac{1}{10})\omega_{cb}$, where ω_{cb} is the desired closed-loop bandwidth.

Once a PID controller is found, the following criterion should be used to validate the solution:

$$ERR = \max_{\omega \in [0, \omega_{cb}]} |\frac{\hat{K}(j\omega) - K(j\omega)}{K(j\omega)}| \leq \epsilon, \tag{9.16}$$

where ϵ is the user-specified fitting error threshold. ϵ is specified according to the desired degree of performance, or accuracy of the SL approximation to the IMC one. Usually ϵ may be set as 3%. If (9.16) holds true, the design is complete.

On the other hand, if the given threshold cannot be met, one can always detune the PID controller by relaxing the IMC specification, i.e., increasing τ. A typical relationship between the tuning parameter τ and the approximation error is shown in Figure 9.3. In general, ERR decreases as τ increases. It provides a simple way to select a minimum τ with respect to the specific accuracy threshold. In practice, however, it is inconvenient to draw such a curve. It is found that the decreasing rate $d(ERR)/d\tau$ is highly influenced by plant dead time L and the right half-plane (RHP) zeros β_i^{-1}, which limit the achievable bandwidth. ω_{cb} is virtually unaffected by the presence of the filter (Rivera et al., 1986) until τ reaches an order of magnitude comparable to L and β_i, respectively. Hence, it is effective and efficient to choose the increment of τ in the PID detuning procedure as the maximum of L and $\mathrm{Re}\,\beta_i$, i.e.,

$$\tau^{k+1} = \tau^k + \eta^k \max(L, \min_i (\mathrm{Re}\,(\beta_i))) \tag{9.17}$$

where k represents the kth iteration, and η is an adjustable factor reflecting the approximation accuracy of the present iteration and is set at $\frac{1}{4}$, $\frac{1}{2}$ and 1, when $3\% < ERR \leq 20\%$, $20\% < ERR \leq 100\%$ and $100\% < ERR$, respectively. The iteration continues until the accuracy bound is fulfilled.

Our detuning rule for τ in (9.17) implicitly assumes that ERR would be sufficiently small when τ is large enough. In this connection, it would be interesting to see if $\lim_{\tau \to \infty} ERR = 0$. Equation (9.5) can be rewritten as

$$\omega_{cb} = \frac{\sqrt{\sqrt[m]{2} - 1}}{\tau}. \tag{9.18}$$

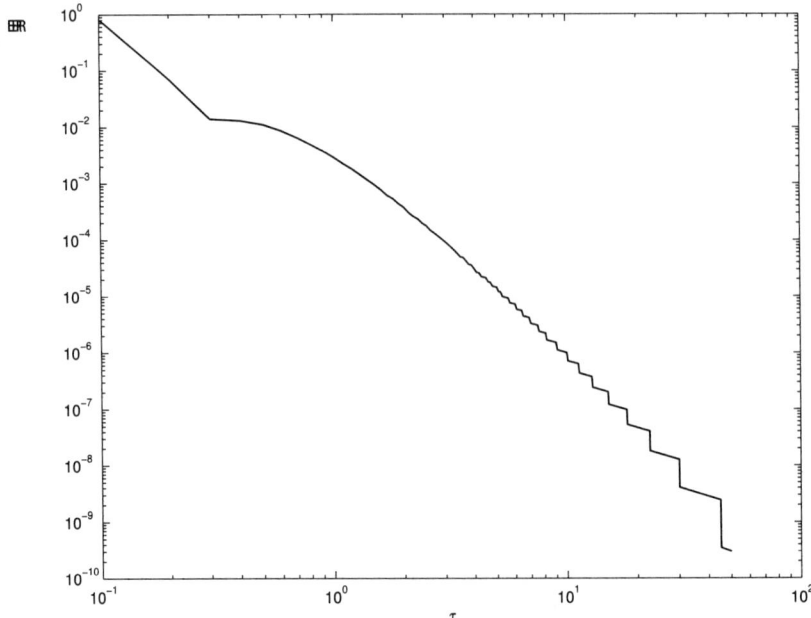

Fig. 9.3. Relationship between filter parameter (τ) and approximation error (ERR)

When τ increases to infinity, it is easy to see from (9.18) that $\lim_{\tau \to \infty} \omega_{cb} = 0$, $G(j\omega)$ can be replaced by $G(0)$ for $\omega \leq \omega_{cb}$, and K becomes $\frac{1}{G_-(0)((\tau s+1)^m-1)}$. For $m = 1$, $K = \frac{1}{sG_-(0)}$ is a pure integrator and can be realized precisely by a PID controller with no error. In general, the Nyquist curve of $K(j\omega)$ for $\omega \in (0, \omega_{cb})$ approaches a straight line as shown in Figure 9.4, when τ tends to infinity. Note that the Nyquist curve for the PID controller is always a vertical straight line, and can match that of $K(j\omega)$ as well as desired for $\tau \to \infty$. One thus expects ERR to converge to 0 as τ approaches infinity.

We now present some simulation examples to demonstrate our PID tuning algorithm and compare it with the original IMC and the PID tuning in Chien (1988). Chien (1988) implemented the following PID form:

$$\tilde{K}_{PID} = K_c(1 + \frac{1}{T_I s} + \frac{T_D s}{\frac{T_D}{N} s + 1}),$$

where the PID settings are given in Table 9.1. The ideal PID controller in (9.14) used for our algorithm development is not physically realizable and thus is replaced by

$$K_{PID} = k_c + \frac{k_i}{s} + \frac{k_d s}{\frac{k_d}{N} s + 1}. \tag{9.19}$$

9.2 PID Controller

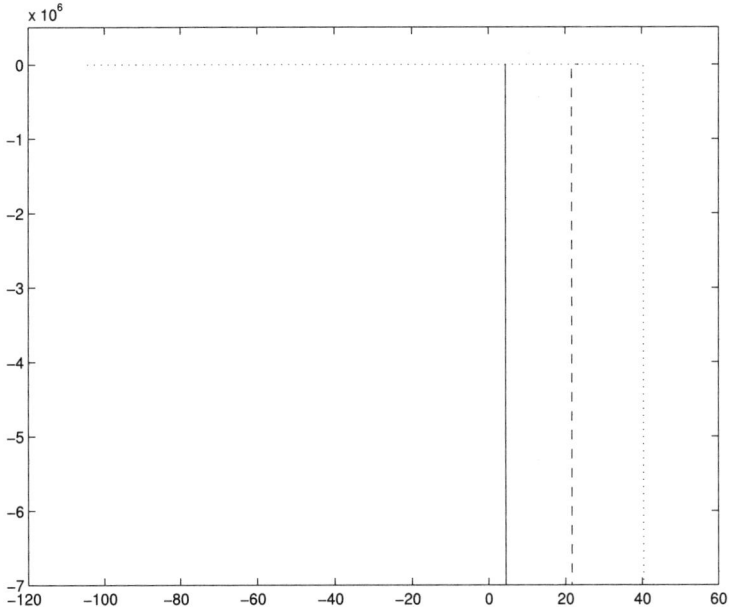

Fig. 9.4. Nyquist curve of K up to ω_{cb} ($m = 3$)
($\cdots \tau = 0.1$, $---\tau = 0.3$, ——— $\tau = 1$)

In both cases, N is suggested to be chosen in the range $[5, 20]$. Simulations are done under the perfect model matching condition, i.e., $\hat{G} = G$ (model mismatch will be considered in Section 9.4). To have fair and comprehensive assessment of controller performance, most performance indices popularly used in process control are measured and they include both time domain ones such as percentage overshoot (M_p), rise time (from 10% to 90%) in seconds (t_r), setting time (to 1%) in seconds (t_s), integral absolute-error ($IAE = \int_0^\infty |r-y|dt$ where the upper limit ∞ may be replaced by T, which is chosen sufficiently large so that $e(t)$ for $t > T$ is negligible); and frequency domain error ERR defined in (9.16). Simulations were made for three typical plants, and the results are shown in Table 9.2.

Example 9.2.1. Consider a first-order plus dead time process:

$$G = \frac{e^{-0.5s}}{s+1}.$$

From (9.9), (9.12) and (9.13), one can readily find $\tau^* = 0.3333$. Then, Chien's formula gives a PI controller,

$$\tilde{K}_{PI} = 1.2029(1 + \frac{1}{s}),$$

Table 9.2. Simulation results for stable processes

Plant	τ	Scheme	$M_p(\%)$	t_r	t_s	U_{max}	$ERR(\%)$	IAE
$G = \dfrac{e^{-0.5s}}{s+1}$	0.3333	Chien's PID	11.7728	0.72	4.15	1.8044	42.59	1.0571
		Proposed PID	0	0.73	2.31	6.3945	1.23	0.8351
		IMC	0	0.73	2.23	3.0184	0	0.8343
$G = \dfrac{(-0.5s+1)e^{-s}}{(s+1)(2s+1)}$	0.8328	Chien's PID	14.0690	1.54	11.66	6.3779	26.33	2.8728
		Proposed PID	0.9339	2.11	8.46	6.1204	2.66	2.8388
		IMC	0	2.17	6.18	4.8031	0	2.8378
	0.6687	Chien				– – –		
		Proposed PID	7.4365	2.67	10.00	5.2860		17.48
		Proposed high-order	0	3.30	8.72	3.5488		0.0045
$G = \dfrac{e^{-2s}}{(s^2+s+1)(s^2+0.6s+1)}$		IMC	0	3.30	8.72	5.0012		0
	1.1687	Chien				– – –		
		Proposed PID	0	5.77	10.00	5.1016		0.39
		IMC	0	0	5.77	10.00	1.0000	0

while the proposed method yields

$$K_{PID} = 1.4005 + \frac{1.2050}{s} + \frac{0.1856s}{\frac{0.1856}{N}s+1},$$

which achieves the specified approximation accuracy $ERR \leq 3\%$. The Ziegler–Nichols step response tuning method (Ziegler and Nichols, 1942) gives

$$K_{PI-ZN} = 2.64(1 + \frac{1}{s} + \frac{0.25s}{\frac{0.25}{N}s+1}),$$

while the Cohen–Coon step response method (Cohen and Coon, 1953) produces

$$K_{PI-CC} = 3.22(1 + \frac{0.9430}{s} + \frac{0.1703s}{\frac{0.1703}{N}s+1}).$$

The closed-loop responses for different designs are shown in Figure 9.5. It

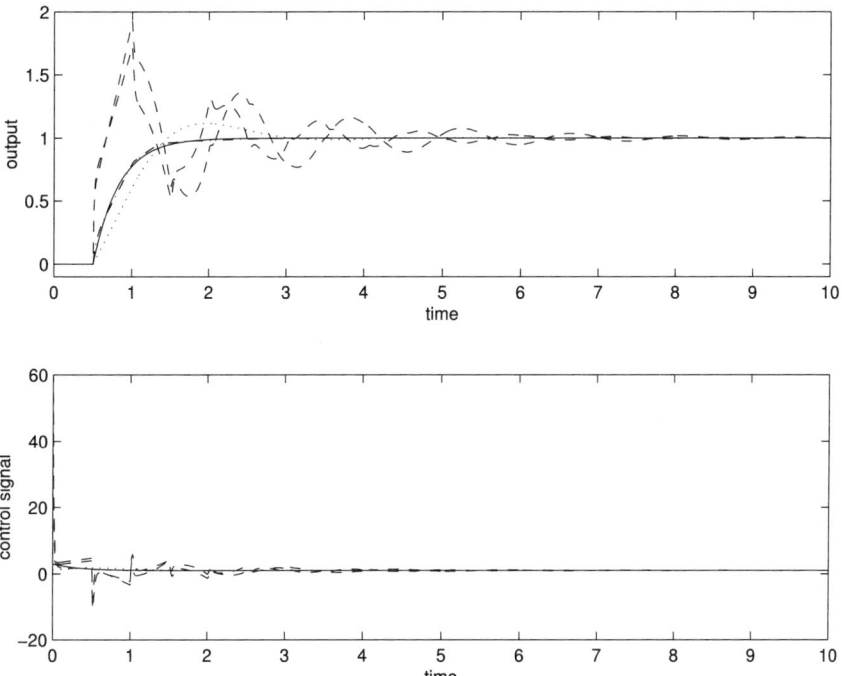

Fig. 9.5. Comparison of set-point responses for $\frac{e^{-0.5s}}{s+1}$
($- \cdot - \cdot -$ proposed PID, $\cdots$ Chien, —— IMC, $- - -$ C-C and Z-N)

can be seen that both Chien's rule and the proposed method show much better performance than the conventional ZN and CC designs. The proposed method is almost identical to the IMC system.

Example 9.2.2. Consider a process with right half-plane zero:
$$G = \frac{(-0.5s+1)e^{-s}}{(s+1)(2s+1)}.$$

From (9.9), (9.12) and (9.13), one gets $\tau^* = 0.6687$, which gives rise to
$$\tilde{K}_{PID} = 1.3779(1 + \frac{0.3111}{s} + \frac{0.8365s}{\frac{0.8365}{N}s+1})$$

by Chien's formula, and
$$K_{PID} = 1.1194 + \frac{0.3569}{s} + \frac{0.9765s}{\frac{0.9765}{N}s+1}$$

by the proposed method with $ERR \leq 3\%$. The closed-loop responses are shown in Figure 9.6.

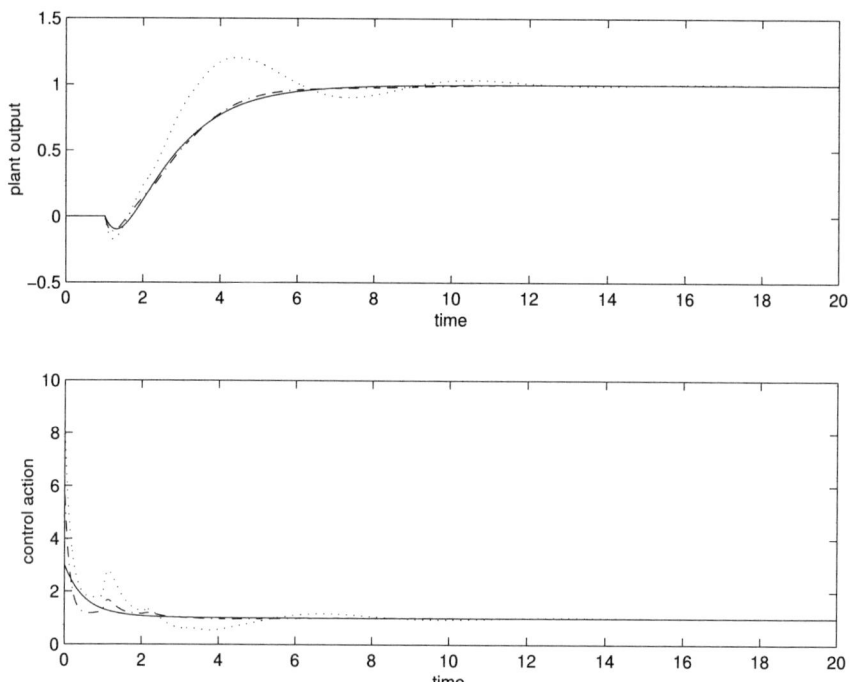

Fig. 9.6. Comparison of set-point responses for $\frac{(-0.5s+1)e^{-s}}{(s+1)(2s+1)}$
($-\cdot-\cdot-$ proposed PID, $\cdots$ Chien, —— IMC)

Example 9.2.3. Consider a high-order and oscillatory process:

$$G = \frac{e^{-2s}}{(s^2 + s + 1)(s^2 + 0.6s + 1)}.$$

Our method produces $\tau^* = 0.6687$ and

$$K_{PID} = 0.2860 + \frac{0.2139}{s} + \frac{0.3962s}{\frac{0.3962}{N}s + 1}. \tag{9.20}$$

This controller has the approximation error $ERR = 17.46\%$, which cannot fulfil the accuracy threshold, and the closed-loop response is very poor, as shown in Figure 9.7. Then τ is adjusted to $\tau^1 = \tau^0 + 0.25L = 0.6687 + 0.5 = 1.1687$

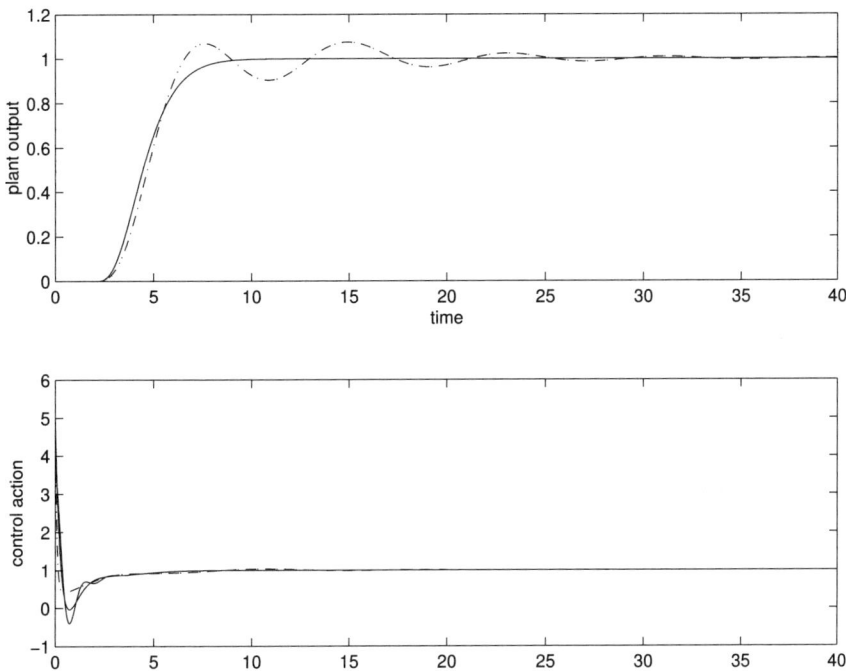

Fig. 9.7. Set-point responses for $\frac{e^{-2s}}{(s^2+s+1)(s^2+0.6s+1)}$ with $\tau = 0.6687$
($- \cdot - \cdot -$ proposed PID, $- - -$ high-order controller, —— IMC)

according to the proposed tuning rule (9.17). The new τ results in

$$K_{PID} = 0.1016 + \frac{0.1498}{s} + \frac{0.1229s}{\frac{0.1229}{N}s + 1}. \tag{9.21}$$

The approximation error ERR of the proposed method has met the specified approximation accuracy $ERR \leq 3\%$. The closed-loop responses are shown in

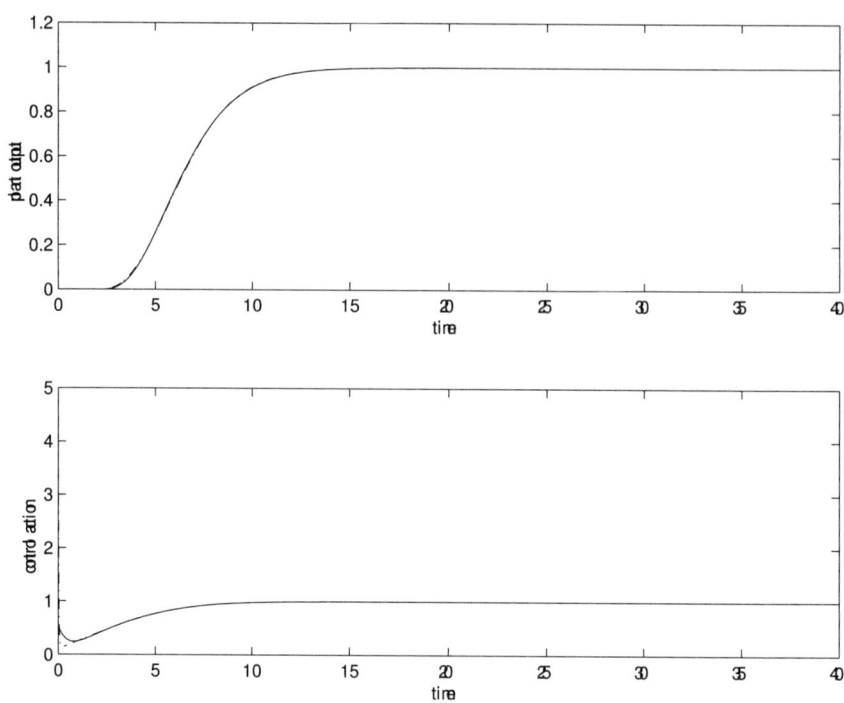

Fig. 9.8. Set-point responses for $\frac{e^{-2s}}{(s^2+s+1)(s^2+0.6s+1)}$ with $\tau = 1.1687$
($-\cdot-\cdot-$ proposed PID, ——— IMC)

Figure 9.8. One observes that the difference between our SL system and the original IMC system is not discernible.

It can be seen from the simulation study in Table 9.2 that the proposed method always yields a PID controller that is a much better approximation to the IMC counterpart than Chien's method, regardless of what τ is chosen. Our experience indicates that for FOPDT and SOPDT processes and a slow closed-loop response requirement of $\tau > \frac{\sqrt{\sqrt[4]{2}-1}}{\omega_{pb}}$, both the proposed IMC-PID method and Chien's rules generate responses close to the IMC counterpart. In particular, the proposed method can always achieve $ERR < 3\%$, and thus the closed-loop performance can be well predicted from the corresponding IMC system. However, when fast closed-loop response, generally $\omega_{cb} > \omega_{pb}$, i.e., $\tau < \frac{\sqrt{\sqrt[4]{2}-1}}{\omega_{pb}}$, is required, the proposed method shows significant improvement over Chien's rules. The improvement is also evident for complex processes with slow responses. Moreover, under the fast response requirement, $\omega_{cb} > \omega_{pb}$, the PID controller derived from Chien's rules may cause large peaks in the manipulated variable, which is harmful to the system. It is, however, noticed that for high-order processes with fast responses, none of the above two IMC-PID methods is able to generate PID systems similar to IMC ones. This implies that a controller in the PID form is insufficient to obtain the desired performance. In this case, a higher-order controller has to be considered for better fitting and performance, which will be discussed in the next section.

9.3 High-order Controller

If the single-loop controller is not limited to PID type and if PID is not adequate to control a given process, we have to consider a general type of proper rational function controller to meet the specifications. The task at hand is then to find an nth-order rational function approximation:

$$\hat{K} = \frac{b_n s^n + b_{n-1} s^{n-1} + ... + b_1 s + b_0}{s^n + a_{n-1} s^{n-1} + ... + a_1 s}, \quad (9.22)$$

with an integrator such that

$$J \triangleq \sum_{i=1}^{M} |W(j\omega_i)(\hat{K}(j\omega_i) - K(j\omega_i))|^2 \quad (9.23)$$

is minimized. The problem can be solved by one of two algorithms for transfer function modelling from frequency response presented in Chapter 7. If the recursive least squares methods (RLS) there is adopted, like the LS algorithm in the preceding section, the frequency range for the RLS is also chosen as $(0.1\omega_{cb}, \omega_{cb})$ with steps of $(\frac{1}{100} \sim \frac{1}{10})\omega_{cb}$.

From the typical relationship of the relative fitting error ERR defined in (9.16) and the rational approximation order n shown in Figure 9.9, we can see that ERR decreases, as n increases. We try to find the minimum n which achieves the approximation bound $ERR < \epsilon$ with a user-specified τ. In general, if faster response is required, a higher-order controller has to be used.

The above algorithm deals with the problem of approximating a given, probably non-rational, transfer function by a rational function. Error bounds for such an approximation have been investigated (Wahlberg and Ljung, 1992; Yan and Lam, 1999). Wahlberg and Ljung (1992) proposed an approach based on weighted least squares estimation, and provided hard frequency-domain transfer function error bounds. However, it is not easy to calculate such a bound, and the convergence of estimation has not been addressed. In our work, we use a maximum likelihood index ERR to evaluate the approximation accuracy,

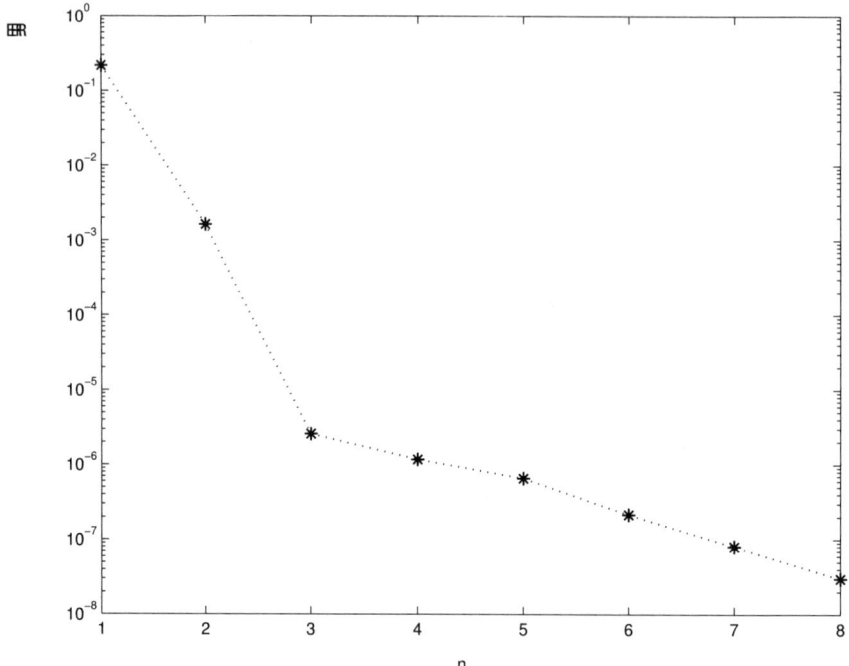

Fig. 9.9. Relationship between order n and approximation error ERR

and assume that the accuracy threshold can be achieved when the controller order is high enough.

When τ is chosen, we first find the PID controller using the standard least squares method and evaluate the corresponding approximation error ERR in (9.16) as described in the preceding section. If ERR cannot achieve the specified approximation accuracy ϵ (usually 3%), we recommend a high-order controller as in (9.22), and start with a controller of order 2 up to the smallest integer n such that $ERR \leq \epsilon$.

Tuning procedure

Step 1. Find the smallest τ^* from (9.9), (9.12) and (9.13), and let $\tau^0 = \tau^*$.

Step 2. Determine the PID controller from (9.15) and evaluate the corresponding approximation error ERR in (9.16). If ERR achieves the specified approximation accuracy ϵ (usually 3%), end the design.

Step 3. Otherwise, we have two ways to solve this problem: if a PID controller is desired, update τ by (9.17), and go to Step 2; else, go to Step 4.

Step 4. Adopt the high-order controller in (9.22), start with a controller of order 2 up to the smallest integer n for which $ERR \leq \epsilon$.

Example 9.2.3 (cont'd). Reconsider

$$G = \frac{e^{-2s}}{(s^2+s+1)(s^2+0.6s+1)},$$

for which $\tau^0 = 0.6687$ and a PID has been obtained with $ERR = 17.46\%$. For a high-order controller, our procedure ends with

$$\hat{K} = \frac{3.5488s^5 + 14.9135s^4 + 23.9669s^3 + 29.6333s^2 + 18.2712s + 9.2024}{s^5 + 4.5140s^4 + 25.8787s^3 + 22.8686s^2 + 43.0211s}, \tag{9.24}$$

with fitting error ERR less than $\epsilon = 3\%$. The closed-loop step responses are shown in Figure 9.7, and their performance indices are also given in Table 9.2. We can see that the new controller $\hat{K}$ restores the IMC performance, while the previous PID controller in (9.20) is not capable of that under such a tight performance specification.

If τ is chosen to be smaller than the value suggested by (9.9), (9.12) and (9.13), this overcomes the limitation of single-loop feedback systems, but then no single-loop controller solution with stability could be found for the corresponding IMC system. For instance, in the above example, choosing τ 50% less than τ^0, i.e., $\tau = 0.3344$, we could not find a controller in the form of (9.22) with ERR less than 3%, which implies that SL controllers are unlikely to achieve a performance tighter than that specified by (9.9), (9.12) and (9.13).

It is observed from our simulation study that usually, the approximation error magnitude of the high-order controller obtained by the RLS is of the order 10^{-4} or less, the controller order is less than 6, and the controller yields a closed-loop response very close to that of the IMC loop provided that τ is set by (9.9), (9.12) and (9.13). The high-order controller does provide significant performance enhancement over PID for complex processes. The proposed method is a simple, effective, and efficient way to design such high-performance controllers.

9.4 Stability Analysis

From the results obtained so far, it is possible to state that the single-loop system with $\hat{K}$ derived using the proposed method has a performance close to the corresponding IMC loop. Thus the stability of the resulting single-loop control system is well related to that of the IMC system. In this section, we consider both nominal stability ($G = \hat{G}$) and robust stability ($G \neq \hat{G}$).

Assume $G = \hat{G}$ in the absence of model uncertainty, the nominal stability of the IMC system automatically guarantees the stability of the feedback system

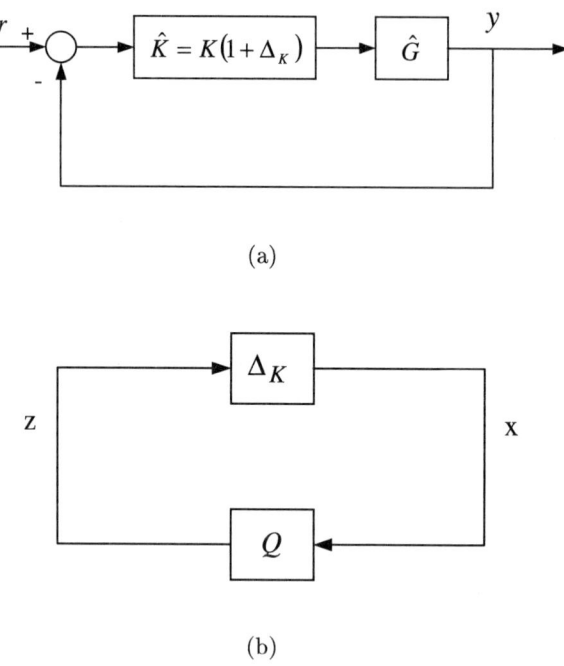

Fig. 9.10. Block diagram of nominal system

in Figure 9.2 with K determined from (9.11). However, the proposed design makes a controller approximation $\hat{K}$ and thus gives the nominal single-loop system shown in Figure 9.10(a), where $\hat{K}(s) = K(s)(1 + \Delta_K(s))$. The system in Figure 9.10(a) can be redrawn as Figure 9.10(b), where

$$Q = -\frac{\hat{G}K}{1 + \hat{G}K}.$$

Using (9.11), $Q(s)$ can be written as $Q = -\hat{G}C$ and is stable. With the standard assumption that $\hat{K}$ has the same number of unstable poles as K, the nominal single-loop feedback system is stable (Green and Limebeer, 1995) if and only if

$$|\hat{G}(j\omega)C(j\omega)\Delta_K(j\omega)| < 1, \quad \forall \omega. \tag{9.25}$$

It follows from (9.1)–(9.3) that $\hat{G}C = \hat{G}_+ f$. But $|\hat{G}_+(j\omega)| = 1, \forall \omega$ and $|\hat{G}_+ f \Delta_K| = |f \Delta_K|$, (9.25) then becomes

$$|f(j\omega)\Delta_K(j\omega)| < 1, \quad \forall \omega. \tag{9.26}$$

Note from (9.4) that $|f(j\omega)|$ decays quickly for $\omega \geq \omega_{cb} = \frac{\sqrt{\sqrt[m]{2}-1}}{\tau}$ and (9.26) is likely to hold for high frequencies. Thus, assume that (9.26) is true for $\omega \geq \omega_{cb}$. One now needs to check (9.26) only for the working frequency range $[0, \omega_{cb}]$, and because $|f(j\omega)| \leq 1$ for all ω, the nominal closed loop is thus stable if

$$|\Delta_K| = |\frac{K(j\omega) - \hat{K}(j\omega)}{K(j\omega)}| < 1, \quad \omega \in [0, \omega_{cb}]. \tag{9.27}$$

In the proposed algorithm, the approximation accuracy has to meet (9.16), where ϵ is usually specified as 3%. The resulting controller $\hat{K}$ then satisfies (9.27) with a large margin and nominal stability of the designed single-loop system is thus expected.

Consider now model uncertainty. Let the actual plant be $G(s) = \hat{G}(s)(1 + \Delta_G(s))$. In the IMC design (Morari and Zafiriou, 1989) to achieve robust stability, the filter parameter τ is chosen big enough to meet the condition:

$$|\hat{G}(j\omega)C(j\omega)\Delta_G(j\omega)| < 1, \quad \text{or} \quad |f(j\omega)\Delta_G(j\omega)| < 1, \quad \forall \omega. \tag{9.28}$$

The single-loop system with process uncertainty is shown in Figure 9.11(a), where $|\Delta_K| \leq \delta_K(\omega)$ and $|\Delta_G| \leq \delta_G(\omega)$, and it can be redrawn into the standard form in Figure 9.11(b), where $\tilde{\Delta}(s)$ is the normalized uncertainty $\tilde{\Delta} = \text{diag}\{\tilde{\Delta}_K, \tilde{\Delta}_G\}$ with $|\tilde{\Delta}_K| \leq 1$ and $|\tilde{\Delta}_G| \leq 1$. The transfer function matrix between z and x has no uncertainty and is given by

$$Q = \begin{bmatrix} \delta_K & 0 \\ 0 & \delta_G \end{bmatrix} \begin{bmatrix} -\hat{G}K & -K \\ \hat{G} & -\hat{G}K \end{bmatrix} (1 + \hat{G}K)^{-1}$$

$$= \begin{bmatrix} \delta_K & 0 \\ 0 & \delta_G \end{bmatrix} \begin{bmatrix} -\hat{G}C & -C \\ \hat{G}(1 - \hat{G}C) & -\hat{G}C \end{bmatrix},$$

whose stability is guaranteed by that of $\hat{G}$ and C. It follows from the stability robustness theorem (Doyle et al., 1982) that the uncertain feedback system remains stable for all $\tilde{\Delta} = \text{diag}\{\tilde{\Delta}_K, \tilde{\Delta}_G\}$ if and only if

$$\|Q\|_\mu < 1, \tag{9.29}$$

where $\|Q\|_\mu = \sup_\omega \mu(Q(j\omega))$ and $\mu(\cdot)$ is the structured singular value with respect to $\tilde{\Delta}$. In our case, the structured singular value $\mu(Q(j\omega))$ can be calculated by

$$\mu(Q(j\omega)) = \mu(DQD^{-1}) = \inf_D \bar{\sigma}(DQD^{-1}),$$

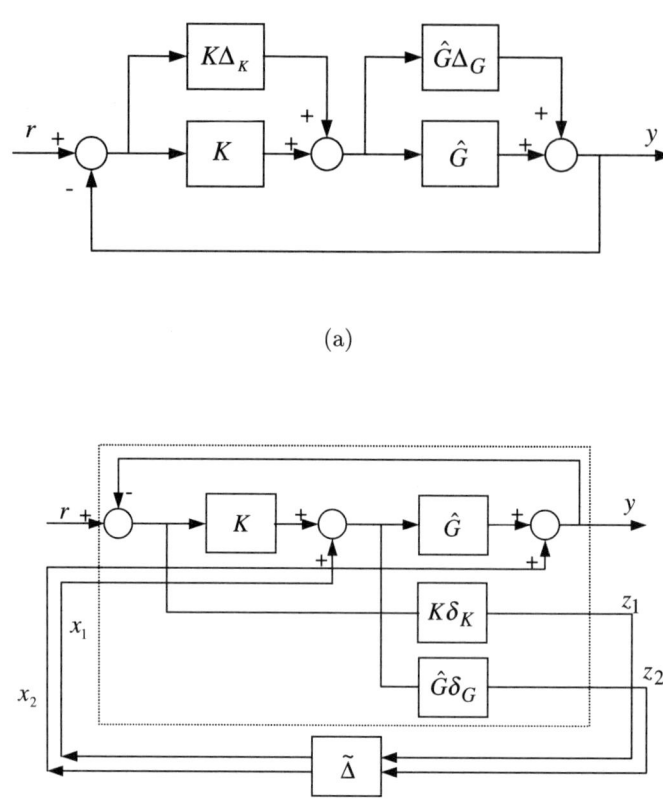

Fig. 9.11. Block diagram of system with process uncertainty

where $D = \text{diag}\{d_1, d_2\}$, $d_1, d_2 > 0$ and $\bar{\sigma}(\cdot)$ represents the largest singular value.

After some calculations, we obtain

$$\mu(Q(j\omega)) = |GC| \cdot$$

$$\cdot \sqrt{\frac{\delta_K^2 + \delta_G^2 + 2\delta_K\delta_G|\frac{1-\hat{G}C}{\hat{G}C}| + \sqrt{(\delta_K^2 + \delta_G^2 + 2\delta_K\delta_G|\frac{1-\hat{G}C}{\hat{G}C}|)^2 - 4\delta_K^2\delta_G^2|\frac{1}{\hat{G}C}|^2}}{2}},$$

(9.30)

and the robust stability condition (9.29) becomes

$$\delta_K^2|\hat{G}C|^2 + \delta_G^2|\hat{G}C|^2 + 2\delta_K\delta_G|(1-\hat{G}C)\hat{G}C| +$$
$$+\sqrt{(\delta_K^2|\hat{G}C|^2 + \delta_G^2|GC|^2 + 2\delta_K\delta_G|(1-\hat{G}C)\hat{G}C|)^2 - 4\delta_K^2\delta_G^2|\hat{G}C|^2}$$
$$< 2, \quad \forall \omega. \tag{9.31}$$

Since $4\delta_K^2\delta_G^2|\hat{G}C|^2 \geq 0$, $\forall \omega$, and $|1 - GC| \leq 1 + |GC| \leq 2$, (9.31) is satisfied if

$$\delta_K^2|\hat{G}C|^2 + \delta_G^2|\hat{G}C|^2 + 4\delta_K\delta_G|\hat{G}C| < 1, \quad \forall \omega, \tag{9.32}$$

i.e.,

$$\delta_K^2(\omega)|f(j\omega)|^2 + \delta_G^2(\omega)|f(j\omega)|^2 + 4\delta_K(\omega)\delta_G(\omega)|f(j\omega)| < 1, \quad \forall \omega.$$

As $|f(j\omega)|$ decays quickly for $\omega \geq \omega_{cb} = \frac{\sqrt{\sqrt[n]{2}-1}}{\tau}$, then (9.32) is likely to hold for high frequencies. Thus, assume that (9.32) is true for $\omega \geq \omega_{cb}$. One now needs to check (9.32) only for the working frequency range $[0, \omega_{cb}]$, and because $|f(j\omega)| \leq 1$ for all ω, the closed loop is robustly stable if

$$\delta_K^2(\omega) + \delta_G^2(\omega) + 4\delta_K(\omega)\delta_G(\omega) < 1, \quad \omega \in [0, \omega_{cb}].$$

In the proposed method, $|\Delta_K|$ is made small, i.e., $\delta_K(\omega) \leq 3\%$. Let $\delta_K = 3\%$, then the robust stability of the closed loop is guaranteed by

$$\delta_G(\omega) \leq 94.13, \quad \omega \in [0, \omega_{cb}]. \tag{9.33}$$

Note that for $\delta_K = 0$, i.e., no controller uncertainty, (9.30) reduces to $\mu(Q(j\omega)) = |\hat{G}C\delta_G|$. Then the robust stability condition (9.29) is simplified to $\sup_\omega |\hat{G}C\delta_G| < 1$, which is equivalent to the robust stability condition (9.28) of the IMC system.

Example 9.2.3 (cont'd). Reconsider

$$G = \frac{\alpha e^{-2s}}{(s^2 + s + 1)(s^2 + 0.6s + 1)},$$

with nominal $\alpha = \alpha_0 = 1$. When $\tau = 0.6687$, the proposed method yields a 5th-order controller in (9.24) and the nominal performance is shown in Figure 9.7. It can be seen that the system is indeed nominally stable. To demonstrate robustness, introduce a 50% perturbation from the nominal gain of α, giving $\alpha = 1.5$. Figure 9.12 shows the resulting performances, and indicates that the single-loop high-order controller K derived using the proposed method exhibits a similar robust performance to the IMC loop.

When $\tau = 1.1687$, the proposed method yields a PID controller as in (9.21) and the nominal performance is shown in Figure 9.8. We also introduce a 50% perturbation in gain, giving $\alpha = 1.5$. Figure 9.13 shows the resulting performance, which is still stable, and more robust than that shown in Figure 9.12 for $\tau = 0.6687$.

260   9. Single-variable Systems

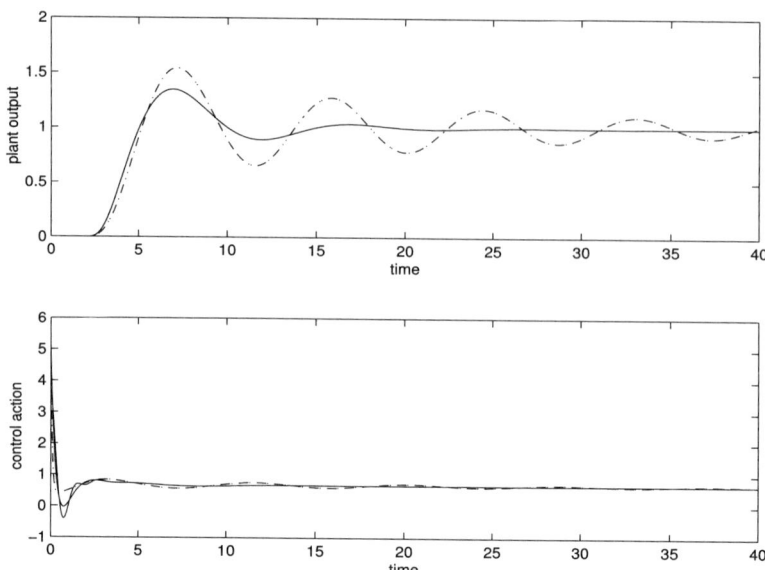

Fig. 9.12. Performance robustness
$(-\cdot-\cdot-$ proposed PID, $---$ high-order controller, — IMC)

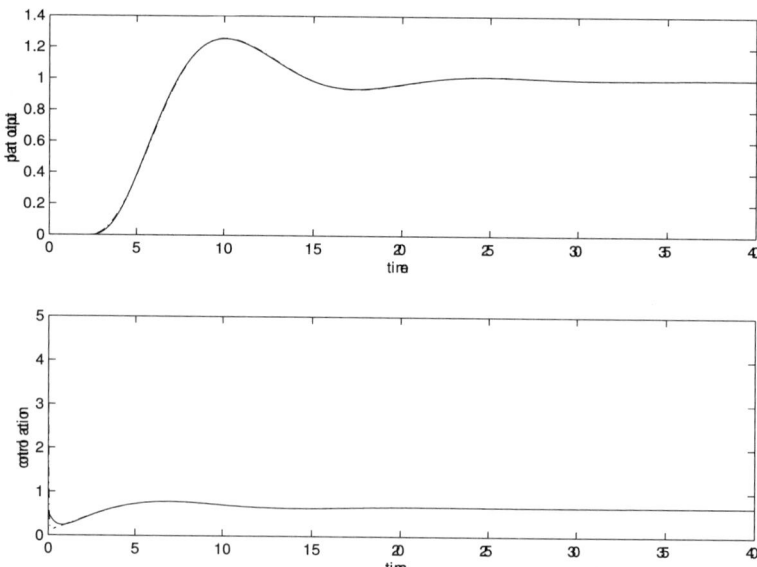

Fig. 9.13. Performances robustness
$(-\cdot-\cdot-$ proposed PID, — IMC)

9.5 Unstable Processes

It is well known that the IMC scheme is internally unstable if the process $G(s)$ is unstable (Morari and Zafiriou, 1989) and is thus not implementable. However, Morari and Zafiriou (1989) suggest that one can still design the controller using the IMC method, and then implement the controller in an equivalent feedback structure as follows. In the nominal case (the uncertain case will be discussed below) of $G(s) = \hat{G}(s)$, the equivalent single-loop feedback system of Figure 9.2 is derived from the IMC system of Figure 9.1 with

$$K = C(1-\hat{G}C)^{-1}. \tag{9.34}$$

The system is internally stable if

$$C \text{ and } (1-\hat{G}C)\hat{G} \text{ are both analytic in the RHP.} \tag{9.35}$$

Generally, the feedback controller K in (9.34) should include an integrator to eliminate the steady-state error, and maintain stability. Thus we require

$$K(\text{after all possible pole–zero cancellations})$$
$$\text{has no pole in the closed RHP except } s = 0. \tag{9.36}$$

Let us consider a class of unstable processes with a single RHP pole only:

$$\hat{G}(s) = \frac{1}{1-Ts}\hat{G}_-(s)e^{-Ls}, \tag{9.37}$$

where $\hat{G}_-(s)$ is rational, stable and of minimum phase. The optimal H_2 IMC controller for step inputs is

$$C = (1-Ts)\hat{G}_-^{-1}f, \tag{9.38}$$

where f is a user-specified low-pass filter and chosen as

$$f(s,\tau) = \frac{\alpha s + 1}{(\tau s + 1)^{m+1}}, \quad \alpha = T(\tau/T+1)^{m+1}e^{L/T}-T, \tag{9.39}$$

where m is an integer large enough to guarantee that the IMC controller C is proper. One can easily verify that (9.35) holds for all $\tau > 0$. Thus, $\tau > 0$ is the only tuning parameter to be selected by the user to meet (9.36) and achieve the appropriate trade-off between performance and robustness, and to keep the manipulated variable within bounds. We now address them separately.

For $\hat{G}$ as in (9.37) and C as in (9.38), K in (9.34) becomes

$$K(s,\tau) = \frac{(1-Ts)(\alpha s + 1)}{\hat{G}_-((\tau s + 1)^{m+1} - (\alpha s + 1)e^{-Ls})}. \tag{9.40}$$

It can be verified that $s = \frac{1}{T}$ is both a zero and pole of $K(s)$. It should be cancelled to form the final $K(s)$ for actual implementation. $s = 0$ is another pole of $K(s)$, which is necessary to eliminate the steady-state error. Equation (9.36) requires that no roots of the denominator of $K(s,\tau)$ lie in the closed RHP, expect $s = \frac{1}{T}$, and $s = 0$. Since $\hat{G}_-$ is of minimum phase, from (9.3) we only need to investigate the root locations of

$$\delta(s,\tau,L) = (\tau s+1)^{m+1} - (\alpha s+1)e^{-Ls}. \tag{9.41}$$

Normalize L, τ, a, s as $\bar{L} = L/T$, $\bar{\tau} = \tau/T$, $\bar{\alpha} = \alpha/T$ and $\bar{s} = sT$, Equation (9.41) then becomes

$$\delta(\bar{s},\bar{\tau},\bar{L}) = (\bar{\tau}\bar{s}+1)^{m+1} - (\bar{\alpha}\bar{s}+1)e^{-\bar{L}\bar{s}}. \tag{9.42}$$

Equation (9.42) is a quasi-polynomial and can be written into a standard form,

$$Q(\bar{s},\bar{L}) = A(\bar{s}) + B(\bar{s})e^{-\bar{L}\bar{s}}, \tag{9.43}$$

where $A(\bar{s})$ and $B(\bar{s})$ are polynomials in $\bar{s}$. Walton and Marshall (1987) proposed a method to study the movement of the roots of (9.43) with respect to a given parameter and this can be employed to determine the minimum $\bar{\tau}$ at which roots of (9.43) lie on the imaginary axis. This $\bar{\tau}_{min}$ should meet

$$\cos(\omega_0 \bar{L}) = \text{Re}\left\{\frac{(j\omega_0 \bar{\tau}_{min} + 1)^{m+1}}{j\omega_0 \bar{\alpha} + 1}\right\}, \tag{9.44}$$

$$\sin(\omega_0 \bar{L}) = \text{Im}\left\{-\frac{(j\omega_0 \bar{\tau}_{min} + 1)^{m+1}}{j\omega_0 \bar{\alpha} + 1}\right\}, \tag{9.45}$$

where

$$\omega_0 = \{\min(\omega_0) | (\omega_0^2 \bar{\tau}_{min}^2 + 1)^{m+1} - (\omega_0^2 \bar{\alpha}^2 + 1) = 0, \omega_0 > 0\}. \tag{9.46}$$

Thus, for a given $\bar{L}$, $\bar{\tau}$ should be chosen to satisfy

$$\bar{\tau} > \bar{\tau}_{min}, \tag{9.47}$$

to ensure stability. For a given m, $\bar{\tau}_{min}$ depends on $\bar{L}$. When $m = 1, 2, 3$, the typical relationship between $\bar{\tau}_{min}$ and $\bar{L}$ is shown in Figure 9.14. It is interesting to note that when $\bar{\tau}_{min}$ tends to infinity, $\bar{L}$ tends to a constant, $\bar{L}_{max}$, which indicates that there is a limitation on tuning τ for stabilizability and that the process is stabilizable only if $\bar{L} \leq \bar{L}_{max}$. $\bar{L}_{max}$ is determined by m only and can be obtained as

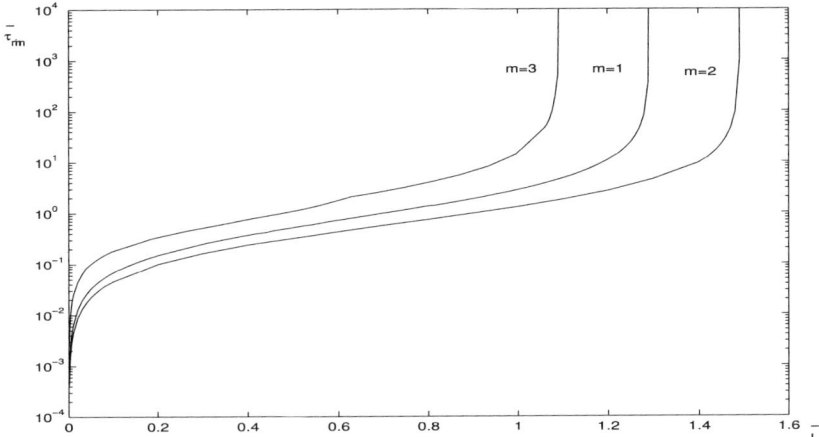

Fig. 9.14. Typical relationship between $\bar{\tau}_{min}$ and $\bar{L}$

$$\bar{L}_{max} = \begin{cases} \{\max(\bar{L}) \mid e^{\bar{L}}\bar{L}^m < (\frac{3\pi}{2})^m\}, m = 4l+1 \\ \{\max(\bar{L}) \mid e^{\bar{L}}\bar{L}^m < (\pi)^m\}, m = 4l+2 \\ \{\max(\bar{L}) \mid e^{\bar{L}}\bar{L}^m < (\frac{\pi}{2})^m\}, m = 4l+3 \\ \{\max(\bar{L}) \mid e^{\bar{L}}\bar{L}^m < (2\pi)^m\}, m = 4l+4 \end{cases} \quad (9.48)$$

where $l = 0, 1, 2 \cdots$.

For system performance, with controller $K(s)$ in (9.40), the closed-loop transfer function is

$$\eta(s) = \frac{G(s)K(s)}{1+G(s)K(s)} = \hat{G}_+(s)f(s) = \frac{\alpha s + 1}{(\tau s + 1)^{m+1}} e^{-Ls}. \quad (9.49)$$

The maximum of the magnitude of η is related directly to $\bar{L}$ and $\bar{\tau}$ as

$$\|\eta\|_\infty = \sqrt{\frac{\frac{1}{m}((\frac{\bar{\alpha}}{\bar{\tau}})^2 - m - 1) + 1}{(\frac{1}{m}(1-(m+1)(\frac{\bar{\tau}}{\bar{\alpha}})^2)+1)^{m+1}}}. \quad (9.50)$$

A typical relation between $\|\eta\|_\infty$ and $\bar{\tau}$ is shown in Figure 9.15. The large amplitude of $\|\eta\|_\infty$ usually produces a peak overshoot in the step response in the time domain perspective (Kuo, 1991). To eliminate the overshoot, a prefilter $F = \frac{1}{\alpha s + 1}$ is added with α given in (9.39), as shown in Figure 9.16.

For system robustness, let the actual process be $G(s) = \hat{G}(s)(1 + \Delta_G(s))$ with $|\Delta_G| \leq \delta_G(\omega)$. In implementation, the presence of dead time in the denominator of $K(s)$ increases the complexity of the controller. Moreover, due to the fact that the denominator of $K(s)$ is not in polynomial form, it is not possible to cancel $s = \frac{1}{T}$ in $K(s)$ explicitly. Thus, model reduction is applied

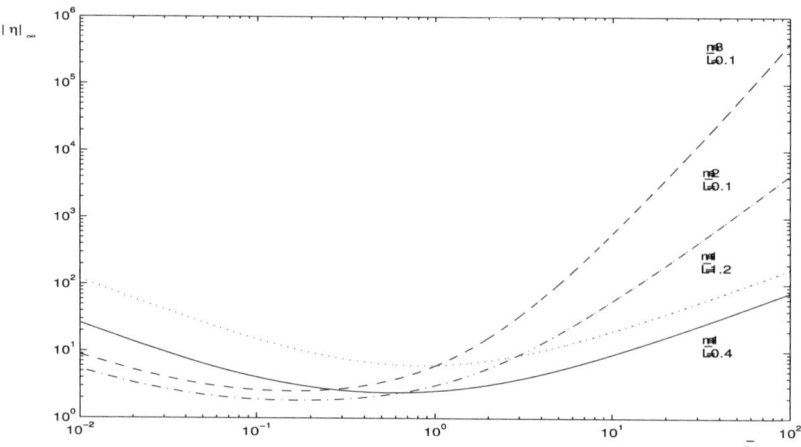

Fig. 9.15. Relation between $\|\eta\|_\infty$ and $\bar{\tau}$

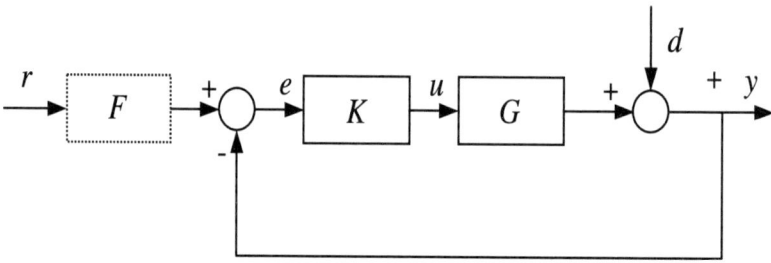

Fig. 9.16. Single-loop control system with pre-filter

to obtain the best approximation $\hat{K}$ to K, where $\hat{K}(s) = K(s)(1 + \Delta_K(s))$ and $|\Delta_K| \leq \delta_K(\omega)$. In our context, $\delta_K(\omega) \leq ERR \leq \epsilon$, where ϵ is the threshold for model reduction error and is specified prior to the design and usually taken as 5%. With the standard assumption that $\hat{K}$ has the same number of unstable poles as K, it follows from the stability robustness theorem (Doyle et al., 1982) and some algebra that the uncertain feedback system remains stable for all $\Delta = \text{diag}\{\Delta_K, \Delta_G\}$ if

$$\delta_K^2(\omega)|\eta(j\omega)|^2 + \delta_G^2(\omega)|\eta(j\omega)|^2 + 4\delta_K(\omega)\delta_G(\omega)|\eta(j\omega)| < 1, \quad \forall \omega. \tag{9.51}$$

It is noted from (9.49) that $|\eta(j\omega)|$ has a peak value at a low frequency ω_p and decays quickly for higher frequencies. Then (9.51) is likely to hold if

$$\delta_K^2(\omega_p)\|\eta\|_\infty^2 + \delta_G^2(\omega_p)\|\eta\|_\infty^2 + 4\delta_K(\omega_p)\delta_G(\omega_p)\|\eta\|_\infty < 1. \tag{9.52}$$

An upper bound on $\|\eta\|_\infty$ can be obtained from (9.52):

$$\|\eta\|_\infty < \left.\frac{2\sqrt{\delta_K^2(\omega) + \delta_G^2(\omega) + 4\delta_K^2(\omega)\delta_G^2(\omega)} - 4\delta_K(\omega)\delta_G(\omega)}{2(\delta_K^2(\omega) + \delta_G^2(\omega))}\right|_{\omega=\omega_p}. \quad (9.53)$$

Combining (9.50) and (9.53) yields

$$\sqrt{\frac{\frac{1}{m}((\frac{\bar{\alpha}}{\bar{\tau}})^2 - m - 1) + 1}{(\frac{1}{m}(1 - (m+1)(\frac{\bar{\tau}}{\bar{\alpha}})^2) + 1)^{m+1}}}$$

$$< \left.\frac{2\sqrt{\delta_K^2(\omega) + \delta_G^2(\omega) + 4\delta_K^2(\omega)\delta_G^2(\omega)} - 4\delta_K(\omega)\delta_G(\omega)}{2(\delta_K^2(\omega) + \delta_G^2(\omega))}\right|_{\omega=\omega_p} \triangleq \delta_p, \quad (9.54)$$

which gives a range of τ to achieve robust stability.

To consider the performance limitation imposed by input constraints, we still use frequency-by-frequency analysis. Assume that at each frequency $|U(j\omega)| \leq \bar{U}$ and the reference signal satisfies $|R(j\omega)| \leq \bar{R}$. The manipulated variable is

$$U(s) = F(s)\frac{K(s,\tau)}{1 + G(s)K(s,\tau)}R(s)$$

$$= \frac{1 - Ts}{(\tau s + 1)^{m+1}}\hat{G}_-^{-1}(s)R(s).$$

It is required that

$$\frac{1 - Tj\omega}{(\tau j\omega + 1)^{m+1}}\hat{G}_-^{-1}(j\omega)R(j\omega) \leq \bar{U}. \quad (9.55)$$

Consider the worst case $|R(j\omega)| = \bar{R}$, which requires

$$\left|\frac{1}{(\tau j\omega + 1)^{m+1}}\right| \leq \left|\frac{1}{1 - Tj\omega}\hat{G}_-(j\omega)\right|\frac{\bar{U}}{\bar{R}}. \quad (9.56)$$

To derive an inequality on τ imposed by the input constraint, let $\omega = \omega_{ob}$, where ω_{ob} is the open-loop bandwidth, and notice that $\left|\frac{1}{1-Tj\omega_{ob}}\hat{G}_-(j\omega_{ob})\right| = \frac{1}{\sqrt{2}}$, we require

$$\left|\frac{1}{(\tau j\omega_{ob} + 1)^{m+1}}\right| \leq \frac{1}{\sqrt{2}}\frac{\bar{U}}{\bar{R}}, \quad (9.57)$$

i.e.,

$$\tau \geq \sqrt{\sqrt[m+1]{\frac{2\bar{R}^2}{\bar{U}^2}} - 1}\bigg/\omega_{ob}. \quad (9.58)$$

Therefore, for open-loop unstable process controller design the tuning parameter τ in the filter (9.39) should be, in general, chosen to meet (9.47), (9.54) and (9.58) simultaneously and this will determine a suitable range, $\tau \in (\tau_{min}, \tau_{max})$.

Once the ideal single-loop controller $K(s)$ has been found, model reduction is again applied to obtain its PID or high-order controller approximation. The procedure is similar to the stable case discussed earlier and we only highlight possible differences and present simulation examples as follows.

9.5.1 PID Controller

Consider a PID controller in the form (9.14). Following the same steps as in Section 9.2, the optimal PID controller K_{PID} can be obtained. If the user-specified fitting error threshold in (9.16) holds true, the design is complete. On the other hand, one can increase τ by the de-tuning rule in (9.17). The iteration continues until the accuracy bound is fulfilled or $\tau^{k+1} > \tau_{max}$. If (9.16) cannot be fulfilled when $\tau^{k+1} > \tau_{max}$, a more complex controller than a PID is necessary. We now present some simulation examples to demonstrate our PID tuning algorithm and the performance is compared with the results of Huang and Chen (1997), Majhi and Atherton (2000), and Park et al. (1998). As in Section 9.2, the ideal PID controller in (9.14) is replaced by version (9.19) for implementation. In simulation examples of this subsection, we consider the nominal case and (9.54) is not used. Normally, (9.58) gives a smaller lower bound on τ and thus only (9.47) is utilized to derive τ_{min}, and we set $\tau^0 = \tau_{min}$ in Examples 9.5.1–9.5.3.

Example 9.5.1. Consider an unstable process (Huang and Chen, 1997)

$$G = \frac{4e^{-2s}}{1-4s}.$$

From (9.47) one obtains $\tau^0 = 1.7$. This results in

$$K_{PID} = 0.6407 + \frac{0.0626}{s} + \frac{0.5633s}{\frac{0.5633}{N}s + 1}. \tag{9.59}$$

This controller has the approximation error $ERR = 1.80\%$, which can meet the accuracy threshold. The PI-PD controller of Majhi and Atherton (2000) are $k_p(1+\frac{1}{T_i s}) = 0.131(1+\frac{1}{2s})$ and $K_f(T_d s + 1) = 0.5(s+1)$. The PID-P controller of Park et al. (1998) has $K_p(1 + \frac{1}{T_i s} + T_d s) = 0.068(1 + \frac{1}{1.885s} + 4.296s)$ and $K_f = 0.350$. The closed-loop responses are shown in Figure 9.17. It can be seen that the proposed method shows much better performance than the other designs.

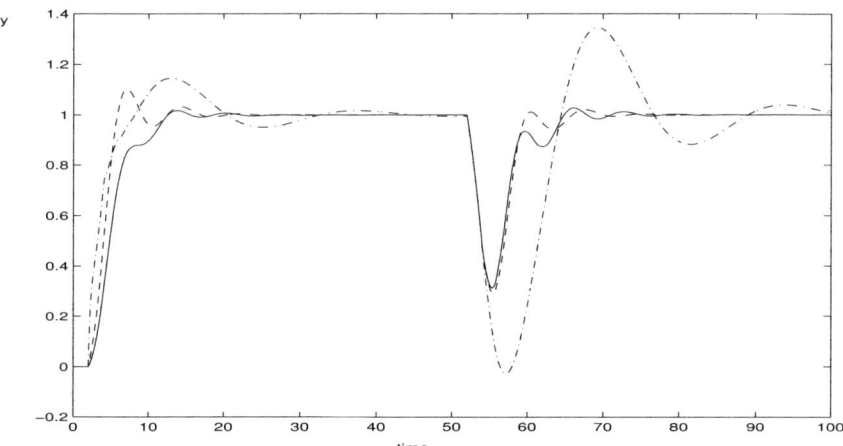

Fig. 9.17. Step response for Example 9.5.1
(—— proposed, − − − Majhi and Atherton, − · − · − Park et al.)

Example 9.5.2. Consider an unstable process (Park et al., 1998)

$$G = \frac{e^{-0.5s}}{(1-2s)(0.5s+1)}.$$

The proposed design gives $\tau^0 = 0.6$ and

$$K_{PID} = 3.1308 + \frac{0.7452}{s} + \frac{1.5651s}{\frac{1.5651}{N}s + 1}, \quad (9.60)$$

with $ERR = 0.47\%$. The PI-PD controller of Majhi and Atherton (2000) are $0.937(1+\frac{1}{1.339s})$ and $2.328(0.53s+1)$. The PID-P controller of Park et al. (1998) are $0.561(1 + \frac{1}{1.165s} + 1.478s)$ and $K_f = 1.687$. The closed-loop responses are shown in Figure 9.18.

Example 9.5.3. Consider an unstable process (Huang and Lin, 1995)

$$G = \frac{e^{-0.5s}}{(1-5s)(2s+1)(0.5s+1)}.$$

It follows that $\tau^0 = 1$ and

$$K_{PID} = 4.3794 + \frac{0.5106}{s} + \frac{7.2571s}{\frac{7.2571}{N}s + 1}, \quad (9.61)$$

with $ERR = 26.07\%$, which cannot meet the accuracy threshold, and the closed-loop response is very poor. Then τ is adjusted to $\tau^1 = \tau^0 + L = 1.5$ according to the proposed tuning rule (9.17). The new τ results in

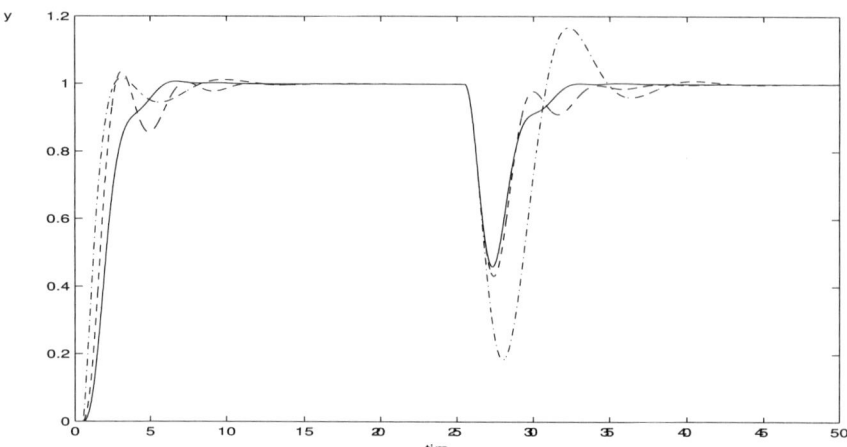

Fig. 9.18. Step response for Example 9.5.2
(—— proposed, − − − Majhi and Atherton, − · − · − Park et al.)

$$K_{PID} = 2.9886 + \frac{0.2335}{s} + \frac{4.6668s}{\frac{4.6668}{N}s + 1}, \tag{9.62}$$

which has $ERR = 2.06$ and meets the specified approximation accuracy $ERR \leq 5\%$. The controller in Huang and Chen (1997) is

$$K_{PID} = 6.1859(1 + \frac{0.1395}{s} + \frac{1.4724s}{\frac{1.4724}{N}s + 1}). \tag{9.63}$$

The closed-loop responses are shown in Figure 9.19. One observes that our design yields great improvement over Huang's method.

It can be seen from the simulation study that the proposed method always yields a PID controller with much better performance than the other methods, regardless of what τ is chosen. Our experience indicates that for FOPDT and SOPDT processes and a slow closed-loop response requirement, the proposed method can always achieve $ERR < 5\%$, and thus the closed-loop performance can be well predicted from the corresponding IMC design. It is, however, noticed that for high-order processes with fast responses, none of the above methods is able to generate PID systems with good performance. This implies that controllers in PID form are insufficient to obtain the desired performance. In this case, a higher-order controller has to be considered for better fitting and performance.

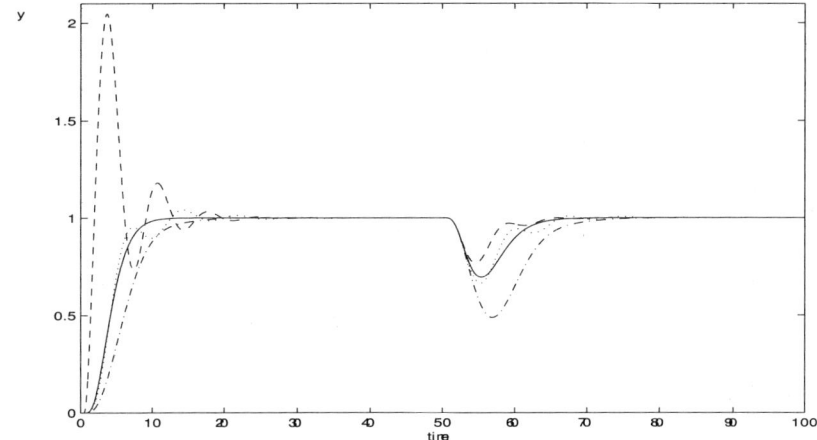

Fig. 9.19. Step response for Example 9.5.3
(······ proposed PID ($\tau = 1$), ─── proposed high order ($\tau = 1$), ─ ─ ─ Huang and Chen, ─ · ─ · ─ proposed PID ($\tau = 1.5$))

9.5.2 High-order Controller

The idea here is the same as in Section 9.3, that is to find the lowest order of controller in the form of (9.22) that can match the ideal controller $K(s)$ as well as possible in the frequency range of interest and can meet a specified approximation accuracy. However, the rules for determining the IMC filter change for the present unstable processes. We thus need to summarize the tuning procedure again.

Tuning Procedure for Unstable Processes

Step 1. Find the smallest τ_{min} from (9.47) and (9.58), and let $\tau^0 = \tau_{min}$. Find the largest τ_{max} from (9.54) if the plant uncertainty δ_G is given.

Step 2. Determine the PID controller from (9.15) and evaluate the corresponding approximation error ERR in (9.16). If ERR achieves the specified approximation accuracy ϵ (usually 5%), end the design.

Step 3. Otherwise, we have two ways to solve this problem: if a PID controller is desired, update τ by (9.17), and go to Step 2 when $\tau < \tau_{max}$; else, go to Step 4.

Step 4. Adopt the high-order controller in (9.22), start with a controller of order 2 up to the smallest integer n for which $ERR \leq \epsilon$.

Example 9.5.3 (cont'd). Reconsider

$$G = \frac{e^{-0.5s}}{(1-5s)(2s+1)(0.5s+1)}.$$

for which $\tau^0 = 1$ and a PID has been obtained with $ERR = 26.07\%$. For a high-order controller, our procedure ends with

$$\hat{K} = \frac{29.8188s^3 + 75.8869s^2 + 39.4556s + 4.3412}{s^3 + 3.2042s^2 + 8.5984s}, \quad (9.64)$$

with fitting error ERR less than $\epsilon = 5\%$. The closed-loop step responses are shown in Figure 9.19. We can see that the new controller $\hat{K}$ restores the IMC performance, while the previous PID controller in (9.61) is not capable of that under such a tight performance specification.

Example 9.5.4. Consider an unstable process

$$G = \frac{e^{-1.2s}}{(1-s)(0.5s+1)}.$$

It follows that $\tau^0 = 2.7$, and

$$K_{PID} = 1.0134 + \frac{0.0063}{s} + \frac{1.0155s}{\frac{1.0155}{N}s + 1}, \quad (9.65)$$

with $ERR = 18.50\%$, which cannot meet the accuracy threshold, and the closed-loop response is very poor. One can de-tune the PID controller by increasing τ. However, this results in a sluggish response. For a high-order controller, our procedure ends with

$$\hat{K} = \frac{5.2221s^4 + 41.6265s^3 + 128.3411s^2 + 131.1146s + 0.7798}{s^4 + 14.0061s^3 + 9.6026s^2 + 123.0987s}, \quad (9.66)$$

with the fitting error ERR less than $\epsilon = 5\%$. The closed-loop step responses are shown in Figure 9.20. We can see that the high-order controller makes a significant improvement over the PID controller. To our knowledge, the PID controller design methods in the literature are not applicable to this example (Huang and Chen, 1997; Majhi and Atherton, 2000; Park et al., 1998).

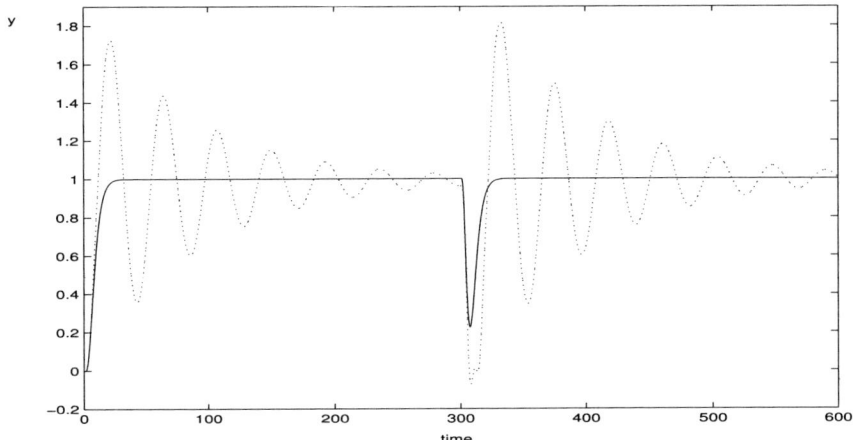

Fig. 9.20. Step response for Example 9.5.4
(······ proposed PID, —— proposed high order)

10. Multivariable Systems

In this chapter, we extend the SISO design methodology of Chapter 9 to the MIMO case. The development of MIMO internal model control (IMC) is not straightforward. In principle, if a multivariable delay model $\hat{G}(s)$ can be factorized as $\hat{G}(s) = \hat{G}_+(s)\hat{G}_-(s)$ such that $\hat{G}_+(s)$ is diagonal and contains all the time delays and non-minimum-phase zeros of $\hat{G}(s)$, then the IMC controller can be designed as $C(s) = \hat{G}_-^{-1}(s)F(s)$, with a possible filter $F(s)$, and the conventional feedback controller is related to the IMC controller, as in the scalar case, via $K(s) = C(s)[I - \hat{G}(s)C(s)]^{-1}$. For a scalar transfer function, it is quite easy to obtain $\hat{G}_+(s)$. However, this becomes much more difficult for a transfer matrix with multi-delays. The factorization is affected not only by the time delays in individual elements but also by their distributions within the transfer function matrix, and a non-minimum-phase zero is not related to that of elements of the transfer matrix at all.

Note also that for the SISO case the output behaviour is completely determined by the delay-free design shifted by the plant time delay, whereas one can hardly tell the MIMO output performance from the delay-free design because the delays in the process could mix up the outputs of the delay-free part to generate a messy actual output response. Decoupling between the loops, in addition to serving its own purpose, can rectify such a problem. Therefore, in this chapter, decoupling will be set as a design objective. The reasons are that (i) decoupling is usually required (Kong, 1995) in practice, at least in the process control industry, for ease of process operations. (ii) According to our experience, poor decoupling in closed loop could, in many cases, lead to poor diagonal loop performance, in other words, good decoupling is helpful for good loop performance. (iii) It should be noted that even if the decoupling requirement is relaxed to limited coupling, it will still lead to near decoupling if the number of inputs/outputs is large. Roughly, if the number is 10, total couplings from all other loops are limited to 30%, then each off-diagonal element will have a relative gain less than 3%, so the system is almost decoupled. (iv) The decoupling also simplifies the design procedure. In fact, it enables elementwise controller design (see Section 10.1). Otherwise, it is impossible and one has to design the controller matrix as a whole if the decoupling is removed or relaxed. It should be emphasized that we are aware that there are cases where decoupling should not be used; instead, couplings are deliberately employed to boost performance. But this is not the topic of the present chapter. We address decoupling the IMC in Section 10.1 and decoupling the conventional feedback scheme in Section 10.2.

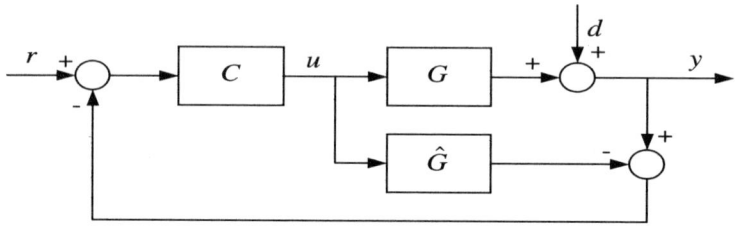

Fig. 10.1. Internal model control

10.1 IMC Scheme

The systems considered here are square, stable, nonsingular MIMO linear systems. In the process industry, most processes are open-loop stable. A process usually has unequal numbers of inputs and outputs, but normally a square subsystem will be selected from it for control purposes. A singular process essentially cannot provide independent control of each output and should not be used. Thus, in this section, we consider stable and nonsingular processes only.

The IMC control system is depicted in Figure 10.1, where G and $\hat{G}$ represent the transfer function matrices of the plant and its model, respectively, and C the controller. $\hat{G}$ is assumed to be identical with G except in our discussion on robust stability:

$$\hat{G}(s) = G(s) = \begin{bmatrix} g_{11}(s) & \cdots & g_{1m}(s) \\ \vdots & \ddots & \vdots \\ g_{m1}(s) & \cdots & g_{mm}(s) \end{bmatrix}, \tag{10.1}$$

where

$$g_{ij}(s) = g_{ij0}(s)e^{-L_{ij}s},$$

and $g_{ij0}(s)$ are strictly proper, stable, and scalar rational functions and L_{ij} are non-negative constants. It is also assumed that a proper input–output pairing has been made for $G(s)$ such that none of the m principal minors of $G(s)$ (the ith principle minor is the determinant of a $(m-1) \times (m-1)$ matrix obtained by deleting the ith row and ith column of $G(s)$) is zero. Our task in this section is to characterize all the realizable controllers C and the resultant closed-loop transfer functions such that the IMC system is internally stable and decoupled.

10.1.1 Decoupling

The closed-loop transfer matrix H between y and r can be derived from Figure 10.1 as

$$H = GC[I + (G - \hat{G})C]^{-1},$$

which becomes

$$H = GC, \quad if \quad \hat{G} = G.$$

Thus, the closed-loop is decoupled, i.e., H is diagonal and nonsingular, if and only if GC is decoupled and the IMC system is internally stable if and only if C is stable. Therefore, the problem becomes to characterize all stable and realizable controllers C and the resulting H such that $GC = H$ is decoupled.

The first question to ask is whether or not a solution exists. For a stable and nonsingular G, one may write

$$G(s) = \frac{N(s)}{d(s)},$$

where $d(s)$ is a least common denominator polynomial of $G(s)$. Now choose $C(s)$ as

$$C(s) = \frac{\text{adj } N(s)}{\sigma(s)},$$

where $\text{adj } N(s)$ is the adjoint matrix of $N(s)$ and $\sigma(s)$ is a stable polynomial with degree high enough to make the rational part of each c_{ij} proper. Then, this $C(s)$ is stable and realizable and decouples $G(s)$ since

$$G(s)C(s) = \frac{|N(s)|}{d(s)\sigma(s)} I_m,$$

where $|\cdot|$ stands for the determinant of a matrix and I_m is the $m \times m$ identity matrix. Thus $C(s)$ is a solution. Nonsingularity of $G(s)$ is obviously also necessary in order for GC to be decoupled.

Theorem 10.1.1. *For a stable, square and multi-delay process $G(s)$, the decoupling problem with stability via the IMC is solvable if and only if $G(s)$ is nonsingular.*

Our task is now to find a $C(s)$ such that $G(s)C(s)$ is decoupled, i.e.,

$$GC = \begin{bmatrix} g_1 \\ g_2 \\ \vdots \\ g_m \end{bmatrix} \begin{bmatrix} c_1 & c_2 & \cdots & c_m \end{bmatrix} = \mathrm{diag}\{q_{ii},\ i=1,2,\ldots,m\}. \tag{10.2}$$

For each column of GC, we have

$$\begin{bmatrix} g_1 \\ g_2 \\ \vdots \\ g_m \end{bmatrix} c_i = \begin{bmatrix} 0 \cdots 0 \ q_{ii} \ 0 \cdots 0 \end{bmatrix}^T, \quad i=1,2,\ldots,m, \tag{10.3}$$

which is equivalent to

$$\begin{bmatrix} g_1 \\ \vdots \\ g_{i-1} \\ g_{i+1} \\ \vdots \\ g_m \end{bmatrix} c_i = 0, \quad i=1,2,\ldots,m, \tag{10.4}$$

and

$$g_i c_i = q_{ii} \neq 0, \quad i=1,2,\cdots,m. \tag{10.5}$$

For any i, we can solve (10.4) to obtain $c_{1,i},\cdots,c_{i-1,i},c_{i+1,i},\cdots,c_{m,i}$, in terms of c_{ii} as

$$\begin{bmatrix} c_{1i} \\ \vdots \\ c_{i-1,i} \\ c_{i+1,i} \\ \vdots \\ c_{mi} \end{bmatrix} = \begin{bmatrix} \psi_{1i} \\ \vdots \\ \psi_{i-1,i} \\ \psi_{i+1,i} \\ \vdots \\ \psi_{mi} \end{bmatrix} c_{ii}, \quad \forall\, i \in \mathbf{m}, \tag{10.6}$$

where $\mathbf{m} = \{1, 2, \cdots, m\}$ and

$$\begin{bmatrix} \psi_{1i} \\ \vdots \\ \psi_{i-1,i} \\ \psi_{i+1,i} \\ \vdots \\ \psi_{mi} \end{bmatrix} \triangleq - \begin{bmatrix} g_{11} & \cdots & g_{1,i-1} & g_{1,i+1} & \cdots & g_{1m} \\ \vdots & \ddots & \vdots & \vdots & \ddots & \vdots \\ g_{i-1,1} & \cdots & g_{i-1,i-1} & g_{i-1,i+1} & \cdots & g_{i-1,m} \\ g_{i+1,1} & \cdots & g_{i+1,i-1} & g_{i+1,i+1} & \cdots & g_{i+1,m} \\ \vdots & \ddots & \vdots & \vdots & \ddots & \vdots \\ g_{m1} & \cdots & g_{m,i-1} & g_{m,i+1} & \cdots & g_{mm} \end{bmatrix}^{-1} \begin{bmatrix} g_{1i} \\ \vdots \\ g_{i-1,i} \\ g_{i+1,i} \\ \vdots \\ g_{mi} \end{bmatrix}.$$
$$\tag{10.7}$$

The ith diagonal element of GC is $\tilde{g}_{ii} c_{ii}$, where

$$\tilde{g}_{ii} = g_{ii} + \sum_{\substack{k=1 \\ k \neq i}}^{m} g_{ik} \psi_{ki}. \tag{10.8}$$

One notes that for a given i, the resulting diagonal element of GC is independent of the controller off-diagonal elements but contains only the controller diagonal element c_{ii}.

To further simplify the above result, let G^{ij} be the cofactor corresponding to g_{ij} in G. It follows from linear algebra (Noble, 1969) that the inverse of G can be written as $G^{-1} = \frac{\mathrm{adj}\, G}{|G|}$, where $\mathrm{adj}\, G = \left[G^{ji}\right]$. This also means

$$\begin{bmatrix} g_{11} & \cdots & g_{1m} \\ \vdots & \ddots & \vdots \\ g_{m1} & \cdots & g_{mm} \end{bmatrix} \begin{bmatrix} G^{11} & \cdots & G^{m1} \\ \vdots & \ddots & \vdots \\ G^{1m} & \cdots & G^{mm} \end{bmatrix} = |G| I_m. \tag{10.9}$$

Equation (10.9) can be equivalently put into the following two relations:

$$\begin{bmatrix} g_{11} & \cdots & g_{1,i-1} & g_{1,i+1} & \cdots & g_{1m} \\ \vdots & \ddots & \vdots & \vdots & \ddots & \vdots \\ g_{i-1,1} & \cdots & g_{i-1,i-1} & g_{i-1,i+1} & \cdots & g_{i-1,m} \\ g_{i+1,1} & \cdots & g_{i+1,i-1} & g_{i+1,i+1} & \cdots & g_{i+1,m} \\ \vdots & \ddots & \vdots & \vdots & \ddots & \vdots \\ g_{m1} & \cdots & g_{m,i-1} & g_{m,i+1} & \cdots & g_{mm} \end{bmatrix} \begin{bmatrix} G^{i1} \\ \vdots \\ G^{i,i-1} \\ G^{i,i+1} \\ \vdots \\ G^{im} \end{bmatrix} = - \begin{bmatrix} g_{1i} \\ \vdots \\ g_{i-1,i} \\ g_{i+1,i} \\ \vdots \\ g_{mi} \end{bmatrix} G^{ii},$$

$$\forall\, i \in \mathbf{m}, \qquad (10.10)$$

and

$$\sum_{k=1}^{m} g_{ik} G^{ik} = |G|, \qquad \forall\, i \in \mathbf{m}.$$

Substituting (10.10) to (10.7) yields

$$\begin{bmatrix} \psi_{1i} \\ \vdots \\ \psi_{i-1,i} \\ \psi_{i+1,i} \\ \vdots \\ \psi_{mi} \end{bmatrix} = - \begin{bmatrix} g_{11} & \cdots & g_{1,i-1} & g_{1,i+1} & \cdots & g_{1m} \\ \vdots & \ddots & \vdots & \vdots & \ddots & \vdots \\ g_{i-1,1} & \cdots & g_{i-1,i-1} & g_{i-1,i+1} & \cdots & g_{i-1,m} \\ g_{i+1,1} & \cdots & g_{i+1,i-1} & g_{i+1,i+1} & \cdots & g_{i+1,m} \\ \vdots & \ddots & \vdots & \vdots & \ddots & \vdots \\ g_{m1} & \cdots & g_{m,i-1} & g_{m,i+1} & \cdots & g_{mm} \end{bmatrix}^{-1} \begin{bmatrix} g_{1i} \\ \vdots \\ g_{i-1,i} \\ g_{i+1,i} \\ \vdots \\ g_{mi} \end{bmatrix}$$

$$= \frac{1}{G^{ii}} \begin{bmatrix} G^{i1} \\ \vdots \\ G^{i,i-1} \\ G^{i,i+1} \\ \vdots \\ G^{im} \end{bmatrix},$$

or

$$\psi_{ji} = \frac{G^{ij}}{G^{ii}}, \qquad \forall\, i, j \in \mathbf{m}, j \neq i. \qquad (10.11)$$

By (10.11), (10.6) and (10.8) become respectively

$$c_{ji} = \frac{G^{ij}}{G^{ii}} c_{ii} = \psi_{ji} c_{ii}, \qquad \forall\, i, j \in \mathbf{m}, j \neq i, \qquad (10.12)$$

and

$$\tilde{g}_{ii} = g_{ii} + \sum_{\substack{k=1 \\ k \neq i}}^{m} g_{ik} \frac{G^{ik}}{G^{ii}} = \frac{1}{G^{ii}} \sum_{k=1}^{m} g_{ik} G^{ik} = \frac{|G|}{G^{ii}}, \quad \forall i \in \mathbf{m}. \quad (10.13)$$

Thus, the decoupled open-loop transfer function matrix is given by

$$GC = \text{diag}\{\tilde{g}_{ii} c_{ii}\} = \text{diag}\left\{\frac{|G|}{G^{ii}} c_{ii}, \quad i = 1, 2, \cdots, m\right\}. \quad (10.14)$$

Before proceeding further, it is interesting to note that $\frac{G^{ii}}{|G|}$ is the (i,i)th element of G^{-1}. And it follows from Bristol (1966) that the (i,i)th element of the relative gain array for G is given by $\lambda_{ii} = g_{ii} \frac{G^{ii}}{|G|}$. Hence, $\tilde{g}_{ii}$ is linked to g_{ii} by $\tilde{g}_{ii} = \lambda_{ii}^{-1} g_{ii}$.

To demonstrate how to find $\tilde{g}_{ii}$ and ψ_{ji}, take the following as an example:

$$G(s) = \begin{bmatrix} \frac{1}{s+2} e^{-2s} & \frac{-1}{s+2} e^{-6s} \\ \frac{s-0.5}{(s+2)^2} e^{-3s} & \frac{(s-0.5)^2}{2(s+2)^3} e^{-8s} \end{bmatrix} \quad (10.15)$$

Simple calculations give

$$|G| = \frac{2(s-0.5)(s+2)e^{-9s} + (s-0.5)^2 e^{-10s}}{2(s+2)^4},$$

$$G^{11} = \frac{(s-0.5)^2 e^{-8s}}{2(s+2)^3}, \quad G^{21} = \frac{e^{-6s}}{s+2},$$

$$G^{22} = \frac{e^{-2s}}{s+2}, \quad G^{12} = -\frac{(s-0.5)e^{-3s}}{(s+2)^2}.$$

It follows from (10.13) that the decoupled loops have their equivalent processes as

$$\tilde{g}_{11} = \frac{2(s+2)e^{-s} + (s-0.5)e^{-2s}}{(s-0.5)(s+2)},$$

$$\tilde{g}_{22} = \frac{2(s-0.5)(s+2)e^{-7s} + (s-0.5)^2 e^{-8s}}{2(s+2)^3},$$

respectively. By (10.11), ψ_{21} and ψ_{12} are found as

$$\psi_{21} = -\frac{2(s+2)}{s-0.5} e^{5s}, \quad \psi_{12} = e^{-4s}.$$

At first glance, each diagonal controller c_{ii} may be designed for $\tilde{g}_{ii}$ according to (10.14) and the off-diagonal controllers can then be determined from (10.12).

Recall that for SISO IMC design, a scalar process $g(s)$ is first factored into $g(s) = g^+(s)g^-(s)$, where $g^+(s)$ contains all the time delay and non-minimum-phase zeros. The controller $c(s)$ is then determined from $g(s)c(s) = g^+(s)f(s)$, where $f(s)$ is the IMC filter. This technique could seemingly be applied to the decoupled equivalent processes $\tilde{g}_{ii}$ to design c_{ii} directly, but it is actually not. The reasons are that in general $\tilde{g}_{ii}$ are not in the usual format of a rational function plus time delay, and more essentially, a realizable c_{ii} derived from (10.14) based on $\tilde{g}_{ii}$ only may result in unrealizable or unstable off-diagonal controllers c_{ji}, $j \neq i$, In other words, realizable c_{ji}, $j \neq i$, may necessarily impose additional time delay and non-minimum-phase zeros to c_{ii}.

10.1.2 Analysis

In what follows, we develop the characterizations of time delays and non-minimum-phase zeros for the decoupling controller and the resulting decoupled loops. For the nonsingular delay process $G(s)$ in (10.1), a general expression for $\tilde{g}_{ii}$ in (10.13) will be

$$\phi(s) = \frac{\sum_{k=0}^{M} n_k(s) e^{-\alpha_k s}}{d_0(s) + \sum_{l=1}^{N} d_l(s) e^{-\beta_l s}}, \tag{10.16}$$

where $n_k(s)$ and $d_l(s)$ are all nonzero scalar polynomials of s, $\alpha_0 < \alpha_1 < \cdots < \alpha_M$ and $0 < \beta_1 < \beta_2 < \cdots < \beta_N$. Define the time delay for nonzero $\phi(s)$ in (10.16) as

$$\gamma(\phi(s)) = \alpha_0.$$

It is easy to verify $\gamma(\phi_1 \phi_2) = \gamma(\phi_1) + \gamma(\phi_2)$ and $\gamma(\phi^{-1}(s)) = -\gamma(\phi(s))$ for any nonzero $\phi_1(s)$, $\phi_2(s)$ and $\phi(s)$. If $\gamma(\phi(s)) \geq 0$, it measures the time required for $\phi(s)$ to have a nonzero output in response to a step input. If $\gamma(\phi(s)) < 0$ then the system output will depend on the future values of the input. It is obvious that for any realizable and nonzero $\phi(s)$, $\gamma(\phi(s))$ cannot be negative. Therefore, a realizable C requires

$$\gamma(c_{ji}) \geq 0, \qquad \forall i \in \mathbf{m}, \ j \in \mathbf{J}_i, \tag{10.17}$$

where $\mathbf{J}_i \triangleq \{j \in \mathbf{m} \mid G^{ij} \neq 0\}$. It follows from (10.12) that

$$\gamma(c_{ji}) = \gamma(\frac{G^{ij}}{G^{ii}} c_{ii}) = \gamma(G^{ij}) - \gamma(G^{ii}) + \gamma(c_{ii}), \qquad \forall i \in \mathbf{m}, \ j \in \mathbf{J}_i.$$

Thus, (10.17) is equivalent to

$$\gamma(c_{ii}) \geq \gamma(G^{ii}) - \gamma(G^{ij}), \qquad \forall\, i \in \mathbf{m},\ j \in \mathbf{J}_i,$$

or

$$\gamma(c_{ii}) \geq \gamma(G^{ii}) - \gamma_i, \qquad \forall\, i \in \mathbf{m}, \tag{10.18}$$

where

$$\gamma_i \triangleq \min_{j \in \mathbf{J}_i} \gamma(G^{ij}). \tag{10.19}$$

Equation (10.18) is a characterization of the controller diagonal elements in terms of their time delays, which indicates the minimum time delay that the ith diagonal controller elements must contain. Consequently, the resulting ith diagonal elements of $H = GC$, $h_{ii} = \tilde{g}_{ii} c_{ii}$, meet

$$\gamma(h_{ii}) = \gamma(\tilde{g}_{ii}) + \gamma(c_{ii}) \geq \gamma(\tilde{g}_{ii}) + \gamma(G^{ii}) - \gamma_i, \qquad \forall\, i \in \mathbf{m},$$

which, by

$$\gamma(\tilde{g}_{ii}) = \gamma\left(\frac{|G|}{G^{ii}}\right) = \gamma(|G|) - \gamma(G^{ii}),$$

becomes

$$\gamma(h_{ii}) \geq \gamma(|G|) - \gamma_i, \qquad \forall\, i \in \mathbf{m}. \tag{10.20}$$

Equation (10.20) is a characterization of the decoupled ith loop transfer function in terms of their time delays, which indicates the minimum time delay that the ith decoupled loop transfer function must contain. One concludes from the above development that for a stable and realizable IMC controller C which decouples G, its diagonal elements are characterized by (10.18) for time delays and they will uniquely determine the off-diagonal elements using (10.12). Equation (10.20) characterizes the time delays of the resultant loops.

For illustration, consider again the example in (10.15). It follows from the definition that the time delays for respective functions are $\gamma(|G|) = 9$, $\gamma(G^{11}) = 8$, $\gamma(G^{12}) = 3$, $\gamma(G^{21}) = 6$ and $\gamma(G^{22}) = 2$. The values for γ_1 and γ_2 can be calculated from (10.19) as

$$\gamma_1 = \min\{\gamma(G^{11}), \gamma(G^{12})\} = \min\{8, 3\} = 3,$$
$$\gamma_2 = \min\{\gamma(G^{21}), \gamma(G^{22})\} = \min\{6, 2\} = 2.$$

It follows from (10.18) that c_{11} and c_{22} are characterized for their time delays by

$$\gamma(c_{11}) \geq \gamma(G^{11}) - \gamma_1 = 8 - 3 = 5,$$
$$\gamma(c_{22}) \geq \gamma(G^{22}) - \gamma_2 = 2 - 2 = 0.$$

Also, h_{11} and h_{22} must meet conditions

$$\gamma(h_{11}) \geq \gamma(|G|) - \gamma_1 = 9 - 3 = 6,$$
$$\gamma(h_{22}) \geq \gamma(|G|) - \gamma_2 = 9 - 2 = 7.$$

Consider now the performance limitation due to the non-minimum-phase zeros. Denote by $\mathbb{C}^+$ the closed right half of the complex plane (RHP). For a nonzero transfer function $\phi(s)$, let $\mathbf{Z}_\phi^+$ be the set of all the RHP zeros of $\phi(s)$, i.e., $\mathbf{Z}_\phi^+ = \{z \in \mathbb{C}^+ \mid \phi(z) = 0\}$. Let $\boldsymbol{\eta}_z(\phi)$ be an integer ν such that $\lim_{s \to z} \phi(s)/(s-z)^\nu$ exists and is nonzero. Thus, $\phi(s)$ has $\boldsymbol{\eta}_z(\phi)$ zeros at $s = z$ if $\boldsymbol{\eta}_z(\phi) > 0$, or $-\boldsymbol{\eta}_z(\phi)$ poles if $\boldsymbol{\eta}_z(\phi) < 0$, or neither poles nor zeros if $\boldsymbol{\eta}_z(\phi) = 0$. It is easy to verify that $\boldsymbol{\eta}_z(\phi_1\phi_2) = \boldsymbol{\eta}_z(\phi_1) + \boldsymbol{\eta}_z(\phi_2)$ and $\boldsymbol{\eta}_z(\phi^{-1}) = -\boldsymbol{\eta}_z(\phi)$ for any nonzero transfer function $\phi_1(s)$, $\phi_2(s)$ and $\phi(s)$. Obviously, a nonzero transfer function $\phi(s)$ is stable if and only if $\boldsymbol{\eta}_z(\phi) \geq 0$, $\forall z \in \mathbb{C}^+$.

A stable C thus requires

$$\boldsymbol{\eta}_z(c_{ji}) \geq 0, \quad \forall i \in \mathbf{m}, j \in \mathbf{J}_i, z \in \mathbb{C}^+. \tag{10.21}$$

It follows from (10.12) that

$$\boldsymbol{\eta}_z(c_{ji}) = \boldsymbol{\eta}_z\left(\frac{G^{ij}}{G^{ii}}c_{ii}\right) = \boldsymbol{\eta}_z(G^{ij}) - \boldsymbol{\eta}_z(G^{ii}) + \boldsymbol{\eta}_z(c_{ii}), \quad \forall i \in \mathbf{m}, j \in \mathbf{J}_i, z \in \mathbb{C}^+.$$

Thus (10.21) is equivalent to

$$\boldsymbol{\eta}_z(c_{ii}) \geq \boldsymbol{\eta}_z(G^{ii}) - \boldsymbol{\eta}_z(G^{ij}), \quad \forall i \in \mathbf{m}, j \in \mathbf{J}_i, z \in \mathbb{C}^+,$$

or

$$\boldsymbol{\eta}_z(c_{ii}) \geq \boldsymbol{\eta}_z(G^{ii}) - \eta_i(z), \quad \forall i \in \mathbf{m}, z \in \mathbb{C}^+, \tag{10.22}$$

where

$$\eta_i(z) \triangleq \min_{j \in \mathbf{J}_i} \boldsymbol{\eta}_z(G^{ij}). \tag{10.23}$$

Since $G(s)$ is stable, so are G^{ij}. One sees that for $\forall i \in \mathbf{m}$ and $\forall z \in \mathbb{C}^+$, $\eta_i(z) \geq 0$. We have, for $\forall i \in \mathbf{m}$ and $\forall z \in \mathbb{C}^+$,

$$\boldsymbol{\eta}_z(G^{ii}) - \eta_i(z) \leq \boldsymbol{\eta}_z(G^{ii}),$$

and

$$\boldsymbol{\eta}_z(G^{ii}) - \eta_i(z) = \boldsymbol{\eta}_z(G^{ii}) - \min_{j \in \mathbf{J}_i} \boldsymbol{\eta}_z(G^{ij}) \geq \boldsymbol{\eta}_z(G^{ii}) - \boldsymbol{\eta}_z(G^{ij})|_{j=i} = 0.$$

Thus, $\boldsymbol{\eta}_z(G^{ii}) - \eta_i(z)$ is bounded by

$$0 \leq \boldsymbol{\eta}_z(G^{ii}) - \eta_i(z) \leq \boldsymbol{\eta}_z(G^{ii}), \qquad \forall\, i \in \mathbf{m},\ z \in \mathbb{C}^+. \tag{10.24}$$

Equation (10.24) also implies that c_{ii} need not have any non-minimum-phase zeros except at $z \in \mathbf{Z}_{G^{ii}}^+$. Therefore, by (10.22), c_{ii} is characterized for its non-minimum-phase zeros by

$$\boldsymbol{\eta}_z(c_{ii}) \geq \boldsymbol{\eta}_z(G^{ii}) - \eta_i(z), \qquad \forall\, i \in \mathbf{m},\ z \in \mathbf{Z}_{G^{ii}}^+, \tag{10.25}$$

with the right side bounded by (10.24).

Consequently, the resulting ith diagonal elements of $H = GC$, $h_{ii} = \tilde{g}_{ii}c_{ii}$, satisfy

$$\boldsymbol{\eta}_z(h_{ii}) = \boldsymbol{\eta}_z(\tilde{g}_{ii}c_{ii}) = \boldsymbol{\eta}_z(\tilde{g}_{ii}) + \boldsymbol{\eta}_z(c_{ii}) \geq \boldsymbol{\eta}_z(\tilde{g}_{ii}) + \boldsymbol{\eta}_z(G^{ii}) - \eta_i(z)$$
$$\forall\, i \in \mathbf{m},\ z \in \mathbb{C}^+,$$

which, by

$$\boldsymbol{\eta}_z(\tilde{g}_{ii}) = \boldsymbol{\eta}_z\left(\frac{|G|}{G^{ii}}\right) = \boldsymbol{\eta}_z(|G|) - \boldsymbol{\eta}_z(G^{ii}),$$

becomes

$$\boldsymbol{\eta}_z(h_{ii}) \geq \boldsymbol{\eta}_z(|G|) - \eta_i(z), \qquad \forall\, i \in \mathbf{m},\ z \in \mathbb{C}^+. \tag{10.26}$$

It is readily seen that for $\forall\, i \in \mathbf{m},\ z \in \mathbb{C}^+$,

$$\boldsymbol{\eta}_z(|G|) - \eta_i(z) \leq \boldsymbol{\eta}_z(|G|),$$

and

$$\boldsymbol{\eta}_z(|G|) - \eta_i(z) = \boldsymbol{\eta}_z\left(\frac{|G|}{(s-z)^{\eta_i(z)}}\right) = \boldsymbol{\eta}_z\left(\frac{1}{(s-z)^{\eta_i(z)}} \sum_{j=1}^m g_{ij} G^{ij}\right)$$
$$= \boldsymbol{\eta}_z\left(\sum_{j=1}^m g_{ij} \frac{G^{ij}}{(s-z)^{\eta_i(z)}}\right) \geq 0.$$

Thus, $\boldsymbol{\eta}_z(|G|) - \eta_i(z)$ is bounded by

$$0 \leq \boldsymbol{\eta}_z(|G|) - \eta_i(z) \leq \boldsymbol{\eta}_z(|G|), \qquad \forall\, i \in \mathbf{m},\ z \in \mathbb{C}^+. \tag{10.27}$$

Equation (10.27) also indicates that the ith closed-loop transfer function h_{ii} need not have any non-minimum-phase zeros except at $z \in \mathbf{Z}^+_{|G|}$. Therefore, their characterizations on non-minimum-phase zeros are given by

$$\eta_z(h_{ii}) \geq \eta_z(|G|) - \eta_i(z), \qquad \forall i \in \mathbf{m}, \; z \in \mathbf{Z}^+_{|G|}, \tag{10.28}$$

with the right side bounded by (10.27).

Consider the previous example again to demonstrate the method. It can be found that $\mathbf{Z}^+_{G_{11}} = \{0.5\}$, $\mathbf{Z}^+_{G_{22}} = \emptyset$, $\mathbf{Z}^+_{|G|} = \{0.5\}$, $\eta_z(|G|)|_{z=0.5} = 1$ and

$$\eta_z(G^{11})|_{z=0.5} = 2, \qquad \eta_z(G^{12})|_{z=0.5} = 1,$$
$$\eta_z(G^{21})|_{z=0.5} = 0, \qquad \eta_z(G^{22})|_{z=0.5} = 0.$$

By (10.23), one sees

$$\eta_1(z)|_{z=0.5} = \min\{\eta_z(G^{11})|_{z=0.5}, \; \eta_z(G^{12})|_{z=0.5}\} = \min\{2,1\} = 1,$$
$$\eta_2(z)|_{z=0.5} = \min\{\eta_z(G^{21})|_{z=0.5}, \; \eta_z(G^{22})|_{z=0.5}\} = \min\{0,0\} = 0.$$

Thus, c_{11} should then satisfy

$$\eta_z(c_{11})|_{z=0.5} \geq \eta_z(G^{11})|_{z=0.5} - \eta_1(z)|_{z=0.5} = 2 - 1 = 1.$$

As $\mathbf{Z}^+_{G_{22}} = \emptyset$, there will be no constraint on c_{22} at its non-minimum-phase zeros. By (10.28), the resultant loops satisfy

$$\eta_z(h_{11})|_{z=0.5} \geq \eta_z(|G|)|_{z=0.5} - \eta_1(z)|_{z=0.5} = 1 - 1 = 0,$$
$$\eta_z(h_{22})|_{z=0.5} \geq \eta_z(|G|)|_{z=0.5} - \eta_2(z)|_{z=0.5} = 1 - 0 = 1,$$

i.e., the second loop must contain a non-minimum-phase zero at $z = 0.5$ of multiplicity one while the first loop need not contain any non-minimum-phase zero.

Theorem 10.1.2. *If the IMC system in Figure 10.1 is decoupled, stable and realizable, then (i) controller C's diagonal elements are characterized by (10.18) on their respective time delays and by (10.25) on their respective RHP zeros, and they uniquely determine the controller's off-diagonal elements by (10.12); (ii) the diagonal elements of GC, $h_{ii} = \tilde{g}_{ii} c_{ii}$, are characterized by (10.20) on their respective time delays and (10.28) on their respective RHP zeros.*

For the example in (10.15), choose the controller diagonal elements as

$$c_{11} = \frac{s - 0.5}{s + \rho} e^{-5s}, \qquad c_{22} = \frac{1}{s + \rho},$$

where $\rho > 0$. It follows from the previous calculations that they satisfy the conditions of Theorem 10.1.2 with the lower bounds in (10.18) and (10.25) exactly met. The off-diagonal controllers are then obtained from (10.12) and the complete controller is

$$C = \begin{bmatrix} \frac{s-0.5}{s+\rho}e^{-5s} & \frac{1}{s+\rho}e^{-4s} \\ \frac{-2(s+2)}{s+\rho} & \frac{1}{s+\rho} \end{bmatrix},$$

which is both realizable and stable, and the resulting GC is calculated as

$$GC = \begin{bmatrix} \frac{2s+4+(s-0.5)e^{-s}}{(s+\rho)(s+2)}e^{-6s} & 0 \\ 0 & (s-0.5)\frac{2s+4+(s-0.5)e^{-s}}{2(s+\rho)(s+2)^3}e^{-7s} \end{bmatrix},$$

whose diagonal elements fall in our previous characterization on time delays and non-minimum-phase zeros, namely, the time delay for the decoupled loops is no less than 6 and 7, respectively, and loop two must contain a non-minimum-phase zero at $s = 0.5$ of multiplicity 1.

However, if we choose the controller diagonal elements as

$$c_{11} = c_{22} = \frac{1}{s+\rho},$$

which violates the conditions of Theorem 10.1.2, this gives rise to

$$c_{21} = \frac{-2(s+2)}{(s+\rho)(s-0.5)}e^{5s},$$

which is neither realizable nor stable. The resultant h_{11} is given by

$$h_{11} = \tilde{g}_{11}c_{11} = \frac{2s+4+(s-0.5)e^{-s}}{(s+\rho)(s+2)(s-0.5)}e^{-s},$$

which is unstable, too. Such c_{ii} should be discarded.

A quite special phenomenon in such a decoupling control is that there might be unstable pole–zero cancellations in forming $\tilde{g}_{ii}c_{ii}$. In contrast to the normal intuition, this will not cause instability. Notice from the previous example that $\tilde{g}_{11}$ has a pole at $s = 0.5$ and c_{11} has at least one zero at the same location as $\eta_z(c_{11})|_{z=0.5} \geq 1$. Hence, an unstable pole–zero cancellation at $z = 0.5$ occurs in forming $\tilde{g}_{11}c_{11}$. However, the resulting closed-loop system is still stable because all the controller diagonal and off-diagonal elements are stable. In fact, as far as G and C are concerned, there is no unstable pole–zero cancellation since both G and C are stable.

10.1.3 Design

In this subsection, the practical design of IMC is considered. We first briefly review the IMC design for the SISO case. Let $g(s)$ be the transfer function model of a stable SISO process and $c(s)$ the controller. The resulting closed-loop transfer function in the case of no process model mismatch is $h(s)$, where $h(s) = g(s)c(s)$. It is quite clear that if $h(s)$ has been specified then $c(s)$ is determined. However $h(s)$ cannot be specified arbitrarily. In IMC design, the process model $g(s)$ is first factored into $g(s) = g^+(s)g^-(s)$, where $g^+(s)$ contains all the time delay and RHP zeros. Then $h(s)$ is chosen as $h(s) = g^+(s)f(s)$, where $f(s)$ is the IMC filter, and consequently $c(s)$ is given by $c(s) = h(s)g^{-1}(s)$. The factorizations which yield $g^+(s)$ are not unique. Holt and Morari (1985a, 1985b) suggested the following form to be advantageous:

$$g^+(s) = e^{-Ls} \prod_{i=1}^{n} \left(\frac{z_i - s}{z_i + s} \right), \tag{10.29}$$

where L is the time delay and $z_1, z_2, \cdots, z_n$ are all the RHP zeros present in $g(s)$.

For multivariable decoupling IMC control, the closed-loop transfer function matrix $H(s)$ in the case of no process model mismatch is

$$H = \text{diag}\{h_{ii}\} = \text{diag}\{\tilde{g}_{ii}c_{ii}\},$$

and thus the key point in design lies in specifying the m decoupled loop transfer function $h_{ii}(s)$. Unlike the SISO case, $h_{ii}(s)$ cannot be chosen based solely on $\tilde{g}_{ii}(s)$ since this may cause an unrealizable and/or unstable controller. Instead, $h_{ii}(s)$ should be chosen according to Theorem 10.1.2. It is also obvious that for best performance of the IMC system and simplicity of the controller it is undesirable to include any more time delays and non-minimum-phase zeros in $h_{ii}(s)$ than necessary, that is, only the minimum time delay in (10.20) and minimum multiplicity of unavoidable non-minimum-phase zeros in (10.28) should be included. This leads to

$$h_{ii} = e^{-(\gamma(|G|)-\gamma_i)s} f_i(s) \prod_{z \in \mathbf{Z}^+_{|G|}} \left(\frac{z-s}{z+s} \right)^{\eta_z(|G|)-\eta_i(z)}, \quad \forall i \in \mathbf{m}, \tag{10.30}$$

where $f_i(s)$ is the ith loop IMC filter. $h_{ii}(s)$ will then determine $c_{ii}(s)$ as

$$c_{ii}(s) = h_{ii}(s)\tilde{g}_{ii}^{-1}(s), \quad i \in \mathbf{m}, \tag{10.31}$$

which in turn yields off-diagonal elements $c_{ji}(s)$ uniquely by (10.12).

Remark 10.1.1. Good control performance of the loops is what we seek apart from the loop-decoupling requirement. However, we do not explicitly define a performance measure and design the controller via the minimization (or maximization) of some measure. These are done in model predictive control, where the "size" of the error between the predicted output and a desired trajectory is minimized; and in LQG control, where a linear quadratic index is minimized to yield an optimal controller. These approaches have a different framework from ours. What we do here is to follow the IMC design approach with a requirement for decoupling the control of multivariable multi-delay processes. Loop performance is simply observed with closed-loop step responses for each output, in terms of traditional performance specifications such as overshoot, rise time and settling time, in comparison with the existing schemes for similar purposes. A link between IMC design theory and quantitative error measurement is that the item with the non-minimum-phase factor $g^+(s)$ in (10.29) used in the IMC closed-loop transfer function in (10.30) can actually minimize the H-2 norm of the error signal in the presence of a step reference change (Morari and Zafiriou, 1989).

Remark 10.1.2. One may note that there is the case where good decoupling and performance may not both be achievable and one is reached at the cost of the other. For example, decoupling should never be used in flight control, instead, couplings are deliberately employed to boost performance. This case is, however, not the topic of the present paper. Here we address applications where decoupling is required. Indeed, decoupling is usually required (Kong, 1995) in the process control industry, to simplify process operations. Poor decoupling in a closed-loop system could, in many cases, lead to poor diagonal loop performance. Thus, the design objective here is to decouple the loops first and then to achieve as good a performance as possible for each loop. Loop performance is considered to be good if best achievable performance is nearly reached among all decoupled and stable loops, subject to limitations imposed by delay, non-minimum-phase zeros and bandwidth. Our design realizes this by selecting the objective loop transfer functions in (10.30) in which each factor reflects respective roles of delay, non-minimum-phase zeros and bandwidth. It might be worth noting that decoupling and loop performance may not always be in conflict. It is shown (Linnemann and Wang, 1993) that for delay-free stable systems some optimal controllers will automatically result in decoupled loops although decoupling is not required in the problem formulation.

For a multivariable multi-delay process, the IMC controller designed by the above procedure is usually in the complicated form (10.16), which is difficult to implement. Model reduction is then employed to find a much simpler rational function plus dead time approximation to the above theoretical controller. The topic of model reduction has been addressed in Chapter 2 in a great detail and two effective algorithms were given there. We can utilize one of them for our present design.

Theoretically, if h_{ii} is specified as in (10.30) so that it contains no more time delays than necessary, then each column of C has at least one element whose time delay is zero. However, if the controller is obtained as a result of model reduction, that element might have non-zero time delay, though it is usually small. In such a case, it is reasonable to subtract it from all elements in the column to speed up the loop response if it is positive or to make the controller realizable if it is negative. This operation will not affect decoupling and stability. The above development is summarized into the following design procedure.

IMC Design Procedure *Seek a controller $C(s)$ given $G(s)$*

(i) *With $i = 1$ for loop one, apply model reduction to $|G|$ and all nonzero G^{ij}. Take time delays and non-minimum-phase zeros (including multiplicities) from their reduced-order models. Find γ_i and $\eta_i(z)$ for all $z \in \mathbf{Z}^+_{|G|}$ from (10.19) and (10.23).*

(ii) *Specify h_{ii} according to (10.30) and determine the diagonal controller c_{ii} from (10.31). Apply model reduction to c_{ii} to obtain $\hat{c}_{ii}$.*

(iii) *Calculate the controller off-diagonal elements c_{ji} from (10.12). Apply model reduction to all c_{ji} to obtain $\hat{c}_{ji}$.*

(iv) *Find the smallest time delay in $\hat{c}_{ji}$ and subtract it from all the time delays of $\hat{c}_{ji}$ if it is nonzero.*

(v) *Repeat the above steps for all other loops.*

The IMC system in Figure 10.1 is referred to as the nominal case if $G = \hat{G}$. It is obvious that the IMC system is nominally stable if both the process G and the controller C are stable, as discussed before. In the real world where the model does not represent the process exactly, nominal stability is not sufficient and robust stability of the IMC system has to be ensured. As a standard assumption in robust analysis, we assume that the nominal IMC system is stable (i.e., both $\hat{G}$ and C are stable). Suppose that the actual process G is also stable and is described by

$$G \in \prod = \{G : \bar{\sigma}\left([G(j\omega) - \hat{G}(j\omega)]\hat{G}^{-1}(j\omega)\right) = \bar{\sigma}\left(\Delta_m(j\omega)\right) \leq \tilde{l}_m(\omega)\},$$

where $\hat{l}_m(\omega)$ is the bound on the multiplicative uncertainty. It follows that the IMC system is robust stable if and only if

$$\bar{\sigma}(\hat{G}(j\omega)C(j\omega)) < l_m^{-1}(j\omega), \quad \forall \omega. \tag{10.32}$$

Let C^* be the ideal controller obtained from (10.31) and (10.12), we have $\hat{G}C^* = \text{diag}\{h_{ii}\}$. Thus (10.32) becomes

$$\bar{\sigma}\left(\text{diag}\{h_{ii}(j\omega)\}C^{*-1}(j\omega)C(j\omega)\right) < l_m^{-1}(j\omega), \quad \forall \omega.$$

One notes from (10.30) that $\text{diag}\{h_{ii}(j\omega)\} = V\text{diag}\{f_i(j\omega)\}$, where V is a unitary matrix. Noticing that $\bar{\sigma}(VA) = \bar{\sigma}(A)$ for a unitary V, a sufficient and necessary robust stability condition is obtained as

$$\bar{\sigma}\left(\text{diag}\{f_i(j\omega)\}C^{*-1}(j\omega)C(j\omega)\right) < l_m^{-1}(j\omega), \quad \forall \omega. \tag{10.33}$$

In practice, we can evaluate and plot the left-hand side of (10.33) and compare it with $l_m^{-1}(j\omega)$ to see whether robust stability is satisfied given the multiplicative uncertainty bound $l_m(j\omega)$. One also notes that if the error between the ideal controller C^* and its reduced-order model C is small, i.e., $C^* \approx C$, then the robust stability condition becomes

$$\bar{\sigma}\left(\text{diag}\{f_i(j\omega)\}\right) < l_m^{-1}(j\omega), \quad \forall \omega,$$

which indicates that the IMC filter should be chosen such that its largest singular value is smaller than the inverse of the uncertainty magnitude for all ω.

10.1.4 Simulation

In this subsection, several simulation examples are given to show the effectiveness of the decoupling IMC design. The effects of time delays and non-minimum-phase zeros are illustrated and control performance as well as its robustness is compared with the existing multivariable Smith predictor schemes. We intend to use commonly cited literature examples with real live background and try to explain technical points of the proposed method as much as possible. The first and third examples are well-known distillation column processes (Wood and Berry, 1973) and (Luyben, 1986). The process in the second example is a modification of the Wood and Berry process where only the delays of its elements have been changed. This modification generates non-minimum-phase zeros for the multivariable process, which is not available in the other two examples, and is used to illustrate relevant design issues.

Example 10.1.1. Consider the well-known Wood/Berry binary distillation column plant (Wood and Berry, 1973):

$$G(s) = \begin{bmatrix} \dfrac{12.8e^{-s}}{16.7s+1} & \dfrac{-18.9e^{-3s}}{21s+1} \\[2mm] \dfrac{6.6e^{-7s}}{10.9s+1} & \dfrac{-19.4e^{-3s}}{14.4s+1} \end{bmatrix}.$$

Step (i) The application of model reduction to $|G|$, G^{11} and G^{12} produces

$$|\hat{G}| = \dfrac{-0.1077s^2 - 4.239s - 0.3881}{s^2 + 0.1031s + 0.0031} e^{-6.3s},$$

$$\hat{G}^{11} = \dfrac{-1.349s - 42.81}{s^2 + 31.85s + 2.207} e^{-3s},$$

$$\hat{G}^{12} = \dfrac{-0.6076s - 17.66}{s^2 + 29.26s + 2.676} e^{-7s}.$$

It is clear that $\gamma(|\hat{G}|) = 6.3$, $\gamma(\hat{G}^{11}) = 3$, and $\gamma(\hat{G}^{12}) = 7$. Thus, by (10.19), one finds $\gamma_1 = \min\{3, 7\} = 3$. Since $|\hat{G}|$ is of minimum phase, $\mathbf{Z}^+_{|\hat{G}|} = \emptyset$ and there is no need to calculate $\eta_1(z)$.

Step (ii) According to (10.30), we have $h_{11} = \dfrac{e^{-3.3s}}{s+1}$ with the filter chosen as $\dfrac{1}{s+1}$. c_{11} is then obtained as $c_{11} = \tilde{g}_{11}^{-1} h_{11}$ from (10.31). The application of model reduction to c_{11} yields

$$\hat{c}_{11} = \dfrac{0.3168s^2 + 0.0351s + 0.0012}{s^2 + 0.1678s + 0.0074} e^{-0.9s}.$$

Step (iii) c_{21} is calculated from (10.12). Applying model reduction to c_{21} yields

$$\hat{c}_{21} = \dfrac{0.1405s^2 + 0.0180s + 0.0007}{s^2 + 0.2013s + 0.0123} e^{-4.9s}.$$

Step (iv) The smallest time delay in $\hat{c}_{11}$ and $\hat{c}_{21}$ is 0.9, which is then subtracted from both elements' time delays to generate the final $\hat{c}_{11}$ and $\hat{c}_{21}$.

Step (v) Loop two is treated similarly, where $h_{22} = \dfrac{e^{-5.3s}}{s+1}$. This results in the overall controller as

$$C = \begin{bmatrix} \dfrac{0.3168s^2 + 0.0351s + 0.0021}{s^2 + 0.1678s + 0.0074} & \dfrac{-0.2130s^2 - 0.0222s - 0.0008}{s^2 + 0.1399s + 0.0051} e^{-2s} \\[2mm] \dfrac{0.1405s^2 + 0.0180s + 0.0007}{s^2 + 0.2031s + 0.0123} e^{-4s} & \dfrac{-0.1798s^2 - 0.0206s - 0.0008}{s^2 + 0.1607s + 0.0073} \end{bmatrix}.$$

The set-point responses of the resulting IMC system (shorten to IMC) are shown in Figure 10.2 as a solid line. The corresponding set-point response of Jerome's multivariable Smith predictor scheme with the controller settings given in Jerome and Ray (1986) is depicted as the dashed line in the figure. The corresponding responses of the ideal IMC system $H(s) = \text{diag}\{h_{ii}\}$ are

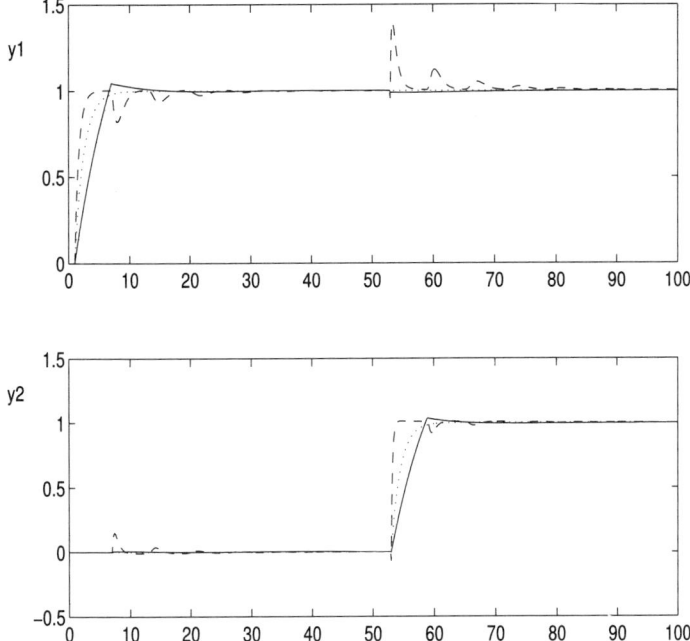

Fig. 10.2. Control performance for Example 10.1.1
(—— IMC; - - - Jerome; ⋯ ideal IMC)

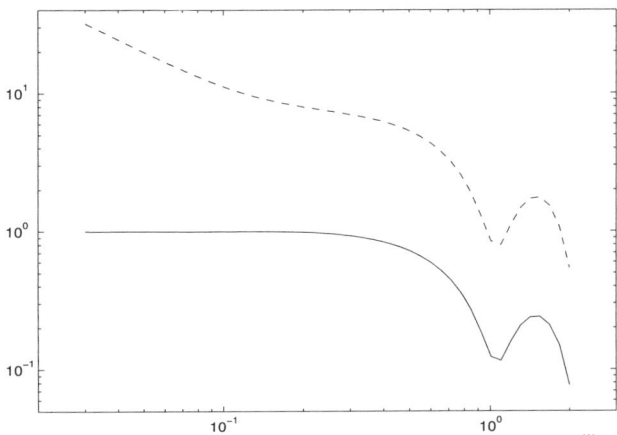

Fig. 10.3. Robustness analysis for Example 10.1.1
(—— left-hand side of (10.33); - - - right-hand side of (10.33))

also shown as the dotted line in the figure. The resulting ISE (Integral Square Error) with equal weighting of the two outputs is calculated as 7.16, 4.73 and 5.01 for the IMC design, Jerome's scheme and the ideal IMC control system, respectively. Although the ISE for Jerome's scheme is smaller than that of the proposed method, Jerome's control system is not robust and exhibits very poor damping and severe oscillation even for the nominal case, while the responses of the IMC are smooth and the decoupling is almost perfect.

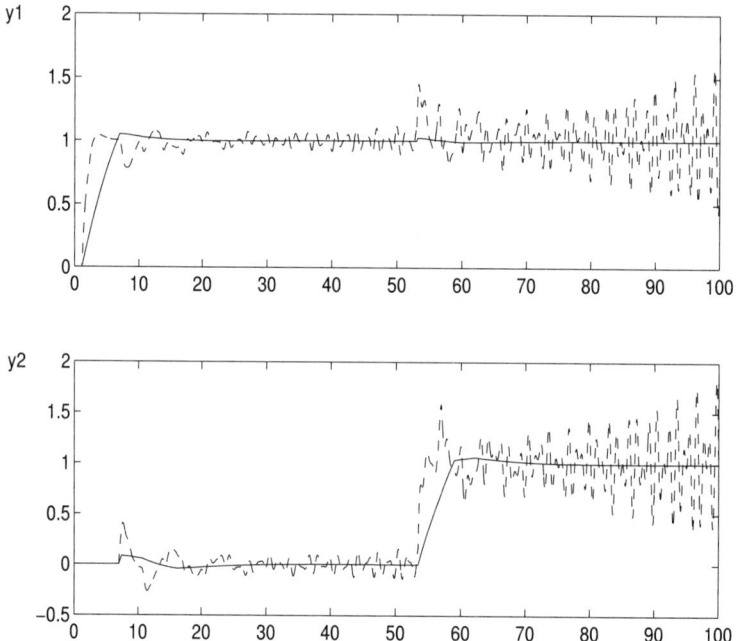

Fig. 10.4. Control system robustness for Example 10.1.1
(——— IMC; - - - Jerome)

In order to demonstrate the robustness of the IMC design, we increase the dead times of the process diagonal elements by 15%. In order to check the stability of the perturbed system, the left-hand side and right-hand side of (10.33) are plotted in Figure 10.3 as a solid line and dashed line respectively. From the figure, we conclude that the IMC system remains stable under the given perturbations. The set-point responses are shown in Figure 10.4, indicating that the performance of the IMC design is far superior to Jerome's multivariable Smith predictor scheme. The resulting ISE is calculated as 7.76 for the IMC method while it is infinite for Jerome's scheme due to instablity.

Example 10.1.2. Consider a variation of the above example:

$$G(s) = \begin{bmatrix} g_{110}(s)e^{-s} & g_{120}(s)e^{-9s} \\ g_{210}(s)e^{-2s} & g_{220}(s)e^{-15s} \end{bmatrix},$$

where the delay free parts $g_{ij0}(s)$ are identical to Example 10.1.1. It follows from the IMC procedure that

$$|\hat{G}| = \frac{0.0374s^2 + 3.72s - 0.5042}{s^2 + 0.1239s + 0.0041} e^{-13.6s},$$

$$\hat{G}^{11} = \frac{-1.3493s - 42.81}{s^2 + 31.85s + 2.207} e^{-15s},$$

$$\hat{G}^{12} = \frac{-0.6072s - 17.66}{s^2 + 29.26s + 2.676} e^{-2s}.$$

One readily sees that $\gamma(|\hat{G}|) = 13.6$, $\gamma(\hat{G}^{11}) = 15$, $\gamma(\hat{G}^{12}) = 2$ and thus $\gamma_1 = \min\{15, 2\} = 2$ from (10.19). A difference from Example 10.1.1 is that $|\hat{G}|$ has a RHP zero at $z = 0.136$ with a multiplicity of one, i.e., $\mathbf{Z}^+_{|\hat{G}|} = \{0.136\}$ and $\eta_z(|\hat{G}|)|_{z=0.136} = 1$. It is easy to find $\eta_z(\hat{G}^{11})|_{z=0.136} = 0$ and $\eta_z(\hat{G}^{12})|_{z=0.136} = 0$ since $\hat{G}^{11}$ and $\hat{G}^{12}$ are both of minimum phase. We thus have $\eta_1(z)|_{z=0.136} = \min\{0, 0\} = 0$ from (10.23) and $h_{11} = \frac{0.136-s}{(0.136+s)(s+1)} e^{-11.6s}$ from (10.30) with the filter of $\frac{1}{s+1}$. For the second loop, with the filter chosen as $\frac{1}{s+1}$, h_{22} is obtained as $h_{22} = \frac{0.136-s}{(0.136+s)(s+1)} e^{-12.6s}$. The rest of design is straightforward and gives

$$C = \begin{bmatrix} \frac{0.3656s^2 + 0.0524s + 0.002}{s^2 + 0.2265s + 0.0124} e^{-13.1s} & \frac{-0.2375s^2 - 0.0294s - 0.0011}{s^2 + 0.1790s + 0.0070} e^{-8s} \\ \frac{0.1551s^2 + 0.0291s + 0.012}{s^2 + 0.2726s + 0.0226} & \frac{-0.2003s^2 - 0.0317s - 0.0014}{s^2 + 0.2215s + 0.0136} \end{bmatrix}.$$

For comparison, Ogunnaike and Ray's multivariable Smith predictor is considered here, where the primary controller given in Ogunnaike and Ray (1979) can still be used since the delay-free part of the process is identical for our case and their case. The set-point responses shown in Figure 10.5 indicate that the performance of the proposed IMC design is significantly better and its response is very close to that of the ideal IMC controller (dotted line). The resulting ISE is calculated as 52.5, 71.16 and 49.4 for the IMC, Ogunnaike and Ray's scheme, and the ideal IMC, respectively.

Example 10.1.3. The process studied by Luyben (1986) has the following transfer function matrix:

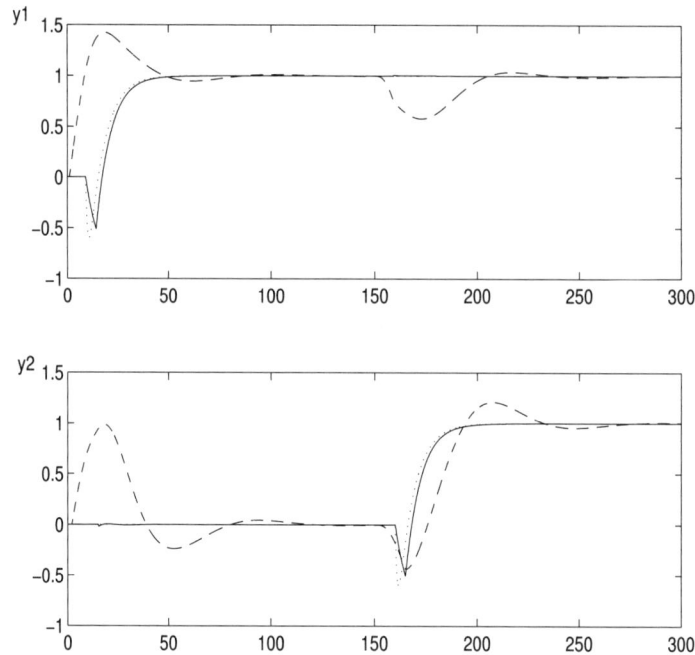

Fig. 10.5. Control performance for Example 10.1.2
(——— IMC; - - - Ogunnaike and Ray; ⋯ ideal IMC)

$$G(s) = \begin{bmatrix} \dfrac{0.126e^{-6s}}{60s+1} & \dfrac{-0.101e^{-12s}}{(45s+1)(48s+1)} \\ \dfrac{0.094e^{-8s}}{38s+1} & \dfrac{-0.12e^{-8s}}{35s+1} \end{bmatrix}.$$

It follows from the design procedure that with filters chosen as $f_1(s) = f_2(s) = \frac{1}{3s+1}$, h_{11} and h_{22} are obtained as $h_{11} = \frac{e^{-6.5s}}{3s+1}$ and $h_{22} = \frac{e^{-8.5s}}{3s+1}$, respectively. The resultant controller is

$$C = \begin{bmatrix} \dfrac{146.8s^2 + 6.355s + 0.0733}{s^2 + 0.3465s + 0.0034} & \dfrac{-6.027s^2 - 0.3416s - 0.0053}{s^2 + 0.0395s + 0.0003}e^{-9.3s} \\ \dfrac{104.5s^2 + 4.719s + 0.0561}{s^2 + 0.34s + 0.0034} & \dfrac{-68.2s^2 - 3.490s - 0.0527}{s^2 + 0.2496s + 0.0024} \end{bmatrix}.$$

Since the multivariable Smith predictor design for this process is not available in the literature, the BLT tuning method (Luyben, 1986) is used for comparison. The set-point responses are shown in Figure 10.6, indicating that significant performance improvement has been achieved with the IMC design and again its response is very close to that of the ideal IMC system. The resulting ISE

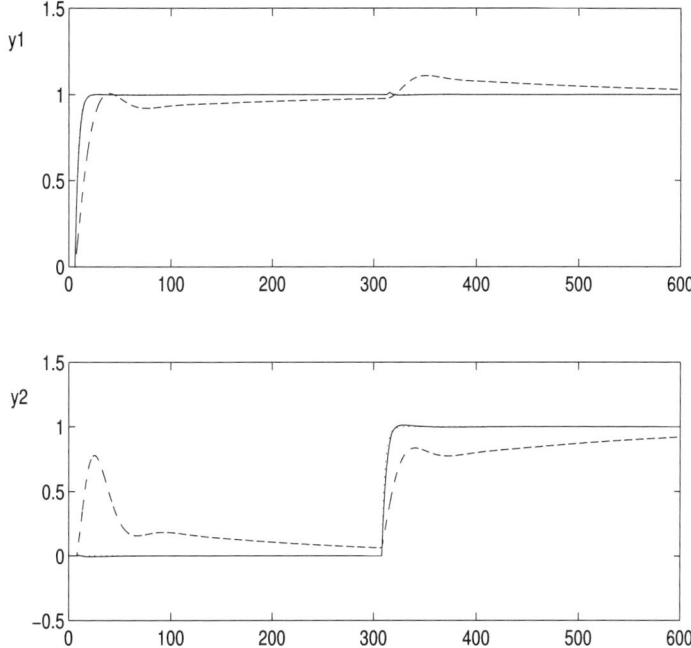

Fig. 10.6. Control performance for Example 10.1.3
(——— IMC; - - - BLT; ··· ideal IMC)

is calculated as 17.7, 51.5 and 17.0 for the IMC, BLT, and ideal IMC system, respectively.

In this section, an approach to the decoupling and stable IMC analysis and design has been presented for multivariable processes with multiple time delays. All the stabilizing controllers which solve this decoupling problem and the resultant closed-loop systems are characterized in terms of their unavoidable time delays and non-minimum-phase zeros. Such delays and zeros can be readily calculated from the process transfer function matrix and clearly quantify performance limitations for any multivariable IMC system which is decoupled and stable. It is interesting to note that a controller may necessarily include some time delays and non-minimum-phase zeros to make itself a solution to the problem. A theoretical control design for the best achievable performance is carried out based on the characterizations. Model reduction is then exploited to simplify both analysis and design. Examples have been given to illustrate our approach and significant performance improvement over existing multivariable Smith predictor control methods is evident.

10.2 Unity Feedback System

In this section, we design a multivariable controller design in the conventional unity output feedback configuration, based on the IMC design of Section 10.1. Experience gained through implementation of this method for various industrial chemical processes reveals that a high-order controller with integrator is more appropriate for complex multivariable processes than the PID controller. This is because a complex process may give rise to very complicated equivalent loops for which PID control is difficult to handle.

10.2.1 Design Methodology

Like the SISO case presented in Chapter 9, we use the IMC controller to derive the single-loop controller, taking into account the performance limitations imposed by the single-loop feedback structure when tuning the filter parameters. Let the closed-loop transfer function between r and y be specified as in (10.30), which is repeated for easy reference as follows:

$$H(s) = \text{diag}\{h_{ii}\},$$

$$h_{ii} = e^{-(\gamma(|G|)-\gamma_i)s} f_i(s) \prod_{z \in \mathbf{Z}^+_{|G|}} \left(\frac{z-s}{z+s}\right)^{\eta_z(|G|)-\eta_i(z)}, \quad \forall\, i \in \mathbf{m}, \quad (10.30)$$

where $f_i(s)$ is the ith loop IMC filter and chosen throughout this section simply as

$$f_i(s) = \frac{1}{\tau_i s + 1}, \quad (10.34)$$

where τ_i is the only tuning parameter for each loop to be selected by the user to achieve the appropriate compromise between performance and robustness and to keep the action of the manipulated variable within bounds. Once $h_{ii}(s)$ are determined, $c_{ii}(s)$ are obtained from (10.31), which in turn yield off-diagonal elements $c_{ji}(s)$ uniquely by (10.12).

The IMC system in Figure 10.1 can be formally redrawn into the equivalent conventional feedback system in Figure 10.7, if the feedback controller K is related to the IMC controller C via

$$K = C[I - GC]^{-1}, \quad (10.35)$$

i.e.,

10.2 Unity Feedback System

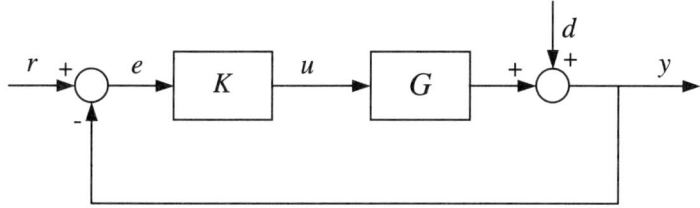

Fig. 10.7. Conventional feedback control

$$k_{ii} = \frac{c_{ii}}{1 - \frac{|G|}{G^{ii}} c_{ii}} = \frac{h_{ii}}{1 - h_{ii}} \frac{G^{ii}}{|G|}, \tag{10.36}$$

$$k_{ji} = \frac{c_{ji}}{1 - \frac{|G|}{G^{ii}} c_{ii}} = \frac{G^{ij}}{G^{ii}} k_{ii}, \quad j \neq i. \tag{10.37}$$

Thus, the open-loop transfer matrix of the conventional feedback system, $Q = \{q_{ij}\} = GK$, must be diagonal:

$$GK = \text{diag}\,\{\frac{|G|}{G^{ii}} k_{ii}\} = \text{diag}\,\left\{\frac{h_{ii}}{1 - h_{ii}}, \quad i = 1, 2, \cdots, m\right\}. \tag{10.38}$$

With h_{ii} specified as in (10.30), the ideal diagonal elements of the controller can be obtained from (10.36). The corresponding decoupling off-diagonal elements $k_{ji}(s)$ of the controller are determined individually by solving (10.37). However, the ideal controller K above, is generally highly complicated and/or unrealizable. We use the same design idea as adopted in the SISO case of Chapter 9 to obtain the best approximation $\hat{K}$ to K to ensure the closed-loop performance. The difference from the SISO case is that both the decoupling performance and the approximation accuracy should be considered in the tuning process. If each loop approximation and/or the overall decoupling performance are not satisfactory, we adjust the IMC controller performance down or increase the controller complexity.

To find a reasonable feedback controller approximation to the corresponding IMC one, it is necessary to obtain a suitable value for τ_i. In SISO applications, performance limitations of single-loop feedback systems due to non-minimum-phase zero and dead time have been considered, and a guideline for tuning IMC filters has been derived, which is also applicable to the present case of decoupled MIMO systems. The gain crossover frequency ω_{ogi} of the ith ideal open-loop transfer function $\frac{h_{ii}}{1-h_{ii}}$, where

$$\left|\frac{h_{ii}(j\omega_{ogi}, \tau_i)}{1 - h_{ii}(j\omega_{og}, \tau_i)}\right| = 1, \tag{10.39}$$

should meet

$$\arg \frac{h_{ii}(j\omega_{ogi}, \tau_i)}{1 - h_{ii}(j\omega_{ogi}, \tau_i)} \geq -180° + \phi_m, \quad (10.40)$$

to have the desired phase margin of ϕ_m. Our studies suggest that $\phi_m = 65°$ will usually be a good choice for most 2 by 2 industrial processes, whereas a larger ϕ_m, such as $\phi_m = 70°$, will be necessary for 3 by 3 or even 4 by 4 processes.

To consider the performance limitation imposed by input constraints, we use a frequency-by-frequency analysis. Assume that at each frequency $|u_i(j\omega)| \leq \bar{u}_i$ and that the model has been appropriately scaled such that the reference signal in each channel satisfies $|r_i(j\omega)| \leq 1$. Now consider the presence of a reference signal in a channel at a time the ideal controlled variable y would be $y = [0 \cdots h_{ii} r_i \cdots 0]^T$, and thus the corresponding manipulated variable becomes

$$u = G^{-1} y = G^{-1} [0 \cdots h_{ii} r_i \cdots 0]^T$$

i.e.,

$$u_j = \frac{G^{ij}}{|G|} h_{ii} r_i.$$

Consider the worst-case references ($|r_i(j\omega)| = 1$); we require

$$\left|\frac{G^{ij}}{|G|} h_{ii}\right| \leq \bar{u}_j, \quad j \in \mathbf{J}_i,$$

or

$$|h_{ii}| \leq \min_{j \in \mathbf{J}_i} \{|\frac{|G|}{G^{ij}}|\bar{u}_j\}.$$

To derive an inequality on τ_i imposed by input constraints, note from (10.30) and (10.34) that $|h_{ii}(j\frac{1}{\tau_i})| = \frac{1}{\sqrt{2}}$ at $\omega = \frac{1}{\tau_i}$ gives

$$\frac{1}{\sqrt{2}} \leq \min_{j \in \mathbf{J}_i} \{|\frac{|G(j\frac{1}{\tau_i})|}{G^{ij}(j\frac{1}{\tau_i})}|\bar{u}_j\}. \quad (10.41)$$

We may simply plot the right hand-side of (10.41) and determine when the inequality holds to find the suitable range of τ_i.

In order to have the fastest response while having a given stability margin and meeting input constraints, the tuning parameter τ_i in the filter (10.34) should be chosen to be the smallest that meets (10.39), (10.40) and (10.41) simultaneously. In the subsequent two subsections, PID and general controllers are considered, respectively.

10.2.2 PID Controller

Consider a multivariable PID controller in the form

$$\hat{K}(s) = \{\hat{k}_{ij}(s)\}, \quad \hat{k}_{ij}(s) = k_{Pij} + k_{Iij}\frac{1}{s} + k_{Dij}s. \tag{10.42}$$

Our task is to find PID parameters so as to match $\hat{k}_{ij}$ to $k_{ij} = c_{ij}[1 - \frac{|G|}{G^{jj}}c_{jj}]^{-1}$ as well as possible. This objective can be realized by minimizing the loss function,

$$\min_{\hat{k}_{ij}} J \triangleq \min_{\hat{k}_{ij}} \sum_{l=1}^{L} |\hat{k}_{ij}(j\omega_l) - k_{ij}(j\omega_l)|^2, \tag{10.43}$$

subject to the same sign for $\hat{k}_{Pij}$, $\hat{k}_{Iij}$, and $\hat{k}_{Dij}$. This problem can be solved by standard non-negative least squares (Lawson and Hanson, 1974) to give the optimal PID parameters. Here, the fitting frequencies ($\omega_1 \sim \omega_L$) are chosen as ($\omega_{bmax}/100, \omega_{bmax}$), where ω_{bmax} is the maximum closed-loop bandwidth among all the loops, i.e., $\omega_{bmax} = \max_i(\omega_{bi}) = \max_i(\frac{1}{\tau_i})$.

In MIMO systems, it might happen that the loop performance is satisfactory while the interactions are too great to be accepted. Let

$$\Omega \triangleq [0, j\omega_L]. \tag{10.44}$$

To ensure both decoupling and loop performance, we require that for $s \in \Omega$

$$ERR_{oi} := \max \left| \frac{\hat{q}_{ii}(s) - q_{ii}(s)}{q_{ii}(s)} \right| \leq \epsilon_{oi}, \quad \forall i \in \mathbf{m}, \tag{10.45}$$

$$ERR_{di} := \max \frac{\sum_{j \neq i} |\hat{q}_{ji}(s)|}{|\hat{q}_{ii}(s)|} \leq \epsilon_{di}, \quad \forall i \in \mathbf{m}, \tag{10.46}$$

where ϵ_{di} and ϵ_{oi} are the specified performance requirements on the loop interactions and loop performance, respectively.

Usually, both ϵ_{oi} and ϵ_{di} may be set as 10%. If both (10.45) and (10.46) hold true, the design is completed. Otherwise, one can always detune PID by relaxing the IMC specification, i.e., increasing τ_i. Our detuning rule is

$$\tau_i^{k+1} = \tau_i^k + \eta_i^k \max\left(\gamma(|G|) - \gamma_i, \min_{z \in \mathbf{Z}_{|G|}^+} (\text{Re}\,(z))\right), \tag{10.47}$$

where k represents the kth iteration, and η_i is an adjustable factor of the ith-loop reflecting both the approximation accuracy and the decoupling performance of the present iteration and is set as

$$\eta_i = \max\{\eta_{oi}, \eta_{di}\},$$

where

$$\eta_{oi} = \begin{cases} 0 & \text{if } ERR_{oi} \leq 10\%, \\ \frac{1}{4} & \text{if } 10\% < ERR_{oi} \leq 50\%, \\ \frac{1}{2} & \text{if } 50\% < ERR_{oi} \leq 100\%, \\ 1 & \text{if } 100\% < ERR_{oi}, \end{cases} \quad \text{and} \quad \eta_{di} = \begin{cases} 0 & \text{if } ERR_{di} \leq 10\%, \\ \frac{1}{2} & \text{if } 10\% < ERR_{di}. \end{cases}$$

The iteration continues until both (10.45) and (10.46) are satisfied.

We now present some simulation examples to demonstrate our PID tuning algorithm and compare it with the multivariable PID controller design in Dong and Brosilow (1997), who used the following PID form:

$$\tilde{K} = \tilde{K}_P + \tilde{K}_I \frac{1}{s} + \frac{\tilde{K}_D s}{\alpha s + 1} \tag{10.48}$$

with $\alpha = \max\{(K_P^{-1} K_D(i,j))/20\}$ if K_p^{-1} exists. The ideal PID controller in (10.42) is not physically realizable and thus is replaced by

$$\hat{k}_{ij}(s) = k_{Pij} + k_{Iij}\frac{1}{s} + \frac{k_{Dij}s}{\frac{|k_{Dij}|}{20}s + 1}.$$

The simulation is done under the perfect model matching condition, i.e., $\hat{G} = G$ (model mismatch will be considered in Section 10.2.4).

Example 10.2.1. Consider the plant in Luyben (1986):

$$G(s) = \begin{bmatrix} \frac{-2.2e^{-s}}{7s+1} & \frac{1.3e^{-0.3s}}{7s+1} \\ \frac{-2.8e^{-1.8s}}{9.5s+1} & \frac{4.3e^{-0.35s}}{9.2s+1} \end{bmatrix}.$$

Following the procedure in Section 10.1, we have $h_{11} = \frac{e^{-1.05s}}{\tau_1 s+1}$ and $h_{22} = \frac{e^{-s}}{\tau_2 s+1}$. From (10.39), (10.40) and (10.41), the initial τ_i are obtained as $\tau_1^0 = 1.0143$ and $\tau_2^0 = 0.6821$. The proposed method results in the overall controller:

$$\hat{K} = \begin{bmatrix} -1.833 - \frac{0.434}{s} - 1.042s & 0.663 + \frac{0.169}{s} + 0.610s \\ -0.646 - \frac{0.306}{s} & 1.398 + \frac{0.265}{s} \end{bmatrix},$$

with $ERR_{o1} = 31.76\%$, $ERR_{o2} = 93.23\%$, $ERR_{d1} = 42.14\%$ and $ERR_{d2} = 43.31\%$. Dong's method generates the PID controller parameters as

$$\tilde{K}_P = \begin{bmatrix} -2.418 & 1.873 \\ -1.312 & 3.151 \end{bmatrix}, \quad \tilde{K}_I = \begin{bmatrix} -0.367 & 0.216 \\ -0.239 & 0.366 \end{bmatrix}, \quad \tilde{K}_D = \begin{bmatrix} 0.188 & -0.558 \\ 1.040 & -1.102 \end{bmatrix}.$$

Since the derivative terms obtained using Dong's method have different signs from other terms in a single PID, we use PI instead of PID controller for this case, as suggested by Dong and Brosilow (1997). The step responses in Figure 10.8(a) show that both the proposed PID controller and Dong's PI controller cannot yield good loop performance. Based on the proposed tuning procedure, τ_i are increased until the approximation error and the decouping index are small enough to satisfy (10.45) and (10.46), respectively. We end with $\tau_1 = 6.0143$, $\tau_2 = 5.9321$ and the overall controller

$$\hat{K} = \begin{bmatrix} -0.669 - \frac{0.106}{s} - 0.507s & 0.270 + \frac{0.033}{s} + 0.428s \\ -0.317 - \frac{0.069}{s} & 0.426 + \frac{0.055}{s} \end{bmatrix},$$

with $ERR_{o1} = 9.05\%$, $ERR_{o2} = 6.38\%$, $ERR_{d1} = 7.70\%$ and $ERR_{d2} = 5.58\%$. Dong's method with the same new τ_i yields

$$\tilde{K}_P = \begin{bmatrix} -0.676 & 0.306 \\ -0.365 & 0.515 \end{bmatrix}, \tilde{K}_I = \begin{bmatrix} -0.105 & 0.036 \\ -0.069 & 0.060 \end{bmatrix}, \tilde{K}_D = \begin{bmatrix} 0.172 & -0.107 \\ 0.362 & -0.206 \end{bmatrix},$$

although Dong and Brosilow (1997) did not suggest any detuning rule there. The closed-loop step responses are shown in Figure 10.8(b). One observes that the performances of both methods are almost identical to the IMC system (not shown for simplicity).

Example 10.2.2. Consider the 3×3 plant presented by Vasnani (1994):

$$G(s) = \begin{bmatrix} \frac{119e^{-5s}}{21.7s+1} & \frac{40e^{-5s}}{337s+1} & \frac{-2.1e^{-5s}}{10s+1} \\ \frac{77e^{-5s}}{50s+1} & \frac{76.7e^{-3s}}{28s+1} & \frac{-5e^{-5s}}{10s+1} \\ \frac{93e^{-5s}}{50s+1} & \frac{-36.7e^{-5s}}{166s+1} & \frac{-103.3e^{-4s}}{23s+1} \end{bmatrix}.$$

It follows that $h_{11} = \frac{e^{-7.3s}}{\tau_1 s+1}$, $h_{22} = \frac{e^{-3.8s}}{\tau_2 s+1}$, and $h_{33} = \frac{e^{-7.9s}}{\tau_3 s+1}$. The guideline for tuning IMC filters gives $\tau_1 = 6.85$, $\tau_2 = 4.90$, $\tau_3 = 6.60$. This results in the proposed multivariable PID controller

$$\hat{K} = 10^{-3} \times \begin{bmatrix} 21.8 + \frac{0.9}{s} + 26.8s & -1.4 - \frac{0.1}{s} - 0.9s & 0.0002 + \frac{0.04}{s} \\ -9.8 - \frac{0.9}{s} & 59.3 + \frac{2.1}{s} + 40.4s & -4.1 - \frac{0.1}{s} \\ 9.3 + \frac{0.7}{s} + 1.5s & -3.2 - \frac{0.3}{s} & -27.8 - \frac{1.1}{s} - 24.7s \end{bmatrix},$$

with $ERR_{o1} = 4.33\%$, $ERR_{o2} = 2.64\%$, $ERR_{o3} = 1.95\%$, $ERR_{d1} = 4.58\%$, $ERR_{d2} = 3.41\%$ and $ERR_{d3} = 5.95\%$. Dong's method yields

302   10. Multivariable Systems

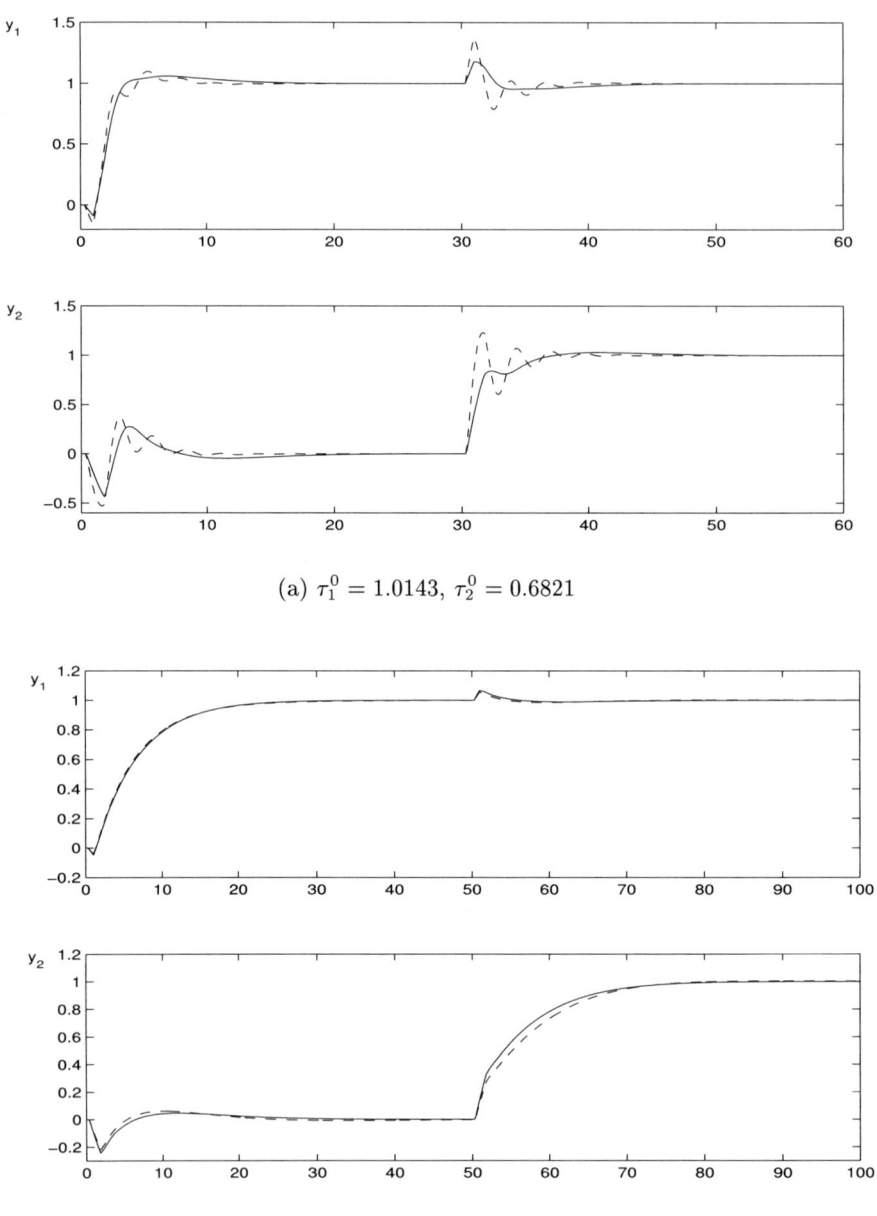

(a) $\tau_1^0 = 1.0143$, $\tau_2^0 = 0.6821$

(b) $\tau_1 = 6.0143$, $\tau_2 = 5.9321$

Fig. 10.8. Step response for Example 10.2.1
(—— proposed PID, – – – Dong's PID)

$$\tilde{K}_P = \begin{bmatrix} -0.144 & 0.365 & -0.010 \\ 0.153 & -0.301 & 0.007 \\ -0.260 & 0.578 & -0.036 \end{bmatrix}, \quad \tilde{K}_I = 10^{-3} \times \begin{bmatrix} 1.1 & -0.8 & 0.01 \\ -1.0 & 2.4 & -0.1 \\ 1.3 & -1.6 & -0.9 \end{bmatrix},$$

and

$$\tilde{K}_D = \begin{bmatrix} 79.312 & -180.405 & 4.583 \\ -76.487 & 173.189 & -4.409 \\ 119.610 & -269.632 & 6.804 \end{bmatrix},$$

which leads to an unstable closed loop. When we eliminate one of the three parameters that has a different sign from the other two elements in $\tilde{k}_{ij}$, a stable step response can be obtained as shown in Figure 10.9. It can be seen that the proposed method shows much better performance than Dong's design.

Generally, an equivalent process after decoupling a significantly interactive multivariable process becomes highly complicated. Though one can always detune the PID controller to a sufficient extent to generate a stable closed loop, a sluggish response must result. This is because a PID controller is too simple to re-shape complicated dynamics satisfactorily over a large frequency range for high performance. In such a case, a more complex controller than PID is necessary.

10.2.3 High-order Controller

In this subsection, we consider controller elements $\hat{k}_{ij}$ in the form of an nth-order rational transfer function plus a time delay L_{ij}:

$$\hat{k}_{ij} = \frac{b_{nij}s^n + b_{n-1ij}s^{n-1} + \ldots + b_{1ij}s + b_{0ij}}{s^n + a_{n-1ij}s^{n-1} + \ldots + a_{1ij}s} e^{-L_{ij}s}, \quad (10.49)$$

which needs to approximate the ideal k_{ij} as well as possible. The recursive least squares (RLS) algorithm for transfer function modelling from the frequency response presented in Chapter 7 can be applied to obtain the solution. Like the LS algorithm, the frequency range for RLS fitting is also chosen as $(\omega_{bmax}/100, \omega_{bmax})$ with steps of $(\frac{1}{100} \sim \frac{1}{10})\omega_{bmax}$. In this range, the RLS yields satisfactory fitting results in the frequency domain. The orders of the rational parts of the controller elements are determined to be the lowest such that for $s \in \Omega$, a set of performance specifications on loop performance (10.45)

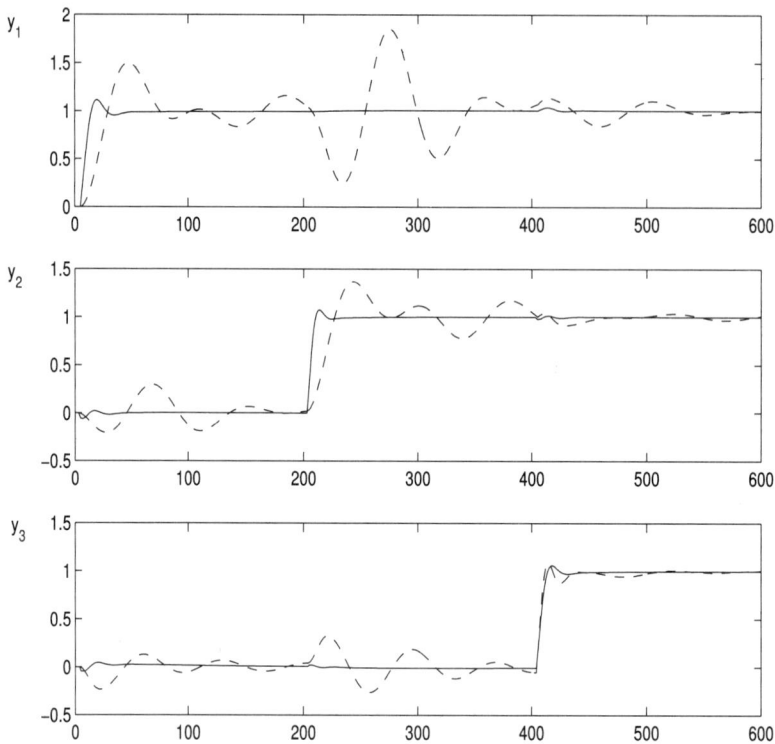

Fig. 10.9. Step response for Example 10.2.2
(——— proposed PID, – – – Dong's PID)

and loop interaction (10.46) are satisfied. The controller design procedure can be summarized as follows.

IMC-based MIMO design procedure

Seek a feedback controller $K(s)$ given $G(s)$:

Step 1. Specify h_{ii} in (10.30) with the method in Section 10.1 and f_i chosen as in (10.34). Determine the smallest τ_i^* for each loop according to (10.39), (10.40) and (10.41), and set $\tau_i^0 = \tau_i^*$.

Step 2. Find the PID controller using the LS method and evaluate ERR_{oi} in (10.45) and ERR_{di} in (10.46). If ERR_{oi} and ERR_{di} meet the specified accuracy ϵ_{oi} and ϵ_{di} (usually 10%) respectively, end the design.

Step 3. Otherwise, we have two options: if a PID controller is desired, update τ_i by (10.47), and go to Step 2; else, go to Step 4.

Step 4. *Adopt the high-order controller in (10.49); for each $\hat{k}_{ij}$, start from a controller order of 2, apply the RLS to obtain $\hat{k}_{ij}$, increase n until the smallest integer n is reached with (10.45) and (10.46) satisfied.*

In this subsection, apart from 2 by 2 processes, examples of 3 by 3 and 4 by 4 processes are also given to show the effectiveness of the proposed design.

Example 10.2.1 (cont'd). Reconsider G in Example 10.2.1. With $\tau_1^0 = 1.0143$ and $\tau_2^0 = 0.6821$, a multivariable PID controller has been obtained there with with $ERR_{o1} = 31.76\%$, $ERR_{o2} = 93.23\%$, $ERR_{d1} = 42.14\%$ and $ERR_{d2} = 43.31\%$. For a high-order controller, our procedure gives

$$K = \begin{bmatrix} \dfrac{-2.054s^2 - 1.400s - 0.148}{s^2 + 0.422s} & \dfrac{1.005s^2 + 0.159s - 0.001}{s^2 + 0.024s} \\ \dfrac{-1.484s - 0.241}{s}e^{-1.5s} & \dfrac{1.226s^2 + 2.468s + 0.260}{s^2 + 1.187s}e^{-0.5s} \end{bmatrix},$$

with $ERR_{o1} = 7.71\%$, $ERR_{o2} = 1.61\%$, $ERR_{d1} = 2.88\%$, and $ERR_{d2} = 3.25\%$. The step response is shown in Figure 10.10. Compared with the PID performance, we can see that the high-order controller gives much improved loop and decoupling performance.

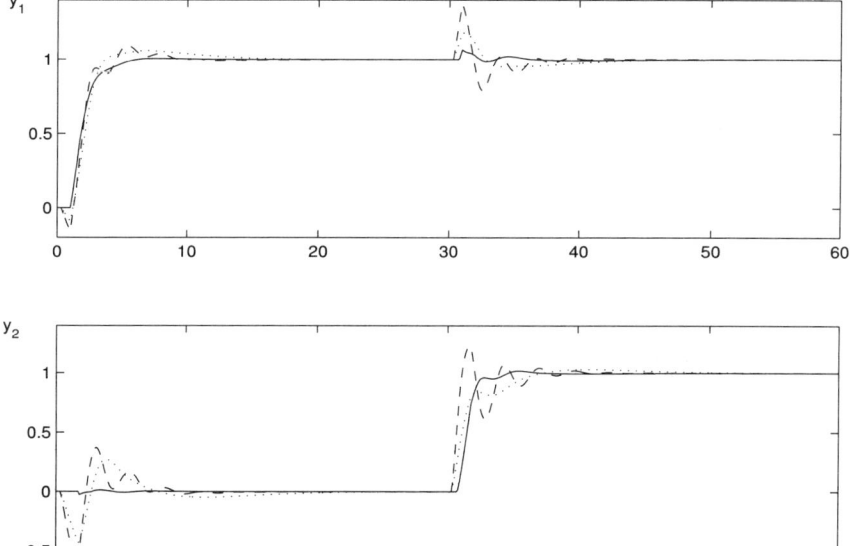

Fig. 10.10. Step response for Example 10.2.1(continued)
(—— proposed high-order, ······ proposed PID, – – – Dong's PID)

Example 10.2.3. The process studied by Luyben (1986) has the following transfer function matrix:

$$G(s) = \begin{bmatrix} \frac{0.126e^{-6s}}{60s+1} & \frac{-0.101e^{-12s}}{(45s+1)(48s+1)} \\ \frac{0.094e^{-8s}}{38s+1} & \frac{-0.12e^{-8s}}{35s+1} \end{bmatrix}.$$

We have $h_{11} = \frac{e^{-6.5s}}{\tau_1 s+1}$ and $h_{22} = \frac{e^{-8.5s}}{\tau_2 s+1}$. The initial parameter values are $\tau_1^0 = 3.84$ and $\tau_2^0 = 5.37$. The proposed IMC-based multivariable PID controller design method results in the overall controller:

$$\hat{K} = \begin{bmatrix} 47.786 + \frac{1.572}{s} + 98.940s & 1.969 + 8.186s \\ 34.617 + \frac{1.212}{s} + 71.481s & -22.903 - \frac{1.0027}{s} - 59.516s \end{bmatrix}, \quad (10.50)$$

with $ERR_{o1} = 13.56\%$, $ERR_{o2} = 25.37\%$, $ERR_{d1} = 0.91\%$ and $ERR_{d2} = 96.83\%$. Dong's method generates

$$\tilde{K}_P = \begin{bmatrix} -19.610 & 94.639 \\ -23.851 & 56.086 \end{bmatrix}, \quad \tilde{K}_I = \begin{bmatrix} 2.168 & -1.343 \\ 1.698 & -1.675 \end{bmatrix},$$

and

$$\tilde{K}_D = 10^3 * \begin{bmatrix} 6.587 & -9.443 \\ 5.444 & -7.948 \end{bmatrix}.$$

Our PID cannot achieve the decoupling performance specified, as shown in Figure 10.11, while Dong's PID controller results in an unstable closed loop in this case. There is little benefit to be gained by increasing τ_1 and τ_2. This is because the Nyquist curve of a PID controller is a vertical line in the complex plane and usually not sufficient to compensate for significant interactions. Therefore, we have to adopt a high-order controller to meet the requirement. With the same τ_i, the resultant controller is

$$\hat{K} = \begin{bmatrix} \frac{87.036s^2+16.852s+0.616}{s^2+0.386s} & \frac{-1.497s^2-0.474s-0.014}{s^2+0.010s}e^{-6.4s} \\ \frac{61.434s^2+11.424s+0.446}{s^2+0.362s} & \frac{-131.604s^3-40.773s^2-1.887s-0.015}{s^3+1.821s^2+0.004s} \end{bmatrix},$$

with $ERR_{o1} = 1.36\%$, $ERR_{o2} = 2.02\%$, $ERR_{d1} = 2.08\%$ and $ERR_{d2} = 0.86\%$. The step response is shown as the solid line in Figure 10.11. We can see that the new controller $\hat{K}$ yields satisfactory performance, while the PID controller in (10.50) is not capable of that.

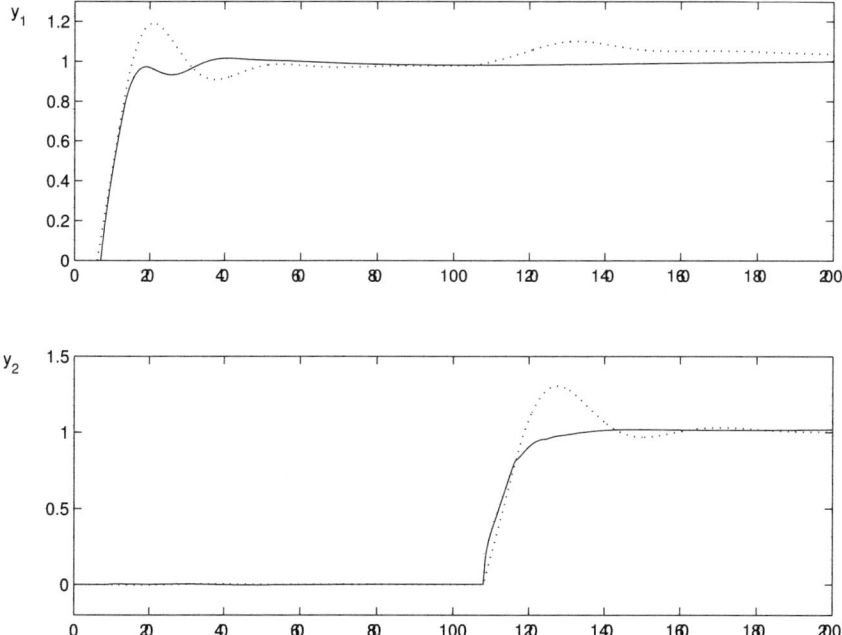

Fig. 10.11. Step response for Example 10.2.3
(——— proposed high-order, ······ proposed PID)

Example 10.2.4. Consider the Tyreus distillation column (Tyreus, 1982):

$$G(s) = \begin{bmatrix} \dfrac{1.986e^{-0.71s}}{66.7s+1} & \dfrac{-5.24e^{-60s}}{400s+1} & \dfrac{-5.984e^{-2.24s}}{14.29s+1} \\ \dfrac{-0.0204e^{-0.59s}}{(7.14s+1)^2} & \dfrac{0.33e^{-0.68s}}{(2.38s+1)^2} & \dfrac{-0.0159e^{-21s}}{10s+1} \\ \dfrac{-0.374e^{-7.75s}}{22.22s+1} & \dfrac{11.3e^{-3.79s}}{(21.74s+1)^2} & \dfrac{9.811e^{-1.59s}}{11.36s+1} \end{bmatrix}.$$

It follows that $h_{11} = \dfrac{(-s+3.0274)}{(s+3.0274)(\tau_1 s+1)} e^{-13.4s}$, $h_{22} = \dfrac{(-s+3.0274)}{(s+3.0274)(\tau_2 s+1)} e^{-8.7s}$, and $h_{33} = \dfrac{(-s+3.0274)}{(s+3.0274)(\tau_3 s+1)} e^{-13.6s}$. The initial τ_i are calculated as $\tau_1^0 = 17.50$, $\tau_2^0 = 11.90$, and $\tau_3^0 = 17.70$. The resultant elements of the high-order controller by the proposed method are

$$\hat{k}_{11} = \frac{1.4990s^4 + 0.1829s^3 + 0.0157s^2 + 0.0002s + 4 \times 10^{-7}}{s^4 + 0.1226s^3 + 0.0129s^2 + 3 \times 10^{-5}s} e^{-13.4s}$$

$$\hat{k}_{21} = \frac{-0.0541s^4 + 0.0112s^3 + 0.0012s^2 + 2 \times 10^{-5}s + 4 \times 10^{-8}}{s^4 + 0.1261s^3 + 0.0237s^2 + 6 \times 10^{-5}s} e^{-6.1s}$$

$$\hat{k}_{31} = \frac{0.0241s^4 + 0.0019s^4 + 0.0002s^3 - 4 \times 10^{-6}s^2 - 1 \times 10^{-7}s - 2 \times 10^{-10}}{s^4 + 0.2674s^3 + 0.0277s^2 + 0.0020s + 4 \times 10^{-6}s} e^{-19.6s}$$

$$\hat{k}_{12} = \frac{0.9397s^5 - 0.2847s^4 - 0.0058s^3 - 0.0061s^2 - 0.0001s - 6 \times 10^{-8}}{s^5 + 0.1793s^4 + 0.0252s^3 + 0.0020s^2 + 5 \times 10^{-6}s} e^{-9.3s}$$

$$\hat{k}_{22} = \frac{1.2078s^2 + 0.1106s + 0.0037}{s^2 + 0.2695s} e^{-8.7s}$$

$$\hat{k}_{32} = \frac{0.0468s^4 - 0.0159s^3 - 0.0007s^2 - 4 \times 10^{-5}s - 7 \times 10^{-8}}{s^4 + 0.0462s^3 + 0.0022s^2 + 4 \times 10^{-6}s} e^{-6.6s}$$

$$\hat{k}_{13} = \frac{0.2547s^3 + 0.0276s^2 + 0.0009s + 3 \times 10^{-5}}{s^3 + 0.0932s^2 + 0.0121s} e^{-12.1s}$$

$$\hat{k}_{23} = \frac{0.3978s^2 + 0.0248s + 0.0007}{s^2 + 0.2283s} e^{-9.6s}$$

$$\hat{k}_{33} = \frac{0.0876s^2 + 0.0064s + 0.0002}{s^2 + 0.4495s} e^{-13.6s}$$

with $ERR_{o1} = 0.31\%$, $ERR_{o2} = 3.41\%$, $ERR_{o3} = 5.78\%$, $ERR_{d1} = 1.25\%$, $ERR_{d2} = 2.96\%$ and $ERR_{d3} = 4.42\%$. The step response is shown in Figure 10.12, and has excellent performance. It should be pointed out that no stabilizing PID controller could be obtained until the closed-loop is so sluggish that it is virtually useless and cannot be shown in the figure.

Example 10.2.5. Consider the 4 by 4 distillation column of Doukas and Luyben (1978):

$$G(s) = \begin{bmatrix} \frac{-11.3e^{-3.79s}}{(21.74s+1)^2} & \frac{0.374e^{-7.75s}}{(22.2s+1)^2} & \frac{-9.811e^{-1.59s}}{11.36s+1} & \frac{-2.37e^{-27.33s}}{33.3s+1} \\ \frac{5.24e^{-60}}{(400s+1)} & \frac{-1.986e^{-0.71s}}{(66.67s+1)^2} & \frac{5.984e^{-2.24s}}{14.29s+1} & \frac{0.422e^{-8.72s}}{(250s+1)^2} \\ \frac{-0.33e^{-0.68s}}{(2.38s+1)^2} & \frac{0.0204e^{-0.59s}}{(7.14s+1)^2} & \frac{2.38e^{-0.42s}}{(1.43s+1)^2} & 0.513e^{-s} \\ \frac{4.48e^{-0.52s}}{(11.11s+1)} & \frac{-0.176e^{-0.48s}}{(6.90s+1)^2} & \frac{-11.67e^{-1.91s}}{(12.19s+1)} & 15.54e^{-s} \end{bmatrix}.$$

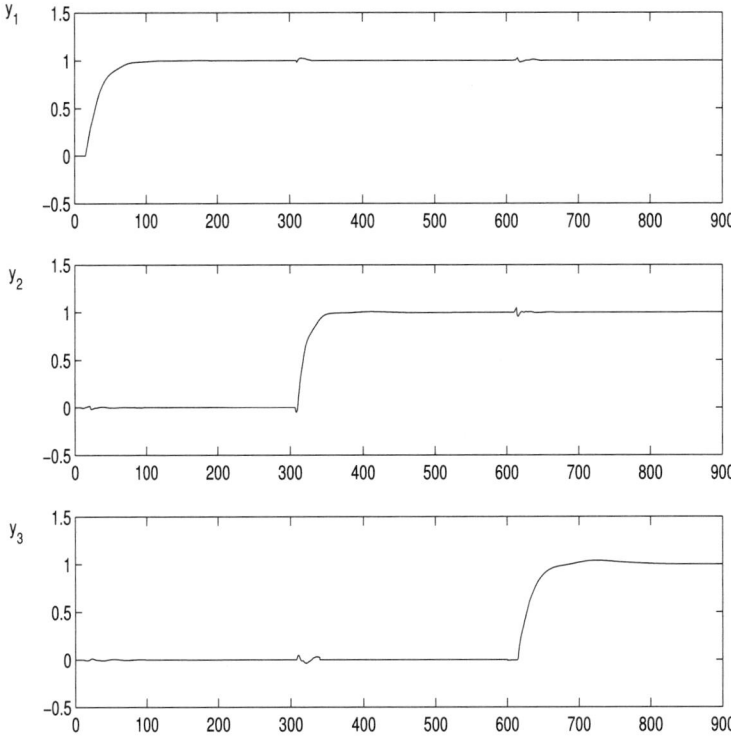

Fig. 10.12. Step response for Example 10.2.4

We have $h_{11} = \frac{1}{(\tau_1 s+1)}e^{-17.78s}$, $h_{22} = \frac{1}{(\tau_2 s+1)}e^{-17.45s}$, $h_{33} = \frac{1}{(\tau_3 s+1)}e^{-17.48s}$, and $h_{44} = \frac{1}{(\tau_4 s+1)}e^{-8.17s}$. The initial τ_i are $\tau_1^0 = 23.90$, $\tau_2^0 = 23.30$, $\tau_3^0 = 23.30$ and $\tau_4^0 = 10.80$. The resultant elements of the high-order controller using the proposed method are

$$\hat{k}_{11} = \frac{0.624s^5 - 0.0528s^4 - 0.0193s^3 - 0.0024s^2 - 0.0001s - 6 \times 10^{-7}}{s^5 + 1.073s^4 + 0.133s^3 + 0.0254s^2 + 0.0003s} e^{-5.9s}$$

$$\hat{k}_{21} = \frac{1.001s^7 + 1.004s^6 + 0.883s^5 - 0.221s^4 - 0.007s^3 - 0.0004s^2 - 7 \times 10^{-6}s - 5 \times 10^{-8}}{s^7 + 1.011s^6 + 0.909s^5 + 0.292s^4 + 0.0292s^3 + 0.003s^2 + 8 \times 10^{-6}s}$$

$$\hat{k}_{31} = \frac{0.571s^5 + 0.0079s^4 - 0.0122s^3 - 0.0014s^2 - 3 \times 10^{-5}s - 4 \times 10^{-8}}{s^5 + 0.974s^4 + 1.069s^3 + 0.155s^2 + 0.0002s} e^{-9.9s}$$

$$\hat{k}_{41} = \frac{-0.0068s^5 + 0.0003s^4 + 0.0001s^3 + 5 \times 10^{-5}s^2 + 6 \times 10^{-6}s + 2 \times 10^{-7}}{s^5 + 0.177s^4 + 0.0467s^3 + 0.0041s^2 + 0.0003s} e^{-5s}$$

$$\hat{k}_{12} = \frac{0.0842s^5 - 0.0398s^4 + 0.0237s^3 - 0.0057s^2 - 0.0002s - 2 \times 10^{-6}}{s^5 + 0.636s^4 + 0.238s^3 + 0.0406s^2 + 0.0035s} e^{-1.1s}$$

$$\hat{k}_{22} = \frac{-11.538s^3 - 0.427s^2 - 0.004s - 1 \times 10^{-5}}{s^3 + 0.231s^2 + 0.0007s} e^{-8.7s}$$

$$\hat{k}_{32} = \frac{0.0594s^5 + 0.0587s^4 - 0.0058s^3 + 0.0022s^2 + 0.0001s + 6 \times 10^{-7}}{s^5 + 1.035s^4 + 1.162s^3 + 0.214s^2 + 0.0124s} e^{-15.3s}$$

$$\hat{k}_{42} = \frac{0.0067s^4 - 0.0019s^3 + 2 \times 10^{-5}s^2 + 3 \times 10^{-6}s + 3 \times 10^{-8}}{s^4 + 0.343s^3 + 0.0282s^2 + 0.0009s} e^{-4.5s}$$

$$\hat{k}_{13} = \frac{0.982s^6 + 0.607s^5 - 0.184s^4 - 0.0132s^3 - 0.0017s^2 - 0.0001s - 2 \times 10^{-7}}{s^6 + 0.980s^5 + 0.422s^4 + 0.0993s^3 + 0.0075s^2 + 2 \times 10^{-5}s} e^{-9.9s}$$

$$\hat{k}_{23} = \frac{9.820s^3 + 0.328s^2 + 0.0018s - 8 \times 10^{-7}}{s^3 + 0.0694s^2 + 2 \times 10^{-5}s} e^{-15.8s}$$

$$\hat{k}_{33} = \frac{0.120s^2 + 0.0201s + 0.0018}{s^2 + 0.166s} e^{-21.2s}$$

$$\hat{k}_{43} = \frac{-0.0365s^3 + 0.0057s^2 + 0.0011s + 0.0001}{s^3 + 0.169s^2 + 0.0067s} e^{-8.9s}$$

$$\hat{k}_{14} = \frac{-0.0439s^4 - 0.0031s^3 - 0.0003s^2 - 4 \times 10^{-5}s - 1 \times 10^{-7}}{s^4 + 0.204s^3 + 0.0256s^2 + 5 \times 10^{-5}s} e^{-5.6s}$$

$$\hat{k}_{24} = \frac{0.529s^3 + 0.0109s^2 + 0.0001s - 2 \times 10^{-7}}{s^3 + 0.0336s^2 + 0.0001s} e^{-6s}$$

$$\hat{k}_{34} = \frac{-0.0023s + 0.0007}{s} e^{-5s}$$

$$\hat{k}_{44} = \frac{-0.0087s^3 + 0.0027s^2 + 0.0003s + 1 \times 10^{-6}}{s^3 + 0.0692s^2 + 0.0003s} e^{-2.1s}$$

with $ERR_{d1} = 0.33\%$, $ERR_{o2} = 9.52\%$, $ERR_{o3} = 3.05\%$, $ERR_{o4} = 0.08\%$, $ERR_{d1} = 7.08\%$, $ERR_{d2} = 4.84\%$, $ERR_{d3} = 8.52\%$ and $ERR_{d4} = 6.53\%$. The step response is shown in Figure 10.13, and the effectiveness of the controller is evident. Like Example 10.2.4, no reasonable PID controller could be found for this example.

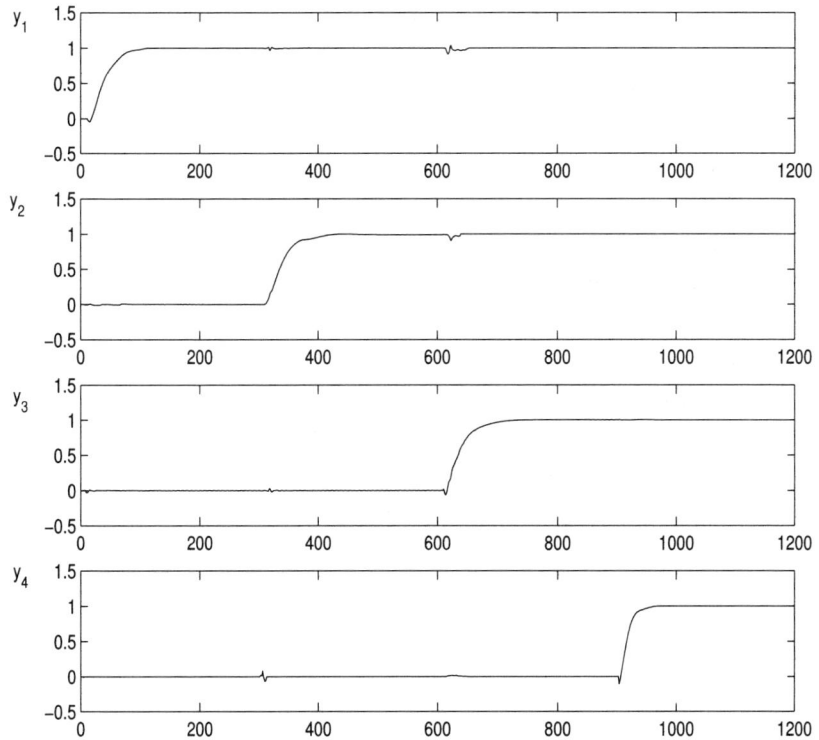

Fig. 10.13. Step response for Example 10.2.5

10.2.4 Stability Analysis

In this subsection, we consider both nominal stability $(G = \hat{G})$ and robust stability $(G \neq \hat{G})$. First, assume $G = \hat{G}$ in the absence of model uncertainty. As the process $G(s)$ is assumed to be stable and the controller $K(s)$ is designed to contain no poles in the right half of the complex plane except the origin, according to the generalized Nyquist theorem (with the Nyquist contour indented to the right around the origin) (Maciejowski, 1989), the closed-loop system is stable if and only if the characteristic loci of $G(s)\hat{K}(s)$, or $\hat{Q}(s)$, taken together, do not encircle the point -1.

Thus, in principle, with the generalized Nyquist theorem, the nominal stability of the system can be determined by calculating the characteristic loci of $\hat{Q}$ and counting their encirclements of the point -1. However, it is not easy to calculate the characteristic loci, especially for processes with many inputs/outputs. Notice that we are adopting a decoupling design, and as a result, the compensated open-loop transfer function $\hat{Q}(s)$ is likely to have a high degree of column dominance, and its Gershgorin bands will be generally narrow. By Gershgorin's theorem (Maciejowski, 1989), we know that the union of the Gershgorin bands "trap" the union of the characteristic loci, and we can assess closed-loop stability by counting the encirclements of -1 by the Gershorin bands, if they exclude -1, since this tells us the number of encirclements made by the characteristic loci. Thus, the closed-loop system is stable if all the Gershgorin bands exclude and make no encirclements of the point -1.

Note that the elements of $G(s)$ are strictly proper and the elements of $\hat{K}(s)$ are designed to be proper. Then, $\hat{q}_{ji}$, the elements of $\hat{Q}(s) = G(s)\hat{K}$, are also strictly proper, i.e., $\lim_{s\to\infty} \hat{q}_{ji}(s) = 0$. Thus, $\sum_{j=1}^{m} |\hat{q}_{ji}(s)| < \epsilon$ (where $0 < \epsilon < 1$) is likely to hold for $s \in \bar{\Omega}$, where $\bar{\Omega}$ denotes the compliment of Ω in the Nyquist contour. We have the following theorem.

Theorem 10.2.1. *(Nominal Stability)* Assume $\sum_{j=1}^{m} |\hat{q}_{ji}(s)| < \epsilon$, for $s \in \bar{\Omega}$ and $i \in \mathbf{m}$, then the multivariable closed-loop system in Figure 10.7 is nominally stable if for $s \in \Omega$ and each $i \in \mathbf{m}$, (10.45) and (10.46) are satisfied, and

$$|h_{ii}|(\epsilon_{oi}+\epsilon_{di}+\epsilon_{oi}\epsilon_{di}) < 1, \quad s \in \Omega, \tag{10.51}$$

or

$$\epsilon_{oi}+\epsilon_{di}+\epsilon_{oi}\epsilon_{di} < 1. \tag{10.52}$$

Proof. By definition, the ith Gershgorin band of $\hat{Q}(s)$ is

$$\mathbf{G}_{er}^{i} := \cup_{s\in D} G_{er}^{i}(s), \tag{10.53}$$

where

$$G_{er}^{i}(s) = \{z \in \mathcal{C} \mid |z-\hat{q}_{ii}| \leq \sum_{j\neq i} |\hat{q}_{ji}(s)|\}. \tag{10.54}$$

Since (10.45) and (10.46) are satisfied for $s \in \Omega$, we have for $s \in \Omega$

$$|\hat{q}_{ii}| = |\hat{q}_{ii}-q_{ii}+q_{ii}| \leq |\hat{q}_{ii}-q_{ii}|+|q_{ii}| \leq (1+\epsilon_{oi})|q_{ii}|,$$

and

$$|\hat{q}_{ii}-q_{ii}|+\sum_{j\neq i}|\hat{q}_{ji}| \leq \epsilon_{oi}|q_{ii}|+\epsilon_{di}|\hat{q}_{ii}| \leq (\epsilon_{oi}+\epsilon_{di}+\epsilon_{oi}\epsilon_{di})|q_{ii}|. \qquad (10.55)$$

Thus for $z \in G_{er}^i$, where $s \in \Omega$, we have

$$|z-q_{ii}| \leq |\hat{q}_{ii}-q_{ii}|+|z-\hat{q}_{ii}| \leq |\hat{q}_{ii}-q_{ii}|+\sum_{j\neq i}|\hat{q}_{ji}| \leq (\epsilon_{oi}+\epsilon_{di}+\epsilon_{oi}\epsilon_{di})|q_{ii}|.$$

If the condition in the theorem holds, then

$$|1+q_{ii}|-|z-q_{ii}| \geq |1+q_{ii}|-(\epsilon_{oi}+\epsilon_{di}+\epsilon_{oi}\epsilon_{di})|q_{ii}| > 0,$$

which means that the distance from q_{ii} to the point -1 is greater than that from q_{ii} to any point in $\mathbf{G}_{er}^i$ so that the Nyquist curve of q_{ii} and the ith Gershgorin band $\mathbf{G}_{er}^i$ have the same encirclements with respect to the point -1.

Since $\sum_{j=1}^m |\hat{q}_{ji}(s)| < \epsilon$, for $s \in \bar{\Omega}$ and $i \in \mathbf{m}$, then for $z \in G_{er}^i(s)$, where $s \in \bar{\Omega}$, we have

$$|z|-|\hat{q}_{ii}| \leq |z-\hat{q}_{ii}| \leq \sum_{j\neq i}|\hat{q}_{ji}|,$$

and

$$|z| \leq |\hat{q}_{ii}|+\sum_{j\neq i}|\hat{q}_{ji}| < \epsilon < 1,$$

which implies that no encirclement can be made by the Gershgorin band. The same can be said for q_{ii}, due to the assumed $|q_{ii}(s)| < 1$ for $s \in \bar{\Omega}$. Thus, the Gershgorin bands make the same number of encirclements of the point -1 as the $Q = \text{diag}\{q_{ii}\}$ does, and the system is stable as $H = \text{diag}\{h_{ii}\}$ is so. Noticing $|h_{ii}(j\omega)| \leq 1, \forall \omega$, (10.51) is true if

$$\epsilon_{oi}+\epsilon_{di}+\epsilon_{oi}\epsilon_{di} < 1. \qquad (10.56)$$

The proof is complete.

In the proposed algorithm, both ϵ_{oi} and ϵ_{di} are usually set as 10%. Then (10.56) can be satisfied with a large margin and nominal stability of the designed multivariable system can be expected.

In the real world where the model does not represent the process precisely, nominal stability is not sufficient, and robust stability of the closed loop has to be ensured. The multivariable control system in Figure 10.7 is referred to as being robustly stable if the closed loop is stable for all members of a family of possible processes. Let the actual process be denoted by $\tilde{G}(s)$ and the nominal process still by $G(s)$. The following assumptions are made.

Assumption 10.1 *The actual process $\tilde{G}(s)$ and the nominal process $G(s)$ do not have any unstable poles. The actual process $\tilde{G}(s)$ and the nominal process $G(s)$ are strictly proper.*

Consider the family of stable processes $\prod$ with norm-bounded uncertainty described by

$$\prod = \{\tilde{G} = [\tilde{g}_{ij}] \mid \left|\frac{\tilde{g}_{ij}(j\omega) - g_{ij}(j\omega)}{g_{ij}(j\omega)}\right| = |\Delta_{ij}(j\omega)| \leq \delta(\omega)\}, \tag{10.57}$$

where $\delta(\omega)$ is the bound on the multiplicative uncertainty Δ_{ij}. Let the perturbed open-loop transfer matrix $\tilde{G}\hat{K}(s)$ be denoted $\tilde{Q}$ and its elements $\tilde{q}_{ji}(s)$. As the elements of $\hat{K}(s)$ are designed to be proper, $\tilde{q}_{ji}$ are thus strictly proper, i.e., $\lim_{s \to \infty} \tilde{q}_{ji}(s) = 0$. Thus, $\sum_{j=1}^{m} |\tilde{q}_{ji}(s)| < \epsilon$ (where $0 < \epsilon < 1$) is likely to hold for $s \in \bar{\Omega}$. We have the following theorem.

Theorem 10.2.2. *(Robust Stability)* Assume $\sum_{j=1}^{m} |\tilde{q}_{ji}(s)| < \epsilon$, for $s \in \bar{\Omega}$ and $i \in \mathbf{m}$, then the multivariable closed-loop system in Figure 10.7 is robustly stable for all processes in $\prod$ if for $s \in \Omega$ and each $i \in \mathbf{m}$, (10.45) and (10.46) are satisfied, and

$$|h_{ii}|(\epsilon_{oi} + \epsilon_{di} + \epsilon_{oi}\epsilon_{di} + \delta(1 + \epsilon_{oi} + \epsilon_{di})) < 1, \quad s \in \Omega. \tag{10.58}$$

Proof. The ith Gershgorin band of the actual open-loop transfer matrix $\tilde{Q}(s)$ is

$$\tilde{\mathbf{G}}_{er}^{i} = \cup_{s \in D} \tilde{G}_{er}^{i}(s),$$

where

$$\tilde{G}_{er}^{i}(s) = \{z \in \mathcal{C} \mid |z - \tilde{q}_{ii}| \leq \sum_{j \neq i} |\tilde{q}_{ji}(s)|\}.$$

Since (10.45) and (10.46) are satisfied for $s \in \Omega$, we have for $s \in \Omega$

$$|z - q_{ii}| = |(\tilde{q}_{ii} - q_{ii}) + (z - \tilde{q}_{ii})|$$
$$\leq |\tilde{q}_{ii} - q_{ii}| + |z - \tilde{q}_{ii}|$$
$$\leq |\tilde{q}_{ii} - q_{ii}| + \sum_{j \neq i} |\tilde{q}_{ji}|$$
$$= |\sum_{l=1}^{m}(1+\Delta_{il})g_{il}\hat{k}_{li} - q_{ii}| + \sum_{j \neq i}|\sum_{l=1}^{m}(1+\Delta_{jl})g_{jl}\hat{k}_{li}|$$
$$\leq |\hat{q}_{ii} - q_{ii}| + |\sum_{l=1}^{m}\Delta_{il}g_{il}\hat{k}_{li}| + \sum_{j \neq i}|\hat{q}_{ji}| + \sum_{j \neq i}|\sum_{l=1}^{m}\Delta_{jl}g_{jl}\hat{k}_{li}|$$
$$\leq \epsilon_{oi}|q_{ii}| + \epsilon_{di}|\hat{q}_{ii}| + \sum_{j,l=1}^{m}|\Delta_{jl}g_{jl}\hat{k}_{li}|$$
$$\leq (\epsilon_{oi} + \epsilon_{di} + \epsilon_{oi}\epsilon_{di})|q_{ii}| + \delta\sum_{j=1}^{m}|\hat{q}_{ji}|$$
$$\leq (\epsilon_{oi} + \epsilon_{di} + \epsilon_{oi}\epsilon_{di})|q_{ii}| + \delta((1+\epsilon_{oi})|q_{ii}| + \epsilon_{di}|q_{ii}|)$$
$$\leq (\epsilon_{oi} + \epsilon_{di} + \epsilon_{oi}\epsilon_{di} + \delta(1 + \epsilon_{oi} + \epsilon_{di}))|q_{ii}|.$$

It follows that

$$|z + 1| = |(1 + q_{ii}) + (z - q_{ii})|$$
$$\geq |1 + q_{ii}| - |z - q_{ii}|$$
$$\geq |1 + q_{ii}| - (\epsilon_{oi} + \epsilon_{di} + \epsilon_{oi}\epsilon_{di} + \delta(1 + \epsilon_{oi} + \epsilon_{di}))|q_{ii}|,$$

which is positive, i.e., the portion of the Gershgorin bands for $s \in \Omega$ exclude the point -1 if (10.58) holds.

Since $\sum_{j=1}^{m}|\hat{q}_{ji}(s)| < \epsilon$, for $s \in \bar{\Omega}$ and $i \in \mathbf{m}$, then for $z \in G_{er}^{i}(s)$, where $s \in \bar{\Omega}$, we have

$$|z| - |\tilde{q}_{ii}| \leq |z - \tilde{q}_{ii}| \leq \sum_{j \neq i}|\tilde{q}_{ji}|,$$

giving

$$|z| \leq |\tilde{q}_{ii}| + \sum_{j \neq i}|\tilde{q}_{ji}| < \epsilon < 1,$$

or

$$|z + 1| > 0.$$

Therefore, the Gershgorin bands will exclude -1 for all s on the Nyquist contour and will not change the number of encirclements of the point -1 compared with the nominal system. The system is robust stable.

Note that $|h_{ii}(j\omega)| < 1$, $\forall \omega$, and it follows that (10.58) is true if

$$\epsilon_{oi} + \epsilon_{di} + \epsilon_{oi}\epsilon_{di} + \delta(\omega)(1+\epsilon_{oi}+\epsilon_{di}) < 1, \quad \forall \omega. \tag{10.59}$$

Let $\epsilon_{oi} = \epsilon_{di} = 10\%$, then the robust stability of the closed-loop can be guaranteed by

$$\delta(\omega) < 64.92\%. \tag{10.60}$$

Note that for $\delta(\omega) = 0$, i.e., in case of no process uncertainty, then the robust stability condition (10.59) is simplified to (10.56).

Example 10.2.2 (cont'd). Reconsider Example 10.2.2. To demonstrate robustness, introduce a 20% perturbation in dead time L_{ij}, and 50% perturbation gain of g_{ij}. It can be easily calculated that $|\Delta_{ij}(j\omega)| \leq 55.98\%$ for $s \in \Omega$. Thus robust stability can be guaranteed by our algorithm with $\epsilon_{oi} = \epsilon_{di} = 10\%$. The closed-loop performance is shown in Figure 10.14, and the column Gershgorin bands are shown in Figure 10.15. One observes that the system is robustly stable, indeed.

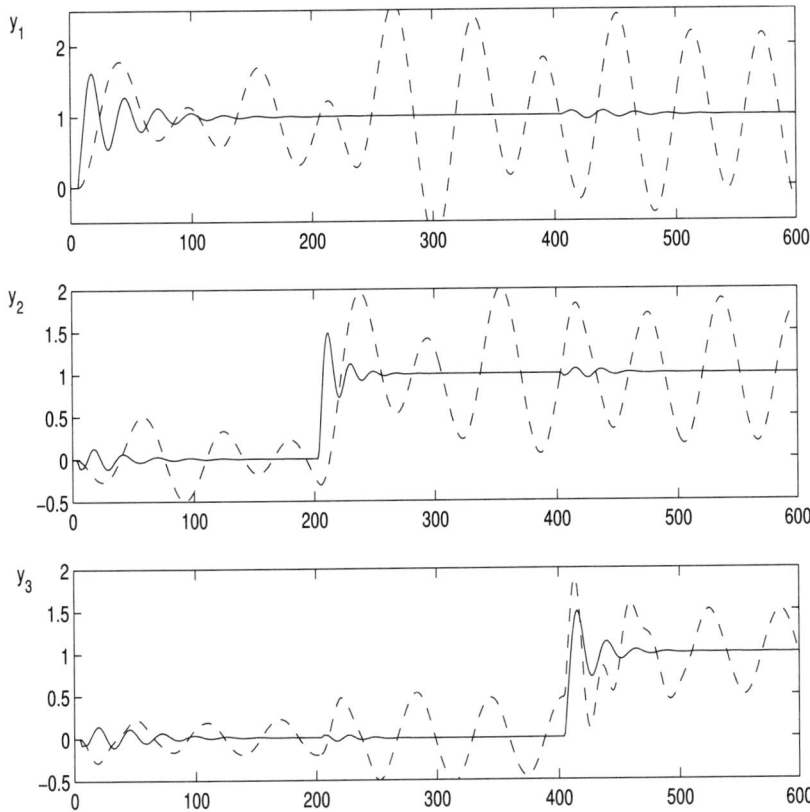

Fig. 10.14. Robust performance for Example 10.2.2(continued)
(——— proposed PID, – – – Dong's PID)

318   10. Multivariable Systems

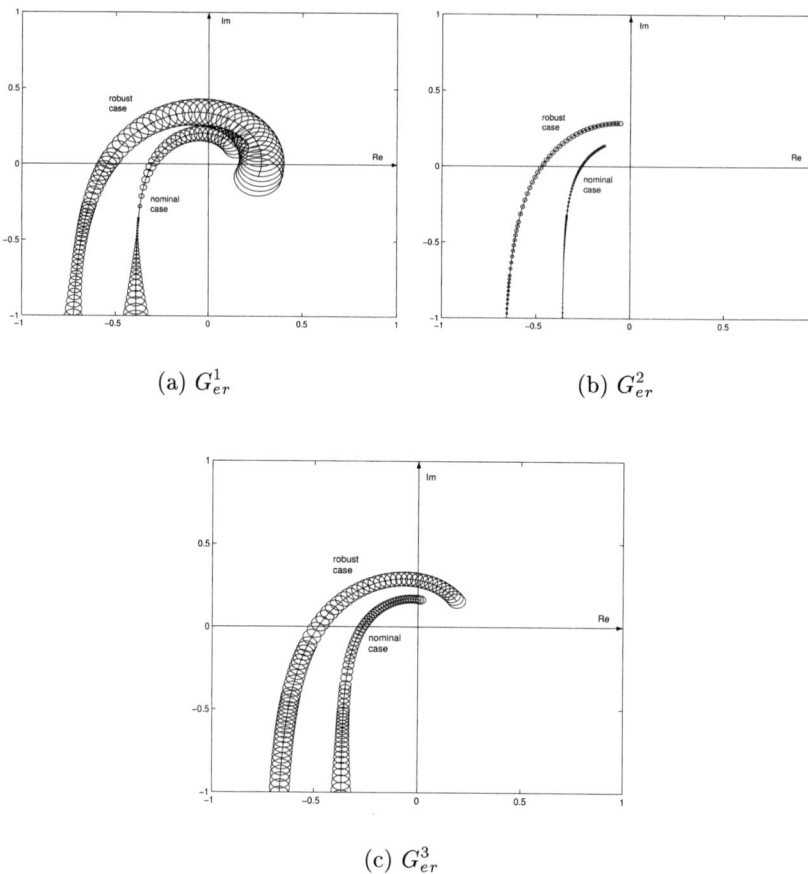

Fig. 10.15. Column Gershgorin bands for Example 10.2.2 (continued)

11. Partial Internal Model Control

For open-loop stable processes, the IMC approach provides a very simple yet powerful parametrization of all stabilizing controllers. The IMC factorization procedure gives valuable insights into the inherent control limitations of particular models. However, when open-loop unstable processes are considered, the original IMC structure cannot be used for control system implementation. This is because under the perfect match situation, the IMC system becomes an open-loop one and unstable processes are thus not stabilized at all. These unstable processes can be pre-stabilized using conventional feedback before the normal IMC structure is applied (Garcia and Morari, 1982). Chapter 9 has presented an alternative IMC-based method to design the conventional *single-loop feedback controller* for the unstable process. The IMC frame is abandoned in implementation in this situation.

In this chapter, a new IMC scheme is proposed to control unstable processes and to provide stability. In this scheme, the process transfer function model is expressed as the sum of two parts. One contains all the unstable poles of the process while the other is the stable part. Unlike the original IMC, in our method, only the stable part of the process model is used as the internal model. In the case of a perfect match, the stable part of the process is cancelled by the internal model. The anti-stable part, which contains all the unstable poles and is usually of low-order dynamics, remains in the closed loop and can easily be stabilized using a low-order controller such as PID. In this scheme, only part of the process model is used as the internal model. Thus, this new approach is named Partial Internal Model Control (PIMC). When a stable process is concerned, the PIMC becomes the IMC. The internal stability analysis of the proposed PIMC scheme shows that the PIMC system is internally stable when the primary controller stabilizes the anti-stable part of the process. Asymptotic tracking and regulation can be achieved by the proposed PIMC system or a Modified PIMC (MPIMC) scheme. The primary controller design is not complicated.

The chapter is organized as follows. The original IMC method is briefly reviewed in Section 11.1 and the PIMC strategy is proposed in Section 11.2. In Section 11.3, the internal stability of the PIMC system is analyzed. In Section 11.4, asymptotic tracking and regulation of the PIMC are addressed. Section 11.5 gives the primary controller design methods for the PIMC and the MPIMC system. Section 11.6 presents the robustness analysis. Several practical issues on the PIMC are discussed in Section 11.7. In Section 11.8, simulation results are given to demonstrate the PIMC scheme. A real-time implementation is presented in Section 11.9.

11.1 Review of the IMC

To understand why the IMC scheme fails for unstable processes and how the problem can be rectified, we briefly review the IMC scheme to motivate our new control scheme. It is well known (Garcia and Morari, 1982; Garcia and Morari, 1985) that an open-loop arrangement represents the best way to achieve fast and accurate set-point tracking. For open-loop scheme, the internal stability problem, which is complicated in feedback systems, is trivial and the controller is easy to design. However, the disadvantages are the sensitivity of the performance to process/model mismatch and the inability to cope with unmeasurable disturbances. To solve these problems, feedback is needed. Based on the idea of taking advantage of both open-loop scheme and feedback strategy, internal model control structure is proposed (Garcia and Morari, 1982). The block diagram of the IMC loop is shown in Figure 11.1, where the process $G(s)$ is assumed to be stable, and $\widehat{G}(s)$ is the process model, which is used as the

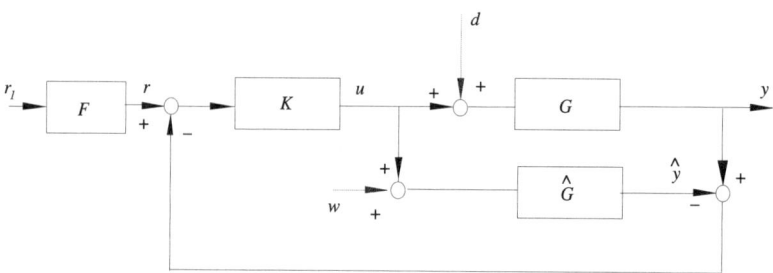

Fig. 11.1. General IMC-class scheme.

internal model. Let r and d be the set-point and load disturbance respectively. It follows from Figure 11.1 that

$$y = \frac{G(s)K(s)}{1+(G(s)-\widehat{G}(s))K(s)}r + \frac{1-\widehat{G}(s)K(s)}{1+(G(s)-\widehat{G}(s))K(s)}G(s)d, \qquad (11.1)$$

which, under the perfect match $G(s) = \widehat{G}(s)$, is reduced to

$$y = \widehat{G}(s)K(s)r + [1-\widehat{G}(s)K(s)]\widehat{G}(s)d. \qquad (11.2)$$

Equation (11.2) shows that 'perfect' control could be achieved by selecting $K(s) = \widehat{G}^{-1}(s)$. However, to ensure internal stability, the controller $K(s)$ should be stable. The process model is then factorized as

$$\widehat{G}(s) = \widehat{G}^+(s)\widehat{G}^-(s), \qquad (11.3)$$

where $\widehat{G}^+(s)$ contains all the time delays and unstable zeros. Consequently, $1/\widehat{G}^-(s)$ is stable and does not involve predictors. The IMC controller is designed as

$$K(s) = f(s)/\widehat{G}^-(s), \qquad (11.4)$$

where $f(s)$ is a low-pass filter which is chosen such that $K(s)$ is bi-proper. With this primary controller, it follows from (11.2) that

$$y = \widehat{G}^+(s)f(s)r + [1-\widehat{G}^+(s)f(s)]G(s)d. \qquad (11.5)$$

The closed-loop transfer function from r to y, $H_{yr} = \widehat{G}^+(s)f(s)$, contains only part of the process, $\widehat{G}^+$. This part limits the achievable control performance, but these inherent limitations cannot be removed by any control system. Thus, this IMC controller will lead to a perfect controller (Garcia and Morari, 1982).

If the model is perfect ($G(s) = \widehat{G}(s)$) and there is no disturbance ($d = 0$), the model output $\widehat{y}$ and the process output y are the same, which causes the feedback signal $(y - \widehat{y})$ to be zero. Thus, when there is no uncertainty, the control system becomes an open loop. This shows very instructively that, for open-loop stable processes, feedback is needed only because of uncertainty.

It is this character of the IMC that makes it unable to control the open-loop unstable system. The reason is simple: if $(G(s) = \widehat{G}(s))$ and $d = 0$, the feedback signal equals zero, which means that the IMC system is open loop and becomes unstable for an unstable process. Morari and Zafiriou (1989) pointed out that in this case we have to abandon the IMC structure for the control system implementation. They then proposed to use the IMC parametrization for the design of a *single-loop controller* to control unstable processes. As they still used the original IMC idea, i.e., $\widehat{G}(s) = G(s)$, the designed controller has to meet some constraints to ensure the single-loop control system is internally stable. Rotstein and Lewin (Rotstein and Lewin, 1991; Rotstein and Lewin, 1992) used the same idea to tune the PID controller for low-order unstable processes. However, when high-order unstable processes are considered, high-order controllers are usually required with these methods. A simple and practical scheme to improve the IMC in this regard is really needed.

11.2 The Proposed PIMC Scheme

Unstable processes are always a challenge to control system design. To deal with such processes, feedback is always needed and the first issue for the feedback design is to ensure the internal stability of the closed-loop system. The idea of using a controller to cancel an unstable process or its unstable poles leads to internal instability of the system and should be avoided (Morari and Zafiriou, 1989). In addition, control performance and controller simplicity are also important for practical applications. Along with these considerations, a general control scheme which includes the IMC as a special case is shown in Figure 11.1, where $F(s)$ is a pre-filter. In the case $\widehat{G} = G$, it becomes the IMC. In general, however, $\widehat{G}$ may not be the same as G. Our goal here is to find the best $\widehat{G}$ for an unstable G in some sense. Recall that the idea of the IMC is choosing the internal model $\widehat{G}$ such that internal stabilization of the compensated process $(G - \widehat{G})$ by the primary controller K is trivial, the design of K to meet performance specifications is easy, and yet the resultant K is simple. This can readily be achieved by setting $\widehat{G} = G$ for a stable G. But for an unstable G, we note that $\widehat{G}$ cannot have any unstable pole of G for internal stability of the closed-loop (this will be shown in Section 11.4). Obviously, we do not want $\widehat{G}$ to have any other unstable poles either. If otherwise, the compensated process $(G - \widehat{G})$ has more unstable poles than the process itself, which further complicates internal stabilization and primary controller design for $(G - \widehat{G})$ and may cause performance deterioration compared with the control design for G. This analysis clearly indicates that only a stable $\widehat{G}$ should be used.

The goal now is to choose a stable $\widehat{G}$ for an unstable G such that $(G - \widehat{G})$ is as simple as possible in some sense for enhancement of control performance and ease of control design and implementation. In the context of control, the order of a process is a major measure of simplicity within the set of rational transfer functions. A low-order process is usually easier to control than a high-order process if other properties such as stability and damping of these two processes are similar. An additional advantage of having a low-order process is that a low-order controller is sufficient to meet control specifications. For instance, we have two processes, $G_1 = \frac{c_1}{s-4}$ and $G_2 = \frac{c_2}{(s-4)(s+1)^3}$, where c_1 and c_2 are constants. In a unity negative feedback configuration, a constant controller can stabilize G_1 and a PI controller can control G_1 very well. But it can be shown (Wang et al., 1997d) that the minimal order of stabilizers for G_2 is 3. The control design for high-order unstable processes like G_2 is not easy. A pole-placement controller of order 4 may be designed for G_2 but it is usually sensitive to modelling errors.

Therefore, our task is now to determine a stable $\widehat{G}$ for a given unstable G such that $(G - \widehat{G})$ has minimal order. To this end, let λ_i^- and λ_j^+ be stable and unstable poles of G, respectively. The partial fraction expansion of G yields

$$G(s) = \sum_{i=1}^{n^-}\sum_{k=1}^{N_{\lambda_i^-}} \frac{b_{ik}^-}{(s-\lambda_i^-)^k} + \sum_{j=1}^{n^+}\sum_{k=1}^{N_{\lambda_j^+}} \frac{b_{jk}^+}{(s-\lambda_j^+)^k}, \quad Re(\lambda_i^-) < 0, \; Re(\lambda_j^+) \geq 0, \tag{11.6}$$

where $N_{\lambda_i^-}$ and $N_{\lambda_j^+}$ are the multiplicities of the poles at $s = \lambda_i^-$ and $s = \lambda_j^+$, respectively. Define

$$G^-(s) = \sum_{i=1}^{n^-}\sum_{k=1}^{N_{\lambda_i^-}} \frac{b_{ik}^-}{(s-\lambda_i^-)^k}, \tag{11.7}$$

and

$$G^+(s) = \sum_{j=1}^{n^+}\sum_{k=1}^{N_{\lambda_j^+}} \frac{b_{jk}^+}{(s-\lambda_j^+)^k}. \tag{11.8}$$

One notes that $(G - \widehat{G})$ always contains unstable poles of G since $\widehat{G}$ has been assumed to be stable. The order of $(G - \widehat{G})$ is thus larger than or equal to the number $(\sum_{j=1}^{n^+} N_{\lambda_j^+})$ of unstable poles of G (multiplicities included) and $\sum_{j=1}^{n^+} N_{\lambda_j^+}$ is thus a lower bound for the order of $(G - \widehat{G})$. As a result, the stable $\widehat{G}$ which minimizes the order of $(G - \widehat{G})$ is given by

$$\widehat{G} = G^-, \tag{11.9}$$

since in this case the order of $G - \widehat{G} = G^+$ is just $\sum_{j=1}^{n^+} N_{\lambda_j^+}$ and thus reaches the lower bound. One sees that G^- is a part but not the whole of the G. Hence, the scheme in Figure 11.1 with $\widehat{G} = G^-$ is called the Partial Internal Model Control, and abbreviated as the PIMC, to distinguish it from the IMC.

As a demonstration, consider an unstable process

$$G(s) = \frac{1}{(s-1)(s+1)^3},$$

which has a high order of 4 and is difficult to control. G can be expressed via the partial fraction expansion as

$$G = \frac{1}{8(s-1)} + \frac{-s^2 - 4s - 7}{8(s+1)^3} = G^+ + G^-.$$

Take $\widehat{G} = G^-$, then the compensated process which the primary controller K faces is $G - \widehat{G} = G^+ = \frac{1}{8(s-1)}$, which is only of first order and is easy to control, as discussed above.

For the control system in Figure 11.1, the closed-loop transfer function is

$$y = \frac{G(s)K(s)}{1 + K(s)(G(s) - \widehat{G}(s))} r + \frac{1 - \widehat{G}(s)K(s)}{1 + K(s)(G(s) - \widehat{G}(s))} G(s)d. \tag{11.10}$$

In the PIMC, $G = G^+ + G^-$, $\widehat{G} = G^-$, and the input–output relationship becomes

$$y = \frac{G(s)K(s)}{1 + G^+(s)K(s)} r + \frac{1 - G^-(s)K(s)}{1 + G^+(s)K(s)} G(s)d. \tag{11.11}$$

Equation (11.11) shows that only the anti-stable part G^+ remains in the loop and needs to be stabilized by the controller $K(s)$. Compared with the original IMC method, the feedback loop here is used not only for process uncertainty but also for stabilization of the process.

We now consider two extreme cases of the PIMC. If G has no stable poles, the partial internal model $\widehat{G} = G^- = 0$, and Figure 11.1 reduces to a unity feedback loop. The controller $K(s)$ is designed with respect to the process G. When G is stable, $G^+ = 0$ and $\widehat{G} = G^- = G$. The PIMC scheme in this case becomes the IMC. Thus, the IMC can be regarded as a special case of the PIMC.

11.3 Internal Stability Analysis

A control system must be stable. In this section, internal stability of the PIMC system is analyzed. In the PIMC, the primary controller K is designed with respect to $G - \widehat{G} = G^+$. Let $G^+ = \frac{b_1}{a^+}$ and $K = \frac{\beta}{\alpha}$ be polynomial fractions. Suppose that K stabilizes $G - \widehat{G} = G^+$, i.e., $q = \alpha a^+ + \beta b_1$ has all its roots in the open left half of the complex plane (LHP) (Chen, 1984). The question is whether the PIMC system is internally stable if K stabilizes $G - \widehat{G} = G^+$.

Recall that a linear time-invariant control system is internally stable if the transfer functions between any two points of the system are stable (Morari and Zafiriou, 1989). Thus, in order to test for internal stability, the transfer functions between all possible inputs and outputs should be examined. Accordingly, for the PIMC system in Figure 11.1 with $\widehat{G} = G^-$, by introduction of the signals w and d, the closed loop is internally stable *if and only if* the transfer matrix H between $[y \ \widehat{y} \ u]^T$ and $[r \ d \ w]^T$ is stable. By a straightforward but tedious analysis, it can be shown that this H is indeed stable under our PIMC scheme if K stabilizes $G - \widehat{G} = G^+$. A simpler proof can be obtained by an easy application of our newly developed theorem on internal stability (Wang et al., 1999e), and is thus presented below. Before stating the theorem, we need the concept of Mason's system determinant. The determinant of the system is defined (Mason, 1956; Franklin et al., 1994) by

$$\Delta = 1 - \sum_i L_{1i} + \sum_j L_{2j} - \sum_k L_{3k} + \ldots, \tag{11.12}$$

where L_{1i} are the loop gains of the system and L_{2j} and L_{3k} are the products of 2 and 3 non-touching loop gains.

Theorem 11.3.1. (Wang et al., 1999e) *A system with m scalar plants $G_i(s) = \frac{b_i}{a_i}$, $i = 1, 2, \ldots, m$, where $\frac{b_i}{a_i}$ are polynomial fractions, is internally stable if and only if the characteristic polynomial defined by*

$$P(s) = \Delta \prod_{i=1}^{m} a_i(s) \tag{11.13}$$

is a stable polynomial, where Δ is the system determinant.

For the PIMC in Figure 11.1 with $\widehat{G} = G^-$, let $G = \frac{b}{a} = G^+ + G^- = \frac{b_1}{a^+} + \frac{b_2}{a^-}$, and $K = \frac{\beta}{\alpha}$. Then, it follows from (11.13) that $P(s)$ for this system is given by

$$P = \left\{ 1 - \left[\frac{\beta b_2}{\alpha a^-} - \frac{\beta}{\alpha} \left(\frac{b_2}{a^-} + \frac{b_1}{a^+} \right) \right] \right\} \alpha a^- a^- a^+$$

$$= (\alpha a^+ + \beta b_1)(a^-)^2, \tag{11.14}$$

which is stable *if and only if* $(\alpha a^+ + \beta b_1)$ is stable, or *if and only if* K stabilizes $G^+ = \frac{b_1}{a^+}$. Hence, we obtain the following theorem on internal stability of the PIMC system.

Theorem 11.3.2. *The PIMC system shown in Figure 11.1 with $\widehat{G} = G^-$ is internally stable if and only if K stabilizes G^+.*

The following claim is made in Section 11.2 and its proof is given here.

Theorem 11.3.3. *The system in Figure 11.1 is internally unstable if $\widehat{G}$ has any unstable pole the same as that of G.*

Proof. Suppose that $\widehat{G}$ and G have a common unstable pole at $s = \lambda^+$, $\lambda^+ \geq 0$, that is $G(s) = \frac{b(s)}{(s-\lambda^+)a_r(s)}$ and $\widehat{G}(s) = \frac{\widehat{b}(s)}{(s-\lambda^+)\widehat{a}_r(s)}$. It follows from (11.13) that for $K = \frac{\beta}{\alpha}$, $P(s)$ in this case is

$$P(s) = [1 - (\frac{\beta \widehat{b}}{\alpha(s-\lambda^+)\widehat{a}_r} - \frac{\beta b}{\alpha(s-\lambda^+)a_r})]\alpha(s-\lambda^+)^2 a_r \widehat{a}_r$$

$$= [\alpha(s-\lambda^+)a_r\widehat{a}_r - \beta \widehat{b} a_r + \beta b \widehat{a}_r](s-\lambda^+), \qquad (11.15)$$

which is always unstable due to the existence of the unstable factor $(s-\lambda^+)$ in $P(s)$. By Theorem 11.3.1, the system in Figure 11.1 is thus internally unstable.

11.4 Asymptotic Tracking and Regulation

Asymptotic tracking and regulation are always desired properties for industrial process control systems. It is noted (Garcia and Morari, 1982) that an implicit integral action is included in the standard IMC to achieve zero offset for step inputs if and only if $K(0) = \widehat{G}^{-1}(0)$. The same can be done for the PIMC. To see this, Figure 11.1 with $\widehat{G} = G^-$ is redrawn into its equivalent in Figure 11.2, where

$$\overline{K} = \frac{K}{1 - G^{-1}K}.$$

To generate an integrator in $\overline{K}$ requires that

$$1 - G^-(0)K(0) = 0,$$

or

$$K(0) = (\bar{G}(0))^{-1}. \qquad (11.16)$$

If (11.16) holds, the output y has zero steady-state error in response to the set-point change r and the load disturbance d of step-type (Garcia and Morari, 1982) provided that the closed loop is stable.

An alternative scheme for tracking/regulation using the PIMC is proposed in Figure 11.3 and termed Modified PIMC (MPIMC) for ease of reference. One will see from below that this MPIMC can automatically achieve asymptotic tracking and regulation for any type of external inputs and removes a

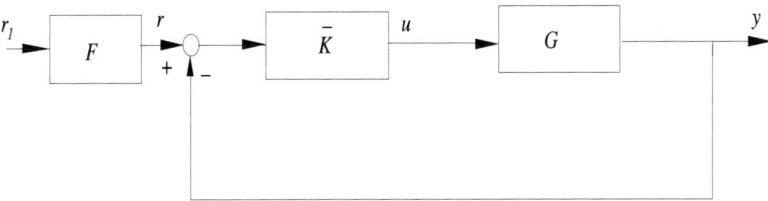

Fig. 11.2. Equivalence of PIMC

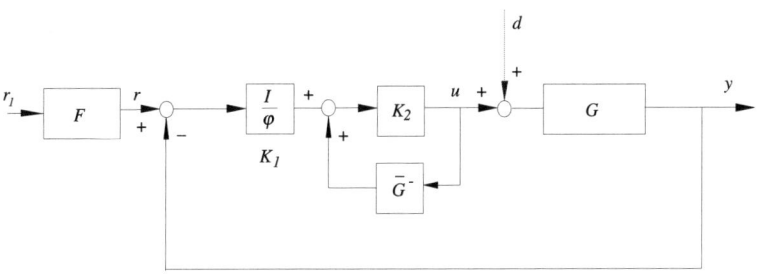

Fig. 11.3. Modified partial internal model control system

(11.16)-like constraint on the controller. For the MPIMC, the controller K is constructed as

$$K = K_1 K_2. \tag{11.17}$$

K_1 is determined by the internal model principle (Chen, 1984) as

$$K_1(s) = \frac{1}{\phi(s)}, \tag{11.18}$$

where $\phi(s)$ is the least common denominator of the anti-stable parts of $r(s)$ and $d(s)$, respectively. For example, if both $r(s)$ and $d(s)$ are of step type, then $r(s) = \frac{1}{s}$ and $d(s) = \frac{1}{s}$, leading to $\phi(s) = s$ and $K_1(s) = \frac{1}{s}$. If either of r and d is changed to ramp type, then $K_1(s) = \frac{1}{s^2}$. We define GK_1 as the generalized process:

$$\overline{G} := GK_1 = \frac{G}{\phi}, \tag{11.19}$$

and expand it as

$$\overline{G} = \overline{G}^- + \overline{G}^+, \tag{11.20}$$

where $\overline{G}^-$ and $\overline{G}^+$ are the stable and anti-stable parts of $\overline{G}$ respectively. Like in the PIMC, only $\overline{G}^-$ is used in MPIMC as the parallel compensator to cancel the stable dynamics of the generalized process, and K_2 is required to stabilize $(\overline{G} - \overline{G}^-) = \overline{G}^+$. We can now establish the following theorem on asymptotic tracking and regulation of the MPIMC.

Theorem 11.4.1. *The MPIMC system in Figure 11.3 has a zero steady-state error provided that $G(s)$ has no zeros coinciding with roots of $\phi(s)$ and K_2 stabilizes $\overline{G}^+$.*

Proof. Under the assumed condition that $G(s)$ has no zeros coinciding with roots of $\phi(s)$, there is no unstable pole–zero cancellation between $G(s)$ and K_1. One may then regard $GK_1 = \overline{G}$ as one single process, and the MPIMC in Figure 11.3 is a PIMC system. It follows from Theorem 11.3.2 that *the MPIMC system is internally stable if K_2 stabilizes $\overline{G} - \overline{G}^- = \overline{G}^+$*. The output error is obtained from Figure 11.3 as

$$e(s) := r(s) - y(s) = G_{er}(s)r(s) + G_{ed}(s)d(s), \tag{11.21}$$

where

$$G_{er}(s) = \frac{[1 - \overline{G}^-(s)K_2(s)]}{\phi(s) + [G(s) - \phi(s)\overline{G}^-(s)]K_2(s)} \phi(s),$$

and

$$G_{ed}(s) = -\frac{[G(s) - G(s)\overline{G}^-(s)K_2(s)]}{\phi(s) + [G(s) - \phi(s)\overline{G}^-(s)]K_2(s)} \phi(s).$$

Since the system has been shown to be stable, the final value theorem is applicable:

$$e(\infty) = \lim_{s \to 0} s(G_{er}(s)r(s) + G_{ed}(s)d(s))$$

$$= \lim_{s \to 0} s \left[\frac{[1 - \overline{G}^-(s)K_2(s)]}{\phi(s) + [G(s) - \phi(s)\overline{G}^-(s)]K_2(s)} \right] [\phi(s)r(s)]$$

$$- \lim_{s \to 0} s \left[\frac{[G(s) - G(s)\overline{G}^-(s)K_2(s)]}{\phi(s) + [G(s) - \phi(s)\overline{G}^-(s)]K_2(s)} \right] [\phi(s)d(s)]$$

$$= \lim_{s \to 0} s \lim_{s \to 0} \left[\frac{[1 - \overline{G}^-(s)K_2(s)]}{\phi(s) + [G(s) - \phi(s)\overline{G}^-(s)]K_2(s)} \right] \lim_{s \to 0} [\phi(s)r(s)]$$

$$- \lim_{s \to 0} s \lim_{s \to 0} \left[\frac{[G(s) - G(s)\overline{G}^-(s)K_2(s)]}{\phi(s) + [G(s) - \phi(s)\overline{G}^-(s)]K_2(s)} \right] \lim_{s \to 0} [\phi(s)d(s)]$$

$$= 0 \cdot c_1 \cdot c_2 - 0 \cdot c_3 \cdot c_4$$

$$= 0,$$

where c_1, c_2, c_3 and c_4 are finite constants. The proof is completed.

11.5 Primary Control Design

The design of the primary controller $K(s)$ in PIMC/MPIMC systems is considered in this section. For the PIMC in Figure 11.1 with $\widehat{G} = G^-$, the primary controller $K(s)$ is determined such that it gives rise to an internally stable system, satisfies (11.16) by having zero steady-state error for external inputs of step-type, and meets the dynamic performance specifications given by the user. For the MPIMC in Figure 11.3, the controller design is the same as for the PIMC except that it does not need to satisfy (11.16). It is noted that the unstable processes encountered most commonly in industry actually have only one unstable pole. Integral processes and chemical reactors are examples. In the rest of this section, we assume that the process is described by

$$G(s) = \frac{b(s)}{(s - \lambda^+)a^-(s)}, \quad \lambda^+ \geq 0, \tag{11.22}$$

where $a^-(s)$ is a stable polynomial. We present an analytical design solution for the primary controller for both PIMC and MPIMC in terms of the desired closed-loop specifications. A similar solution for the process with two unstable poles is given in Appendix A of this chapter.

11.5.1 PIMC Primary Controller Design

Consider Figure 11.1 with $\widehat{G} = G^-$. G in (11.22) can be expanded into

$$G(s) = \frac{b_1}{s - \lambda^+} + \frac{b_2(s)}{a^-(s)}. \tag{11.23}$$

Choose $K(s)$ as

$$K(s) = \frac{k_2 s + 1}{k_1 s + G^-(0)}. \tag{11.24}$$

Note that $K(0) = (G^-(0))^{-1}$ and that such a $K(s)$ satisfies (11.16). Assume that the desired closed-loop specifications are given by the following stable second-order dynamics:

$$P_d(s) = s^2 + 2\xi\omega_0 s + \omega_0^2. \tag{11.25}$$

With $G - G^- = G^+ = \frac{b_1}{s-\lambda^+}$ and $K(s)$ in (11.24), the actual characteristic polynomial $P(s)$ for Figure 11.1 with $\hat{G} = G^-$ is determined using (11.14) as

$$P(s) = (a^-(s))^2 k_1 [s^2 + \frac{(G^-(0) - k_1 \lambda^+ + k_2 b_1)}{k_1} s + \frac{b_1 - \lambda^+ G^-(0)}{k_1}]$$
$$:= (a^-(s))^2 k_1 P_c(s).$$

We match the adjustable part $P_c(s)$ in $P(s)$ to $P_d(s)$, that is $P_c(s) = P_d(s)$. This yields the solution for the remaining controller parameters as

$$\begin{bmatrix} k_1 \\ k_2 \end{bmatrix} = \begin{bmatrix} \frac{b_1 - \lambda^+ G^-(0)}{\omega_0^2} \\ \frac{1}{b_1}\{-G^-(0) + \frac{(2\xi\omega_0 + \lambda^+)[b_1 - \lambda^+ G^-(0)]}{\omega_0^2}\} \end{bmatrix}. \tag{11.26}$$

ξ and ω_0 are important factors for satisfactory behaviour of a system. ξ is the resultant system damping factor and is usually chosen as 0.707. A large value of ω_0 results in a fast system response but may require much control effort. The selection of ω_0 is usually related to the real part of the unstable pole λ^+.

11.5.2 MPIMC Primary Controller Design

Consider Figure 11.3 with G in (11.22) and $\phi(s) = s$, which is the most common case. The generalized process $\overline{G} = \frac{G}{s}$ is expanded into

$$\overline{G}(s) = \frac{G(s)}{s} = \frac{\mu_1 s + \mu_2}{s(s - \lambda^+)} + \frac{b_2(s)}{a^-(s)}, \tag{11.27}$$

where a^- is a stable polynomial. Formulas for determining μ_1, μ_2 and b_2 for typical unstable processes can be found in Appendix B of this chapter. $K_2(s)$ is taken as $K_2(s) = k_1 s + k_2$. In this case, the complete primary controller is

$$K = K_1 K_2 = \frac{k_1 s + k_2}{s}, \tag{11.28}$$

which is exactly a PI controller. It follows from (11.13) that the closed-loop characteristic polynomial for the MPIMC can be obtained as

$$P(s) = (a^-(s))^2(1+k_1\mu_1)(s^2 + \frac{k_1\mu_2 - \lambda^+ + k_2\mu_1}{1+k_1\mu_1}s + \frac{k_2\mu_2}{1+k_1\mu_1})$$
$$:= (a^-(s))^2(1+k_1\mu_1)P_c(s), \qquad (11.29)$$

where only P_c is adjustable. Suppose that the desired closed loop is still specified by (11.25). Then, we make

$$s^2 + \frac{k_1\mu_2 - \lambda^+ + k_2\mu_1}{1+k_1\mu_1}s + \frac{k_2\mu_2}{1+k_1\mu_1} = s^2 + 2\xi\omega_0 s + \omega_0^2, \qquad (11.30)$$

whose solution is

$$\begin{bmatrix} k_1 \\ k_2 \end{bmatrix} = \begin{bmatrix} \frac{2\xi\omega_0\mu_2 + \lambda^+\mu_2 - \omega_0^2\mu_1}{\mu_1^2\omega_0^2 + \mu_2^2 - 2\xi\omega_0\mu_1\mu_2} \\ \frac{\omega_0^2(\mu_2 + \lambda^+\mu_1)}{\mu_1^2\omega_0^2 + \mu_2^2 - 2\xi\omega_0\mu_1\mu_2} \end{bmatrix}. \qquad (11.31)$$

11.6 Robustness Analysis

In the real world where the model does not represent the process exactly, nominal stability analysis is not sufficient. Stability robustness of the proposed PIMC scheme has to be investigated.

11.6.1 Robust Stability

Suppose that an actual process is described by G, which is unknown to us. What we know is a nominal process (or a model for it), $\widetilde{G} = \widetilde{G}^+ + \widetilde{G}^-(s)$. Since the MPIMC is a special case of PIMC, we need only to consider the PIMC system in Figure 11.4 with the internal model $\widetilde{G}^-$ and the process described by

$$\prod = \{G(s): \ |G(s) - \widetilde{G}(s)| \leq \gamma(s)\}, \qquad (11.32)$$

where $\gamma(s)$ is a stable bound function on the additive uncertainty $G(s) - \widetilde{G}$.

As the standard assumptions in robust analysis, we assume that a primary controller K has been designed such that the nominal closed-loop system is stable, i.e., K stabilizes $\widetilde{G}^+$, and that G has the same number of unstable poles as $\widetilde{G}$. The PIMC system in Figure 11.4 can be re-drawn into Figure 11.5, where $\widetilde{K} = \frac{K}{1+\widetilde{G}^+K}$ is stable. It then follows from the the robust stability theorem (Doyle et al., 1982) that the system in Figure 11.5 is robustly stable *if and only if* $\|\widetilde{K}(G - \widetilde{G})\|_\infty < 1$, or $\|\widetilde{K}\gamma\|_\infty < 1$ holds. Therefore, we obtain the following result on robust stability of PIMC systems.

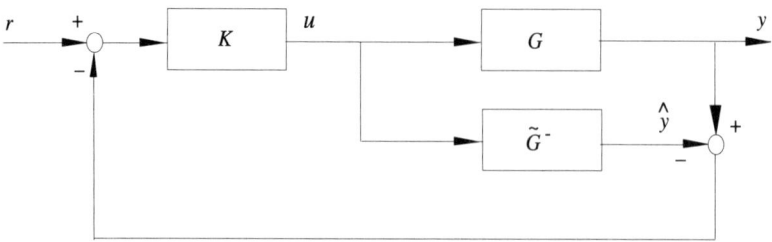

Fig. 11.4. PIMC robust stability analysis

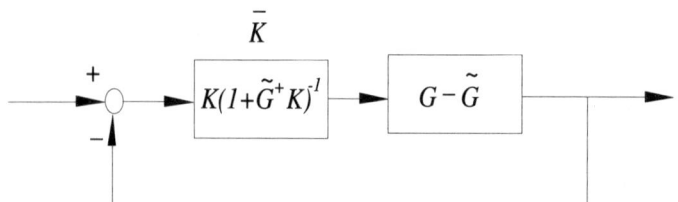

Fig. 11.5. Equivalent robust stability analysis scheme for PIMC

Theorem 11.6.1. *A PIMC system is robustly stable if and only if*

$$\|\widetilde{K}(s)\gamma(s)\|_\infty < 1. \tag{11.33}$$

11.6.2 Practical Stability

Palmor (1980) pointed out that there is a 'practical unstable' problem in some Smith Predictor control system in the sense that a nominally stable Smith system may become unstable when there is an infinitesimal mismatch. The practical stability concept (Yamanaka and Shimemura, 1987; Wang et al., 1998b) addresses this phenomenon. The closed loop is said to be practically stable *if and only if* there exist positive numbers ω_M and σ such that the closed loop is stable for every process satisfying

$$|G-\widetilde{G}| \leq \sigma, \quad 0 \leq \omega \leq \omega_M. \tag{11.34}$$

As such, practical stability analysis concerns system stability during an infinitesimal perturbation of the process, or an infinitesimal mismatch between the process and the model, and thus is a special case of robust stability.

If ω is finite, $\widetilde{K}(j\omega)$ is also finite. We can thus choose sufficiently large ω_M and sufficiently small σ such that $\gamma(j\omega)$ can be made arbitrarily small and

$$|\widetilde{K}(j\omega)\gamma(j\omega)| < \sigma, \quad 0 \leq \omega \leq \omega_M. \tag{11.35}$$

When ω approaches infinity, one notes that

$$\lim_{\omega \to \infty} |\widetilde{K}(j\omega)| = |\lim_{\omega \to \infty} \frac{K(j\omega)}{1 + G^+(j\omega)K(j\omega)}| < \infty, \tag{11.36}$$

since $K(s)$ is proper and $G^+(s)$ is strictly proper, and that

$$\lim_{\omega \to \infty} |G(j\omega) - \widetilde{G}(j\omega)| = 0,$$

or

$$\lim_{\omega \to \infty} \gamma(j\omega) = 0, \tag{11.37}$$

if $G(s)$ and $\widetilde{G}(s)$ are both strictly proper. Equations (11.36) and (11.37) imply that

$$\lim_{\omega \to \infty} |\widetilde{K}(j\omega)\gamma(j\omega)| = 0, \tag{11.38}$$

for strictly proper $G(s)$ and $\widetilde{G}(s)$. And (11.35) and (11.38) together show that $\|\widetilde{K}\gamma\|_\infty < 1$ and the system is stable for every $G(s)$ satisfying (11.34).

Theorem 11.6.2. *The PIMC system of Figure 11.4 is practically stable if both the process G and its model $\widetilde{G}$ are strictly proper.*

11.7 Practical Aspects

In this section, several issues concerning the use of PIMC/MPIMC systems are discussed.

11.7.1 Pre-filter Design

It is noted that severe overshoot and undershoot may sometimes exist in a PIMC for a set-point change. This is caused by improper zeros/poles in the closed-loop transfer function. In order to reduce this, a pre-filter $F(s)$ can be placed in front of the closed loop, as shown in Figure 11.1.

PIMC Case. Consider the most common case in which the process has only one unstable pole at $s = \lambda^+$ and is described by

$$G(s) = \frac{b(s)}{(s-\lambda^+)a^-(s)} = \frac{b_1}{s-\lambda^+} + \frac{b_2(s)}{a^-(s)} = G^+(s) + G^-(s),$$

where a^- is a stable polynomial. Suppose that the PIMC primary controller K in (11.24) has been obtained. It follows from (11.11) that the closed-loop transfer function from r to y is

$$\begin{aligned}G_{yr}(s) &= \frac{G(s)K(s)}{1+G^+(s)K(s)}\\ &= \frac{b(s)(k_2 s + 1)}{a^-(s)[(s-\lambda^+)(k_1 s + G^-(0)) + b_1(k_2 s + 1)]}\\ &= \frac{b^+(s)b^-(s)(k_2 s + 1)}{a^-(s)[(s-\lambda^+)(k_1 s + G^-(0)) + b_1(k_2 s + 1)]},\end{aligned}$$

where $b^+(s)$ contains all the unstable zeros of $b(s)$.

The ideal design of the pre-filter $F(s)$ is such that the transfer function between r_1 and y in Figure 11.1 has exactly the desired second-order dynamics, that is,

$$G_{yr}(s)F(s) = \frac{\omega_o^2 b^+(s)}{s^2 + 2\xi\omega_0 s + \omega^2}.$$

This requires

$$F(s) = \frac{b^-(0)a^-(s)}{a^-(0)b^-(s)(k_2 s + 1)}.$$

But this $F(s)$ may not be proper, and we may thus take $F(s)$ as

$$F(s) = \frac{b^-(0)a^-(s)}{a^-(0)b^-(s)(k_2 s + 1)} \cdot \frac{1}{(\tau s + 1)^m}, \tag{11.39}$$

where $\frac{1}{(\tau s+1)^m}$ is chosen such that $F(s)$ is proper. It may be noted that if the stable poles of the process are much smaller than ω_0, the pre-filter in (11.39) will result in a faster set-point response but with much greater control effort. To avoid this, the pre-filter can be chosen as

$$F(s) = \frac{b^-(0)}{b^-(s)(k_2 s + 1)}.$$

This rule is also applicable to the following MPIMC case.

11.7 Practical Aspects

MPIMC Case. Consider the process with one unstable pole at $s = \lambda^+$:

$$G(s) = \frac{b(s)}{(s-\lambda^+)a^-(s)}$$

and

$$\overline{G}(s) = \frac{G(s)}{s} = \overline{G}^+(s) + \overline{G}^-(s) = \frac{\mu_1 s + \mu_2}{s(s-\lambda^+)} + \frac{b_2(s)}{a^-(s)}.$$

It follows from (11.11) and (11.21) that the closed-loop transfer function from r to y is obtained as

$$G_{yr}(s) = \frac{G(s)K(s)}{1+\overline{G}^+(s)K_2(s)}$$

$$= \frac{b(s)(k_1 s + k_2)}{a^-(s)[(s-\lambda^+)s + (\mu_1 s + \mu_2)(k_1 s + k_2)]}.$$

Like the PIMC case, $F(s)$ is chosen as

$$F(s) = \frac{b^-(0)k_2 a^-(s)}{a^-(0)b^-(s)(k_1 s + k_2)} \cdot \frac{1}{(\tau s + 1)^m}. \qquad (11.40)$$

11.7.2 Determination of G^-

In order to implement the PIMC, one has to obtain G^- from G. For a complex or high-order unstable process, the computation of G^- using *the residual theorem* is quite involved, especially for the case of multiple poles. Actually, this can be avoided. As we know, practical industrial processes usually have only one unstable pole. This means that

$$G(s) = \frac{b_1}{s-\lambda^+} + G^-(s), \quad \lambda^+ \geq 0.$$

Then we have

$$b_1 = \lim_{s \to \lambda^+} (s-\lambda^+)G(s), \qquad (11.41)$$

and

$$G^-(s) = G(s) - \frac{b_1}{s-\lambda^+}, \quad \lambda^+ \geq 0. \qquad (11.42)$$

The decomposition in (11.41) and (11.42) is sufficient for the PIMC. But, for the MPIMC, the generalized process is

$$\overline{G}(s) = \frac{G(s)}{s}.$$

If $\lambda^+ \neq 0$, then

$$\overline{G}(s) = \frac{\mu_1}{s - \lambda^+} + \frac{\mu_2}{s} + \overline{G}^-(s). \tag{11.43}$$

μ_1 and μ_2 can still be calculated simply by the residual theorem and $\overline{G}^-$ is thus obtained from (11.43). If $\lambda^+ = 0$, then $\overline{G}$ has a double pole at the origin. However, a simple solution still exists for this case. We express $\overline{G}$ as

$$\overline{G}(s) = \frac{\mu_1}{s^2} + \frac{\mu_2}{s} + \overline{G}^-(s) := \frac{\mu_1}{s^2} + Z(s),$$

where μ_1 is determined by $\mu_1 = \lim_{s \to 0} s^2 \overline{G}(s)$ and $Z(s) = \overline{G}(s) - \frac{\mu_1}{s^2}$. $Z(s) = \frac{\mu_2}{s} + \overline{G}^-(s)$ has only a single unstable pole at the origin, and μ_2 can be calculated accordingly. Thus, we obtain

$$\overline{G}^+(s) = \frac{\mu_1}{s^2} + \frac{\mu_2}{s},$$

and

$$\overline{G}^-(s) = \overline{(G)} - (\frac{\mu_1}{s^2} + \frac{\mu_2}{s}).$$

11.7.3 Dead Time

We have assumed so far that the process has a rational transfer function $G(s)$. If the given unstable process has small dead time, it can be approximated as

$$e^{-\tau s} = \frac{\sum_{i=1}^{n} \frac{1}{i!} (\frac{-\tau s}{2})^i}{\sum_{i=1}^{n} \frac{1}{i!} (\frac{\tau s}{2})^i}, \tag{11.44}$$

and it contributes additional stable poles to $G(s)$. Partial function expansion can then be applied to it. If the process dead time is large, the Finite Spectrum Assignment method may be used for dead time compensation (Wang et al., 1998b).

11.8 Simulation Results

In this section, several simulation examples of different kinds of unstable processes are presented to illustrate the PIMC/MPIMC methods.

Example 11.8.1. Consider the unstable batch reactor of (Rotstein and Lewin, 1992):

$$G(s) = \frac{-745}{(-2442s + 1)(372s + 1)}.$$

For the PIMC scheme, the process is decomposed in the form of (11.6) into the stable and anti-stable parts as

$$G(s) = G^+(s) + G^-(s) = \frac{-646.514}{-2442s + 1} + \frac{-98.486}{372s + 1}.$$

With ξ and ω_0 in (11.25) specified as 0.707 and 0.02 respectively, the primary controller K in (11.24) is designed for G^+ as

$$K(s) = \frac{453.843s + 1.000}{762.7s - 98.486}.$$

The pre-filter is also designed using (11.39) to reduce the overshoot in response to the set-point change. The pre-filter in (11.39) is designed as

$$F(s) = \frac{372s + 1}{(453.843s + 1)(10s + 1)}.$$

The response to a step change in the set-point and a step load disturbance is shown in Figure 11.6. The process can also be controlled using the MPIMC.

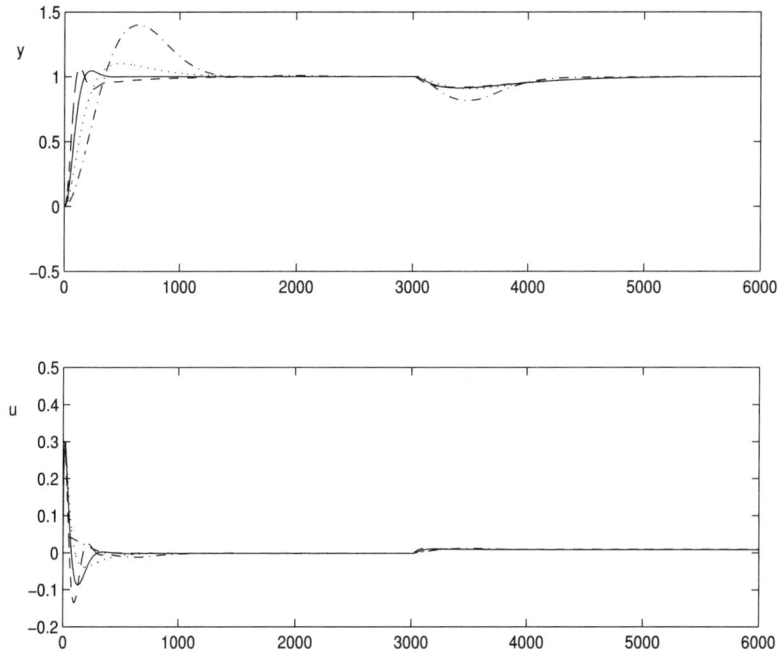

Fig. 11.6. PIMC and MPIMC performance
(solid lines: PIMC/MPIMC; dotted lines: PIMC with pole shift; dashed lines: PIMC with gain variation; dash dotted lines: Rotstein and Lewin method)

In the case that the set-point and disturbance is of step-type, the generalized process in the form of (11.19) is expressed as

$$\overline{G}(s) := \frac{G(s)}{s} = \overline{G}^+(s) + \overline{G}^-(s) = \frac{2.405 \times 10^5 s - 745}{s(-2442s + 1)} + \frac{3.664 \times 10^4}{372s + 1}.$$

The primary controller K is designed with respect to $\overline{G}^+$. With ξ and ω_0 in (11.25) specified as 0.707 and 0.02 respectively, the controller in the form of (11.28) is computed from (11.31) as

$$K(s) = 0.010 + \frac{2.200 \times 10^{-5}}{s}.$$

The pre-filter in (11.40) is calculated as

$$F(s) = \frac{372s + 1}{(453.843s + 1)(10s + 1)}.$$

The closed-loop response is shown in Figure 11.6, which is the same as the response of the system designed using the PIMC scheme.

For comparison, the PID tuning method using IMC parametrization (Rotstein and Lewin, 1992) is adopted here, the PID controller from that method being

$$K(s) = 0.0502(1 + \frac{1}{612s} + 101.97s).$$

The system response is shown in Figure 11.6 for comparison. Our method has a better performance.

Performance robustness of the proposed method under process perturbations is also considered. The PID settings are still the same as above. In the first case, we change the process gain from 745 to 745 × 2. In the second case, the unstable time constant is varied from 2442 to 2442 × 2. The resultant responses under these two circumstances are shown in Figure 11.6. The responses are quite satisfactory even under process perturbations.

Example 11.8.2. Consider the high-order unstable process in Shafiei and Shenton (1994):

$$G(s) = \frac{27}{(s-1)(s+2.8)^3}.$$

Partial fraction expansion gives

$$G(s) = G^+(s) + G^-(s) = \frac{0.4921}{s-1} + \frac{-0.4921s^2 - 4.6253s - 16.1984}{(s+2.8)^3}.$$

11.8 Simulation Results

With ξ and ω_0 in (11.25) specified as 0.707 and 1.5, the resultant primary controller is computed as

$$K(s) = \frac{4.943s + 1.000}{0.5466s - 0.7379}$$

and the pre-filter as

$$F(s) = \frac{(s+2.8)^3}{2.8^3(4.943s+1)(0.2s+1)^2}.$$

With these settings, the closed-loop response of the PIMC is illustrated in Figure 11.7. For the MPIMC scheme, the generalized process $\frac{G(s)}{s}$ is decomposed into

$$\bar{G} = \frac{G(s)}{s} = \frac{-0.7379s + 0.4921}{s(s-1)} + \frac{0.7379s^2 + 5.7063s + 12.7301}{(s+2.8)^3}.$$

With the same ξ and ω_0 as above, the primary controller is

$$K(s) = 1.179 + \frac{0.238}{s},$$

and the pre-filter is

$$F(s) = \frac{(s+2.8)^3}{2.8^3(4.943s+1)(0.2s+1)^2}.$$

The MPIMC performance is shown in Figure 11.7. The tuning procedure described by Shafiei and Shenton (1994) is quite involved and a pre-filter is also required there. The performance of his method is also given in Figure 11.7, which shows that the proposed method has a better response.

Example 11.8.3. Consider an oscillatory process with one unstable pole:

$$G(s) = \frac{0.25}{(s^2+s+0.4)(s-1.5)}e^{-0.3s}.$$

With (11.44), the dead time is approximated as

$$e^{-0.3s} \approx \frac{1-0.15s}{1+0.15s}.$$

The process then becomes

$$G(s) \approx \frac{0.25(1-0.15s)}{(s^2+s+0.4)(s-1.5)(1+0.15s)}.$$

In the case of the PIMC scheme, the process is decomposed into

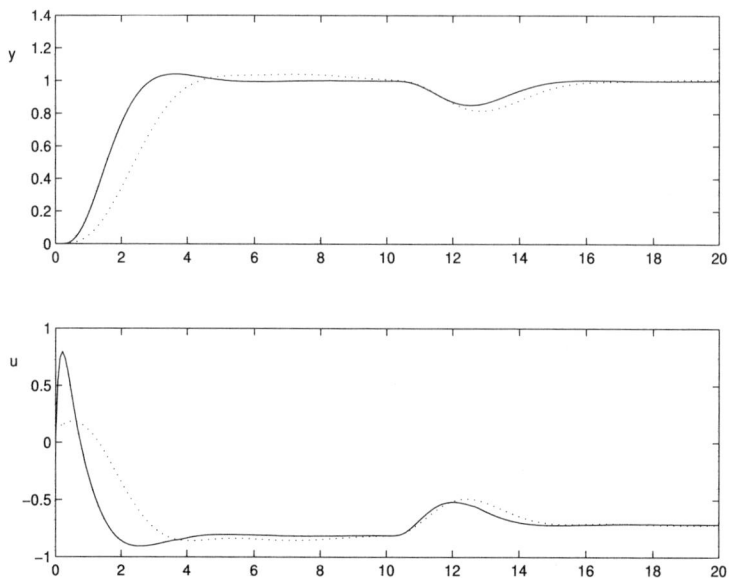

Fig. 11.7. PIMC and MPIMC performance
(solid lines: PIMC/MPIMC; dashed lines: Shafiei method)

$$G(s) \approx \frac{0.25(1 - 0.15s)}{(s^2 + s + 0.4)(s - 1.5)(1 + 0.15s)}$$
$$= \frac{0.0381}{s - 1.5} + \frac{-0.0318s^2 - 0.3943s - 1.0436}{(s^2 + s + 0.4)(s + 6.667)}.$$

When ξ and ω_0 are specified as 0.7 and 3, respectively, the primary controller is

$$K(s) = \frac{20.652s + 1.000}{0.069s - 0.3913},$$

and the pre-filter is

$$F(s) = \frac{0.15s + 1}{20.652s + 1}.$$

With these settings, the closed-loop response of the PIMC is illustrated in Figure 11.8. When the MPIMC scheme is adopted, the generalized process $\frac{G(s)}{s}$ is first expanded into

$$\overline{G} = \frac{G(s)}{s} = \frac{-0.391s + 0.625}{s(s - 1.5)} + \frac{0.391s^2 + 2.961s + 2.415}{(s^2 + s + 0.4)(s + 6.667)}.$$

With the same ξ and ω_0 as above, the primary controller K is determined as

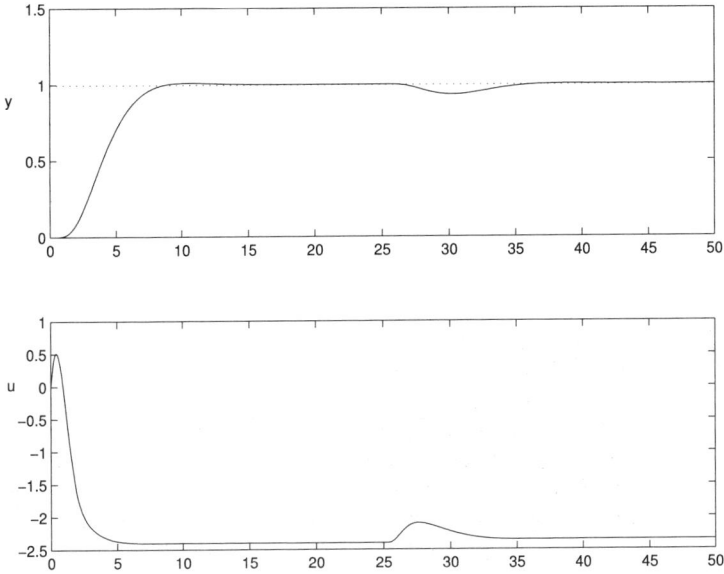

Fig. 11.8. PIMC and MPIMC performance

$$K(s) = 2.536 + \frac{0.124}{s}.$$

The pre-filter is designed as

$$F(s) = \frac{0.0186s + 0.124}{2.536s + 0.124}.$$

The control performance using our method is shown in Figure 11.8. Satisfactory performance is obtained using the proposed PIMC/MPIMC scheme.

11.9 Real-time Implementation

In this section, the proposed PIMC scheme is applied to a motor position control pilot-scale system. The system comprises an L. J. Electronics DC motor servo apparatus and a PC-based data acquisition system, as shown in Figure 11.9. The DC motor apparatus diagram is presented in Figure 11.10. It is obvious that the transfer function between the control voltage and the position in Figure 11.10 is unstable. It is true that a simple controller can be designed to realize motor position control due to the simplicity of this motor apparatus. Our intention in utilizing this motor apparatus is to demonstrate that our scheme is capable of being applied to a real process.

Fig. 11.9. A motor servo pilot system

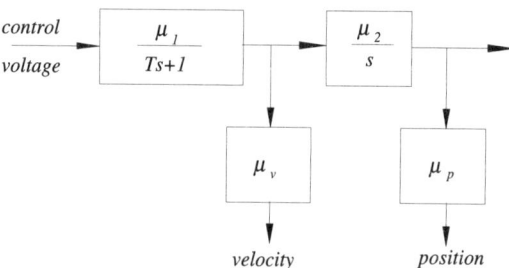

Fig. 11.10. Diagram of the motor apparatus

The motor apparatus is identified using both open-loop and closed-loop tests. First, a step set-point change of control voltage is applied to the open-loop motor apparatus. The time constant T is estimated by measuring the time for the motor velocity to reach 63.2% of its steady speed (Franklin *et al.*, 1994). Then, a closed loop is formed by feeding back the position signal to the reference via a constant controller. When a step set-point change in control voltage occurs, the time for the motor to reach its first peak is measured, from which we can calculate $\mu_1\mu_2$ (Franklin et al., 1994). The transfer function between the reference and the position output is then estimated as

$$G(s) = \frac{8.72}{(0.149s+1)s}.$$

In the proposed PIMC scheme, $G(s)$ is decomposed into

$$G(s) = G^+(s) + G^-(s) = \frac{8.72}{s} - \frac{8.72}{s+6.71}.$$

Only $G^-(s)$ is used as the internal model. With ξ and ω_0 specified as 0.7 and 7, respectively, the primary controller is computed as

$$K(s) = \frac{0.349s + 1.000}{0.178s - 1.299},$$

and the pre-filter as

$$F(s) = \frac{0.149s + 1}{0.349s + 1}.$$

The experimental result on the position control performance is shown in Figure 11.11 and is quite good.

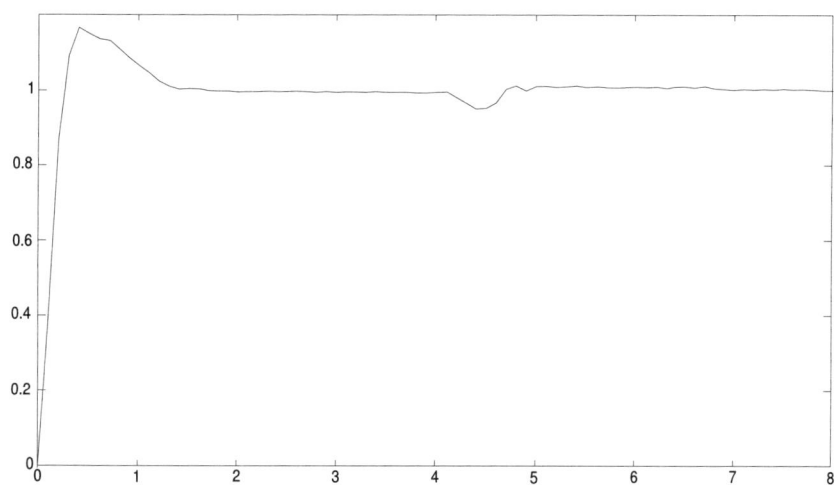

Fig. 11.11. Position control performance under PIMC

Appendix A: Controller Design for Processes with Two Unstable Poles

Let the process be represented by

$$G(s) = \frac{b(s)}{(s^2 + \nu_1 s + \nu_0)a^-(s)} = G^+(s) + G^-(s),$$

where $b(s)$ is a polynomial and $a^-(s)$ is a stable polynomial,

$$G^+(s) = \frac{\mu_1 s + \mu_0}{s^2 + \nu_1 s + \nu_0},$$

and

$$G^-(s) = \frac{b_2(s)}{a^-(s)}.$$

(i) PIMC Design

Let

$$K = \frac{k_2 s^2 + k_1 s + 1}{k_4 s^2 + k_3 s + G^-(0)}.$$

Then $K(0) = (G(0))^{-1}$. Such a controller K automatically satisfies (11.16) and has zero offset under step inputs for a stable closed loop. Suppose that the desired characteristic polynomial is specified as

$$P_d(s) = s^4 + \alpha_3 s^3 + \alpha_2 s^2 + \alpha_1 s + \alpha_0. \tag{11.45}$$

The characteristic polynomial of the actual PIMC system is

$$P = a^- k_4 [s^4 + (\nu_1 + \frac{k_3 + \mu_1 k_2}{k_4})s^3 + (\nu_0 + \frac{\nu_1 k_3 + \mu_1 k_1 + \mu_0 k_2 + G^-(0)}{k_4})s^2$$
$$+ \frac{\nu_0 k_3 + \mu_1 + \mu_0 k_1 + \nu_1 G^-(0)}{k_4} s + \frac{\mu_0 + \nu_0 G^-(0)}{k_4}]$$
$$:= a^- k_4 P_c.$$

Setting $P_c = P_d$ yields

$$\begin{bmatrix} 0 & \mu_1 & 1 & \nu_1 - \alpha_3 \\ \mu_1 & \mu_0 & \nu_1 & \nu_0 - \alpha_2 \\ \mu_0 & 0 & \nu_0 & -\alpha_1 \\ 0 & 0 & 0 & \alpha_0 \end{bmatrix} \begin{bmatrix} k_1 \\ k_2 \\ k_3 \\ k_4 \end{bmatrix} = \begin{bmatrix} 0 \\ -G^-(0) \\ -\mu_1 - \nu_1 G^-(0) \\ \mu_0 + \nu_0 G^-(0) \end{bmatrix},$$

from which the controller parameters $[k_1\ k_2\ k_3\ k_4]^T$ can be determined. The pre-filter F is designed in the same way as in Section 11.7.2, and given by

$$F(s) = \frac{b^-(0)a^-(s)}{a^-(0)b^-(s)(k_2 s^2 + k_1 s + 1)} \cdot \frac{1}{(\tau s + 1)^m}.$$

(ii) MPIMC Design

Suppose that the primary controller in a MPIMC system is given by

$$K = K_1 K_2,$$

where $K_1 = \frac{1}{s}$, $K_2 = \frac{k_2 s^2 + k_2 s + k_0}{k_3 s + 1}$. The generalized process is decomposed as

$$\overline{G} = \frac{G}{s} = \frac{\mu_2 s^2 + \mu_1 s + \mu_0}{s(s^2 + \nu_1 s + \nu_0)} + \frac{b_2}{a^-}$$
$$= \overline{G}^+ + \overline{G}^-.$$

Suppose that the desired characteristic polynomial is specified the same as (11.45) and is matched by the characteristic polynomial of the actual MPIMC, leading to

$$\begin{bmatrix} 0 & \mu_2 & \mu_1 - \mu_2 \alpha_3 & \nu_1 - \alpha_3 \\ \mu_2 & \mu_1 & \mu_0 - \mu_2 \alpha_2 & \nu_0 - \alpha_2 \\ \mu_1 & \mu_0 & -\mu_2 \alpha_1 & -\alpha_1 \\ \mu_0 & 0 & -\mu_2 \alpha_0 & \alpha_0 \end{bmatrix} \begin{bmatrix} k_0 \\ k_1 \\ k_2 \\ k_3 \end{bmatrix} = \begin{bmatrix} -1 \\ -\nu_1 \\ -\nu_0 \\ 0 \end{bmatrix},$$

where the controller parameters $[k_0 \; k_1 \; k_2 \; k_3]^T$ can be computed. The pre-filter in this case is

$$F(s) = \frac{b^-(0) k_0 a^-(s)}{a^-(0) b^-(s)(k_2 s^2 + k_1 s + k_0)} \cdot \frac{1}{(\tau s + 1)^m}.$$

Appendix B: Formulas for Decomposition of some Typical Unstable Processes

(i) If the unstable process is given by

$$G(s) = \frac{1}{s - \lambda^+} \cdot \frac{\mu_1 s + \mu_0}{s + \lambda^-}, \quad \lambda^+ > 0 \text{ and } \lambda^- > 0,$$

then we have

$$\overline{G}(s) = \frac{G(s)}{s} = \frac{b_1(s)}{s(s - \lambda^+)} + \frac{b_2}{s + \lambda^-},$$

where

$$b_1(s) = \frac{\mu_1 \lambda^- - \mu_0}{\lambda^-(\lambda^+ + \lambda^-)} s + \frac{\mu_0}{\lambda^-},$$

and
$$b_2(s) = \frac{-\mu_1 \lambda^- + \mu_0}{\lambda^-(\lambda^+ + \lambda^-)}.$$

(ii) If the unstable process is given by
$$G(s) = \frac{1}{s - \lambda^+} \cdot \frac{\mu_2 s^2 + \mu_1 s + \mu_0}{s^2 + \nu_1 s + \nu_0}, \quad \lambda^+ > 0, \ \nu_1 > 0 \ and \ \nu_0 > 0,$$

then it follows that
$$\overline{G}(s) = \frac{G(s)}{s} = \frac{b_1(s)}{s(s - \lambda^+)} + \frac{b_2(s)}{s^2 + \nu_1 s + \nu_0},$$

where
$$b_1(s) = -\theta s + \frac{\mu_0}{\nu_0},$$
$$b_2(s) = -\theta s + (\mu_2 - \theta \lambda^+ - \theta \nu_1 - \frac{\mu_0}{\nu_0}),$$

and
$$\theta = \frac{-\mu_0 \lambda^+ - \mu_0 \nu_1 + \mu_2 \lambda^+ \nu_0 + \mu_1 \nu_0}{\nu_0(\lambda^{+2} + \nu_1 \lambda^+ + \nu_0)}.$$

12. Decentralized Control

Processes with inherently more than one variable at the output to be controlled are frequently encountered in industry and are known as multivariable or multi-input multi-output (MIMO) processes. Interactions usually exist between control loops, which account for the renowned difficulty in their control compared with single-input single-output (SISO) processes. Depending on the application and requirement, multivariable controllers or multi-loop controllers can be adopted for MIMO processes. Multi-loop controllers, sometimes known as decentralized controllers, have much simpler structures and fewer tuning parameters to handle. In addition, in the case of actuator or sensor failure, they are relatively easy to stabilize manually, because only one loop is directly affected by the failure (Palmor *et al.*, 1996; Skogestad and Morari, 1989). Hence for processes with modest interactions, multi-loop controllers are often more favoured than multivariable controllers.

Many multi-loop controller design methods have been reported in the literature. In the BLT (Biggest Log Modulus Tuning) method presented by Luyben (1986), the familiar Ziegler–Nichols rule is modified with the inclusion of a detuning factor, which determines the trade-off between stability and performance of the system. Individual controllers are designed for their respective loops by first ignoring all interactions. The calculated controller gains are then scaled by the detuning factor to guarantee stability. Despite the simple computations involved, the design regards interactions as elements obstructing system stability and attempts to dispose of them rather than control them to speed up individual loops. It is hence too conservative to exploit process structures and characteristics for the best achievable performance. In the sequential loop closing method, loops are closed one after another, with those previously closed fitted with appropriate controllers. The interactions are well taken care of only if the loops are of considerably different bandwidths and the closing sequence starts from the fastest loop. This assumption of high gains in the loops that have already been closed can rarely be justified, except at low frequencies (Maciejowski, 1989). There are some other existing methods which need iterations. The designer changes the trade-off between conflicting constraints by adjusting certain inequalities or objective functions (Maciejowski, 1989; Zakian, 1979) until satisfactory response is obtained. As with all other iterative designs, the major drawback with these methods is the initial estimate and convergence problem.

In this chapter, a simple independent design method for multi-loop controllers is presented which exploits process interactions for the improvement of loop performance. Unlike many other methods that emphasize suppression of interactions for decoupling purposes, our method is developed to channel the effect of interactions to individual loops to speed up loop responses. This is achieved by regarding each loop together with its corresponding interactions from all other loops as an equivalent single-input single-output (SISO) plant, and designing an independent SISO controller for it. Once an objective transfer function is specified for each of these equivalent processes, a set of simultaneous equations is formed and separated into independent ones, each of which contains one controller element only. They are then solved to obtain exact solutions, which are usually irrational. The exact solutions can be well approximated by rational functions. The popular multi-loop PID controllers can naturally be obtained as a special case of rational approximation, and they give a reasonable trade-off between loop and decoupling performance. Simulation examples are provided to show the effectiveness of the proposed method, and comparisons are made with the BLT method.

This chapter is organized as follows. In Section 12.1, the proposed method is presented. A detailed discussion on the choice of solutions to the controller gain equations is given in Section 12.2. The rational function approximation method follows in Section 12.3. In Section 12.4, the reduction of controller structure to the PID format is investigated. Section 12.5 focuses on stability analysis of the feedback system. Extension of the proposed method to the m by m case is provided in Section 12.6. Simulation examples are included in the various sections for illustrations.

12.1 The Proposed Independent Design Strategy

Consider a stable 2×2 process described via the transfer function matrix $G(s)$ as $Y(s) = G(s)U(s)$ or

$$\begin{bmatrix} y_1(s) \\ y_2(s) \end{bmatrix} = \begin{bmatrix} g_{11}(s) & g_{12}(s) \\ g_{21}(s) & g_{22}(s) \end{bmatrix} \begin{bmatrix} u_1(s) \\ u_2(s) \end{bmatrix}. \tag{12.1}$$

12.1 The Proposed Independent Design Strategy

Assume that proper input–output pairing has been made for the process. If the process is inherently poorly paired, the RGA (Relative Gain Array) method (Maciejowski, 1989) can be employed to make the necessary arrangement. The process is to be controlled by a multi-loop controller with the structure:

$$K(s) = \begin{bmatrix} k_1(s) & 0 \\ 0 & k_2(s) \end{bmatrix}. \tag{12.2}$$

The resultant control system is shown in Figure 12.1. The controller design objective is to find $k_1(s)$ and $k_2(s)$ such that both loops achieve satisfactory performance. The boxed portion in Figure 12.1 can be viewed as an individual

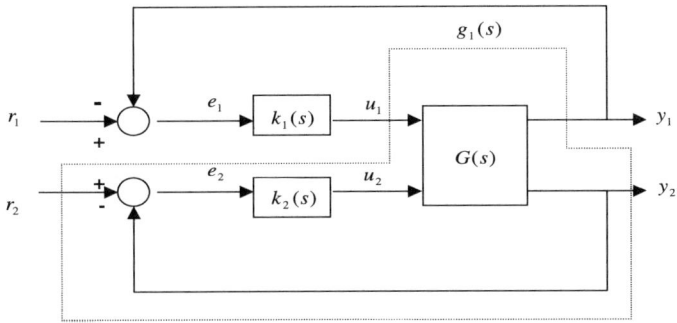

Fig. 12.1. Equivalent $g_1(s)$ for a 2×2 plant

SISO process with an equivalent transfer function $g_1(s)$ between input u_1 and output y_1. It follows that $g_1(s)$ can be obtained (Maciejowski, 1989) as

$$g_1 = g_{11} - \frac{k_2 g_{12} g_{21}}{1 + k_2 g_{22}}. \tag{12.3}$$

Similarly, the equivalent plant between u_2 and y_2 is given by

$$g_2 = g_{22} - \frac{k_1 g_{21} g_{12}}{1 + k_1 g_{11}}. \tag{12.4}$$

In principle, $k_1(s)$ and $k_2(s)$ should be designed for $g_1(s)$ and $g_2(s)$ respectively. But the generalized process g_1 is a function of k_2 as seen from (12.3), and therefore we cannot design the controller k_1 for it unless k_2 is known beforehand. Similarly, to design k_2 requires k_1 to be known. Hence, these conditions fall into a loop and are the main difficulty in multi-loop controller deign. A common way of solving this problem is by iterations (Maciejowski, 1989). Here, we introduce a new method which needs no iterations and therefore does not suffer the convergence problem. The performance specifications on individual closed loops may be formulated in terms of the desired closed-loop transfer functions:

$$h_{d_i} = \frac{\omega_i^2 e^{-sL_i}}{s^2 + 2\xi_i \omega_i s + \omega_i^2}, \quad i = 1, 2. \tag{12.5}$$

The corresponding desired open-loop transfer functions are

$$q_{d_i} = \frac{h_{d_i}}{1 - h_{d_i}}, \quad i = 1, 2. \tag{12.6}$$

The k_i are designed to match the actual open loops $g_i k_i$ to the desired open-loops q_{d_i}:

$$g_1 k_1 = \left(g_{11} - \frac{k_2 g_{12} g_{21}}{1 + k_2 g_{22}} \right) k_1 = q_{d_1} = \frac{\omega_1^2 e^{-sL_1}}{s^2 + 2\xi_1 \omega_1 s + \omega_1^2 (1 - e^{-sL_1})}, \tag{12.7}$$

$$g_2 k_2 = \left(g_{22} - \frac{k_1 g_{12} g_{21}}{1 + k_1 g_{11}} \right) k_2 = q_{d_2} = \frac{\omega_2^2 e^{-sL_2}}{s^2 + 2\xi_2 \omega_2 s + \omega_2^2 (1 - e^{-sL_2})}. \tag{12.8}$$

This pair of coupling equations are solved to obtain the following two decoupled equations:

$$[(1 + q_{d2})g_{11}\Delta]k_1^2 + [(1 - q_{d1}q_{d2})g_{11}g_{22} + (q_{d2} - q_{d1})\Delta]k_1$$
$$+ [-(1 + q_{d2})q_{d1}g_{22}] = 0, \tag{12.9}$$

$$[(1 + q_{d1})g_{22}\Delta]k_2^2 + [(1 - q_{d1}q_{d2})g_{11}g_{22} + (q_{d1} - q_{d2})\Delta]k_2$$
$$+ [-(1 + q_{d1})q_{d2}g_{11}] = 0, \tag{12.10}$$

where

$$\Delta = g_{11}g_{22} - g_{12}g_{21}. \tag{12.11}$$

Therefore, each gain k_i can be determined separately by solving the corresponding single quadratic equation.

The values of $\{\omega_1, \omega_2\}$, $\{L_1, L_2\}$ and $\{\xi_1, \xi_2\}$ have to be determined before (12.9) and (12.10) are solved for $k_1(s)$ and $k_2(s)$. It is found from simulations that in general suitable values of ω_i lie between 0.6 and 0.8 times the critical frequencies $\omega_{ci}(\angle g_{ii}(j\omega_{ci}) = -\pi)$ of the respective diagonal elements $g_{ii}(s)$. As a guideline, this is taken to be 0.7 times as default. The value of L_i is chosen to be the equivalent dead time τ_i in the element g_{ii}. For g_{ii} of order higher than one, τ_i can be approximated by fitting g_{ii} into the first-order plus dead time model $g_{ii}(s) = \frac{a_i}{1+sT_i} e^{-sT_i}$ at the point $s = 0$ and $s = j\omega_{ci}$, or obtained equivalently from the equation:

$$\tau_i = \frac{\pi - \tan^{-1}\sqrt{|\frac{g_{ii}(0)}{g_{ii}(j\omega_{ci})}|^2 - 1}}{\omega_{ci}}. \tag{12.12}$$

The damping ratios $\{\xi_1, \xi_2\}$ are all set to 0.707 for a fair trade-off between speed and the amount of overshoot. Hence, we recommend

$$\left.\begin{array}{l} \omega_i = 0.7\omega_{ci}, \\ \xi_i = 0.707, \\ L_i = \tau_i. \end{array}\right\} \quad i = 1, 2. \tag{12.13}$$

12.2 Choice of Solutions to Controller Gain Equations

In general, (12.9) or (12.10) has multiple or two solutions at each frequency point $s = j\omega_j$, $j = 1, 2, \ldots, N$. Every solution satisfies the corresponding equation from a mathematical point of view. However, when stability is concerned, one solution has to be discarded. In order to address this problem, typical solutions to (12.9) are plotted in the complex plane as shown in Figure 12.2. Similar treatment can be applied to (12.10). Since $k_1(s)$ has two solutions, two branches are shown in Figure 12.2. According to the Nyquist stability criterion, for a stable open-loop system there has to be no encirclement of the point -1 by the Nyquist plot of the loop gain as $j\omega$ travels along the imaginary axis, to ensure closed-loop stability. To suppress the steady-state error in response to step inputs, integrators have to be present in k_1 and k_2. Owing to these explicitly introduced poles in the controllers at the origin, the Nyquist path is modified to bypass the origin by making a semi-circular turn of infinitesimal radius to the right of the origin so that the contour engulfs the entire right half complex plane except the origin. Since $\lim_{s\to\infty}[g_1(s)k_1(s)] = 0$, $\lim_{s\to 0+,real}[g_1(s)k_1(s)]$ has to be positive in order not to have any encirclement of the point -1. It follows that if $\lim_{s\to 0+,real}[g_1(s)] > 0$, then $\lim_{s\to 0+,real}[k_1(s)]$ has to be positive, or equivalently $k_1(s)$ has to start from negative infinity in the complex plane. Owing to the presence of an integrator in $k_2(s)$, we have $\lim_{s\to 0+,real}[k_2(s)] = \infty$. It can then be readily evaluated from (12.3) that $\lim_{s\to 0+,real}[g_1(s)]$ equals $g_{11}(0) - \frac{g_{12}(0)g_{21}(0)}{g_{22}(0)}$. Conversely, if $\lim_{s\to 0+,real}[g_2(s)] = g_{22}(0) - \frac{g_{21}(0)g_{12}(0)}{g_{11}(0)} < 0$, then $\lim_{s\to 0+,real}[k_1(s)]$ has to be negative, and thus $k_1(s)$ has to start from positive infinity in the complex plane.

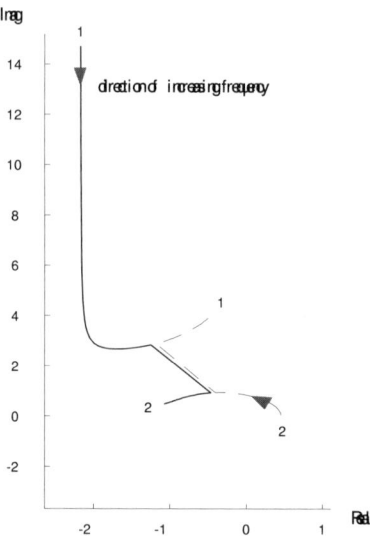

Fig. 12.2. Nyquist plot of controller gain $k_1(s)$

Suppose $\lim_{s \to 0^+, real}[g_1(s)] = g_{11}(0) - \frac{g_{12}(0)g_{21}(0)}{g_{22}(0)} < 0$. Based on the above argument, branch 1 in Figure 12.2 should be chosen as the solution. Unlike in the above discussion on the choice of branches for the stability of the closed loop, we now shift our focus to the stability of the controller itself. In order to have all poles and zeros of the controller $k_1(s)$ in the left half complex plane, the Nyquist plot of the controller $k_1(s)$ should have no encirclement of the origin. Following the earlier argument that $\lim_{s \to 0^+, real}[g_1(s)] < 0$ requires $\lim_{s \to 0^+, real}[k_1(s)] < 0$, $k_1(s)$ should cut the negative real axis at some high frequency for the plot not to encircle the origin. As observed from Figure 12.2, branch 1 heads towards the real axis initially but detours on the way, and shows no sign of cutting the negative real axis as frequency advances. On the contrary, branch 2 turns towards the negative real axis in its late stage. As a result, if we break branch 1 halfway and join it to branch 2, the resulting hybrid branch will make the correct crossing and ensure stable controller poles and zeros. To minimize the gap caused by branch jumping, the break point is chosen to be the point where the two branches are nearest to each other. The re-combined solution branch is shown as a solid line in Figure 12.2, which starts with branch 1 and ends with 2. It forms the ultimate solution to (12.9) where stability is guaranteed. It is noted from (12.7) and (12.8) that $k_2(s)$ is dependent on $k_1(s)$.

Hence whenever there is branch jumping in $k_1(s)$, a corresponding branch jump should also be performed on $k_2(s)$.

This analysis leads to the following rules for the solution selection:

(i) For each i, if $g_i(0) > 0$, the solution should be chosen so that $k_i(j\omega)$ for a very small ω has an imaginary part at negative infinity. It should cross the positive real axis as the frequency approaches infinity. If that is not satisfied, branch hopping has to be carried out at the point where the chosen branch is nearest to some other branch that cuts the real positive axis at infinite frequency.

(ii) Likewise, if $g_i(0) < 0$, the imaginary part of $k_i(j\omega)$ for very small ω should be located at positive infinity and should intersect the negative real axis by possible switching among the solution branches.

In the above discussion, we base solution selection on the assumption of having a stable open loop (stable equivalent process $g_i(s)$ and controller $k_i(s)$), as a result of which there must be no encirclement of the point -1 for closed-loop stability according to the Nyquist stability criterion. For the 2×2 case, there are two solutions for each $k_1(s)$ and $k_2(s)$. If the alternative solutions are selected or no branch jumping is performed, there will be encirclements of the point -1. However, it is noticed that closed-loop stability can still be guaranteed in such cases if the open loop makes the same number of encirclements of the origin. If the equivalent processes are all open-loop stable, these encirclements must be contributed by unstable poles and zeros of the controllers. This in practice is not desirable although closed-loop stability is still achieved, because it results in only a conditionally stable system, where an unconditionally stable system is usually preferred if possible. Hence, in our method, only stable controllers are considered.

To further illustrate how the solutions can be selected, we now look at some popularly used examples. In all the examples, the objective transfer functions q_{d1} and q_{d2} are obtained according to the default settings given in (12.13).

Example 12.2.1. Consider the eight-tray + reboiler distillation column separating methanol and water studied by Wood and Berry (Luyben, 1986). The transfer function matrix is given by

$$G(s) = \begin{bmatrix} \frac{12.8e^{-s}}{16s+1} & \frac{-18.9e^{-3s}}{21s+1} \\ \frac{6.6e^{-7s}}{10.9s+1} & \frac{-19.4e^{-3s}}{14.4s+1} \end{bmatrix}.$$

The elements g_{11} and g_{22} have critical frequencies of 1.62 rad/s and 0.58 rad/s and equivalent dead times of 1 s and 3 s respectively. The objective transfer functions are thus set according to (12.13) as

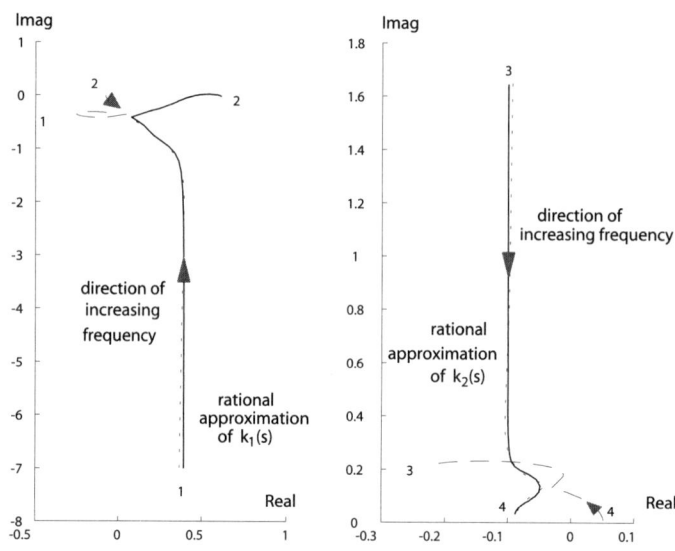

Fig. 12.3. Approximation of controller gains by rational functions in Example 12.2.1

$$q_{d1} = \frac{\omega_1^2 e^{-sL_1}}{s^2 + 2\xi_1\omega_1 s + \omega_1^2(1-e^{-sL_1})} = \frac{1.134^2 e^{-s}}{s^2 + 1.414 \times 1.134s + 1.134^2(1-e^{-s})},$$

$$q_{d2} = \frac{\omega_2^2 e^{-sL_2}}{s^2 + 2\xi_2\omega_2 s + \omega_2^2(1-e^{-sL_2})} = \frac{0.404^2 e^{-3s}}{s^2 + 1.414 \times 0.406s + 0.406^2(1-e^{-3s})}.$$

The values of $k_1(s)$ and $k_2(s)$ are then calculated at $s = j\omega$ using (12.9) and (12.10) and are plotted in Figure 12.3. There are two roots for each of them, corresponding to the two branches shown in dashed lines and dotted lines in the figure. Since $\lim_{s \to 0+, real}[g_1(s)] = g_{11}(0) - \frac{g_{12}(0)g_{21}(0)}{g_{22}(0)} = 6.4 > 0$ and $\lim_{s \to 0+, real}[g_2(s)] = g_{22}(0) - \frac{g_{21}(0)g_{12}(0)}{g_{11}(0)} = -9.7 < 0$, branches 1 and 3 are chosen as the solutions at low frequency. However, they do not move towards the real axis for intersection at high frequency. Therefore, the solutions have to hop to branches 2 and 4 at the nearest points. The resulting Nyquist plots for the final solution are given in solid lines.

Example 12.2.2. The 24-tray tower separating methanol and water examined by Vinante and Luyben (Luyben, 1986) has the following transfer function matrix:

$$G(s) = \begin{bmatrix} \frac{-2.2e^{-s}}{7s+1} & \frac{-1.3e^{-0.3s}}{7s+1} \\ \frac{-2.8e^{-1.8s}}{9.5s+1} & \frac{4.3e^{-0.35s}}{9.2s+1} \end{bmatrix}.$$

Following the rules, we obtain $k_1(j\omega)$ and $k_2(j\omega)$ as shown in Figure 12.4.

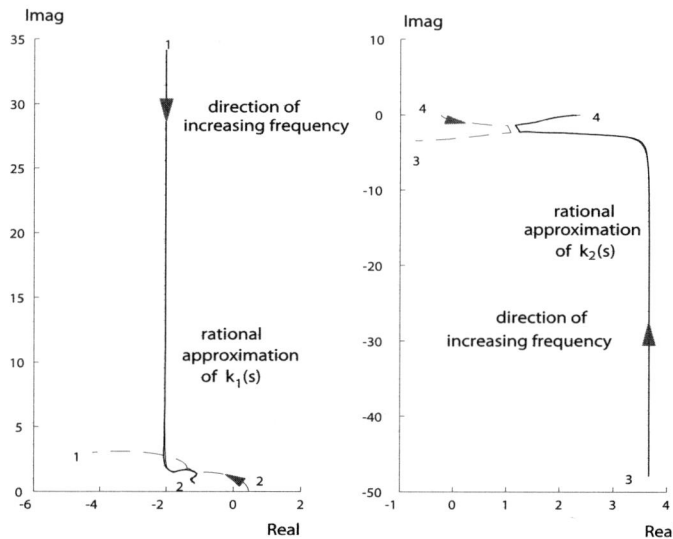

Fig. 12.4. Approximation of controller gains by rational functions in Example 12.2.2

Example 12.2.3. The process studied by Wardle and Wood (Luyben, 1986) has the following transfer function matrix:

$$G(s) = \begin{bmatrix} \frac{0.126e^{-6s}}{60s+1} & \frac{-0.101e^{-12s}}{(48s+1)(45s+1)} \\ \frac{0.094e^{-8s}}{38s+1} & \frac{-0.12e^{-8s}}{35s+1} \end{bmatrix}.$$

$k_1(j\omega)$ and $k_2(j\omega)$ are computed and the selected solutions are shown in Figure 12.5.

12.3 Rational Approximation of the Irrational Solutions

It should be noted that the solutions obtained in Section 12.2 and selected in Section 12.3 are irrational in general, i.e., such a solution $k(j\omega)$ may not have a rational function realization. For practical implementation, a rational function approximation $\widehat{k}(s)$ is required to approximate a possible irrational $k(s)$. The problem at hand is to find a rational function:

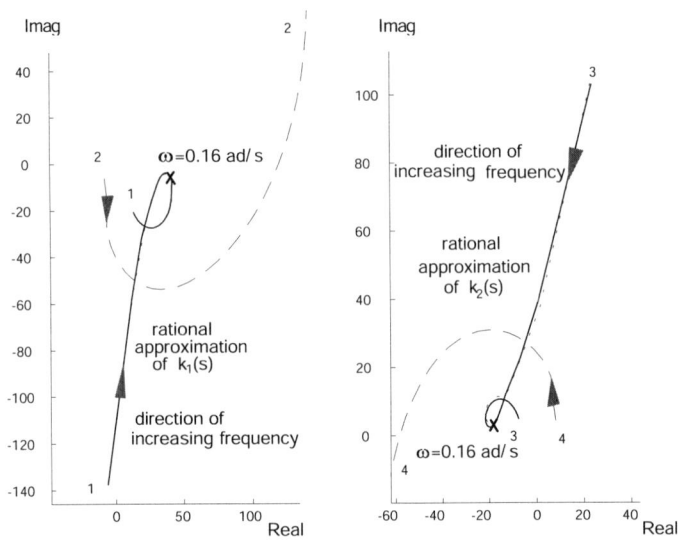

Fig. 12.5. Approximation of controller gains by rational functions in Example 12.2.3

$$\widehat{k}(s) = \frac{b_m s^m + b_{m-1} s^{m-1} + \ldots + b_0}{s^n + a_{n-1} s^{n-1} + \ldots + a_1 s}, \qquad (12.14)$$

where an integrator has been imposed in $\widehat{k}(s)$ to match that of $k(s)$ such that the approximation error e defined as

$$e = \sqrt{\frac{1}{N} \sum_{j=1}^{N} |k(j\omega_j) - \widehat{k}(j\omega_j)|^2}, \qquad (12.15)$$

meets

$$|e \cdot g(0)| < \varepsilon, \qquad (12.16)$$

where $g(0)$ is the equivalent process $g(s)$ at zero frequency and ε is a specified threshold and usually takes the value 0.1 ~ 0.15. We do not set a direct constraint on the error e because processes with high gains are usually fitted with controllers having relatively small gains, which naturally lead to smaller error e. Hence, the absolute error is scaled by the static gain of the equivalent process for a more reasonable evaluation of the approximation. The recursive least squares algorithm presented in Chapter 7 can then be used to solve this rational approximation problem. The data points in k_1 and k_2 used for the

12.3 Rational Approximation of the Irrational Solutions

approximation should stretch from positive/negative infinity to somewhere in close proximity to the real axis so as to capture the general trends at high frequencies. Usually these points can be conveniently taken from the frequency range $(0, \min\{\omega_{c1}, \omega_{c1}\})$, where ω_{ci} as defined earlier corresponds to the critical frequency of the element g_{ii} of the process. The range is usually sufficient unless k_1 and k_2 make sharp turns at high frequencies.

This method provides a general framework for the design of rational function controllers. Note that PID controllers can be viewed as a special case of rational function controllers with the order of numerator and denominator being two and one respectively. It is noted that although nonlinear least squares and stochastic (or statistical) methods may also be used, they demand far more computational resources and hence are not recommended.

We shall now apply the approximation method to demonstrate its effectiveness. Comparisons will be made with the well-known BLT design introduced by Luyben (1986) to show how our method can improve control performance. A threshold of 0.1 is set in (12.16) for the determination of controller orders.

Example 12.2.1 (cont'd). The recursive least squares method is then used to compute the coefficients of the rational function. Since $\min\{\omega_{c1}, \omega_{c2}\} = 0.58$ rad/s, and the general trend at high frequencies is well exhibited in the range $(0, 0.58)$ rad/s, the points in this range are used for the approximation. This results in the multi-loop controller:

$$K(s) = \mathrm{diag}\left\{\frac{n_1}{d_1}, \frac{-0.0904s^3 - 0.0196s^2 - 0.00262s - 0.000257}{s^3 + 0.0640s^2 + 0.0156s}\right\},$$

where

$$n_1 = 0.485s^7 + 0.318s^6 + 0.117s^5 + 0.0249s^4 + 0.00391s^3 + 0.000469s^2$$
$$+ 0.0000340s + 0.00000229,$$

$$d_1 = s^7 + 0.453s^6 + 0.218s^5 + 0.0264s^4 + 0.0054s^3 + 0.000301s^2 + 0.0000327s.$$

The error–gain products $|e \cdot g(0)|$ as defined in (12.16) are 0.054 and 0.049 for $k_1(s)$ and $k_2(s)$, respectively. Their Nyquist plots are shown in Figure 12.3 and close fitting is achieved. The step response of the feedback system using the above controller is shown in Figure 12.6. For comparison, the step response using the BLT method with

$$K_{BLT}(s) = \mathrm{diag}\left[0.375(1 + \frac{1}{8.29s}), -0.075(1 + \frac{1}{23.6s})\right]$$

is also given in the figure. It is observed that the proposed method has significant improvement in the loop performance. To ease the assessment of controller performance, the performance index ISE (Integral Square Error) with

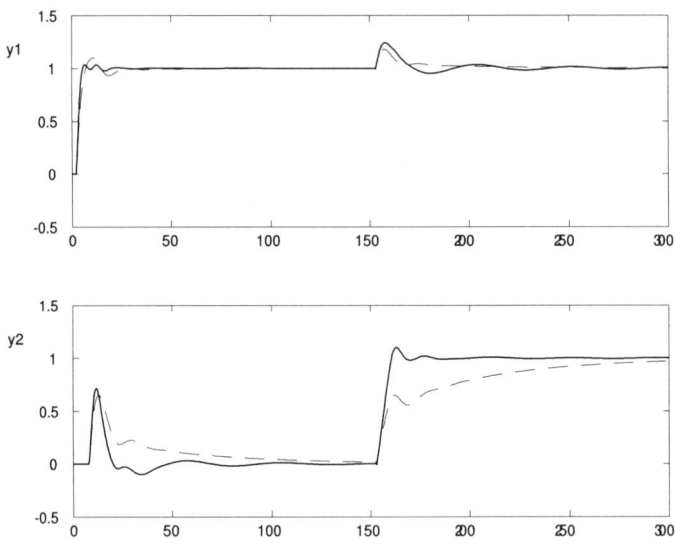

Fig. 12.6. Step response for Example 12.2.1
(solid lines: proposed method; dashed lines: BLT method)

equal weighting of the two outputs are calculated for the proposed and BLT methods as 10.2 and 18.22, respectively. Figure 12.7 shows the Nyquist plots of the equivalent open-loop gains. It is seen that they are reasonably close to the desired values. The dashed lines show those of the corresponding $g_1 k_{BLT_1}$ and $g_2 k_{BLT_2}$.

Example 12.2.2 (cont'd). The solid lines in Figure 12.4 show the Nyquist plots of the controller gains obtained using rational function approximation, and they are extremely close to the exact gains given by the dashed lines. The controller is

$$K(s) = \mathrm{diag}\left\{\frac{-1.03s^5 - 1.50s^4 - 0.735s^3 - 0.453s^2 - 0.0951s - 0.00651}{s^5 + 0.664s^4 + 0.391s^3 + 0.166s^2 + 0.0191s}, \frac{n_2}{d_2}\right\}.$$

where

$n_2 = 2.76s^7 + 1.76s^6 + 2.92s^5 + 1.14s^4 + 0.963s^3 + 0.217s^2 + 0.101s + 0.0100,$
$d_2 = s^7 + 0.696s^6 + 0.872s^5 + 0.380s^4 + 0.240s^3 + 0.0519s^2 + 0.0210s.$

12.3 Rational Approximation of the Irrational Solutions

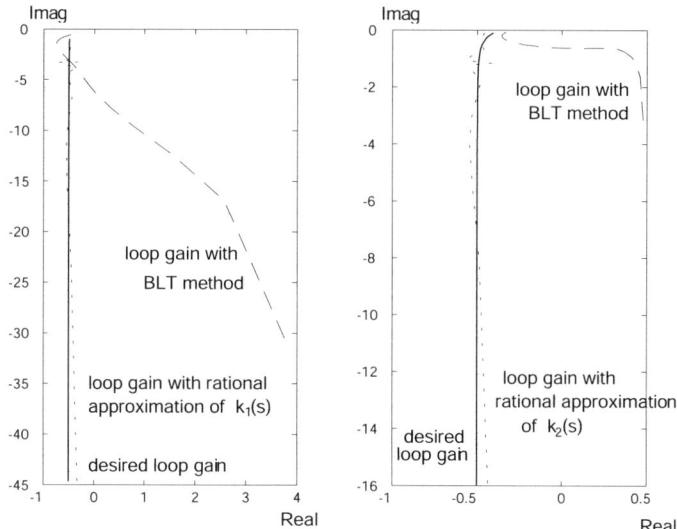

Fig. 12.7. Nyquist plot of equivalent open-loop gains in Example 12.2.1

The step response of the feedback system is shown in Figure 12.8. As seen from the figure, the proposed controller gives very much better loop performance than the BLT method with

$$K_{BLT}(s) = \text{diag}\left[-1.07(1+\frac{1}{7.1s}), 1.97(1+\frac{1}{2.58s})\right],$$

and the performance index ISE of the proposed and BLT methods is calculated as 3.09 and 3.86, respectively. The corresponding Nyquist plots of the equivalent open-loop gains are given in Figure 12.9.

Example 12.2.3 (cont'd). Following the same procedure, the controller designed is

$$K(s) = \text{diag}\left\{\frac{11.3s^5 + 18.4s^4 + 15.1s^3 + 1.21s^2 + 0.0414s + 0.000473}{s^5 + 0.730s^4 + 0.377s^3 + 0.0301s^2 + 0.000280s}\right.$$
$$\left.\frac{-8.55s^4 - 7.16s^3 - 2.42s^2 - 0.0995s - 0.00216}{s^4 + 0.348s^3 + 0.131s^2 + 0.00161s}\right\}.$$

The points used for the approximation lie in the frequency range $\{0, 0.8\}$ since there is a sharp turn in both k_1 and k_2 at frequencies above $min\{\omega_{c1}, \omega_{c2}\} = 0.16$ rad/s. Close fitting of the approximate gains to the exact gains is seen in Figure 12.5. The step response of the feedback system is shown in Figure 12.10.

360   12. Decentralized Control

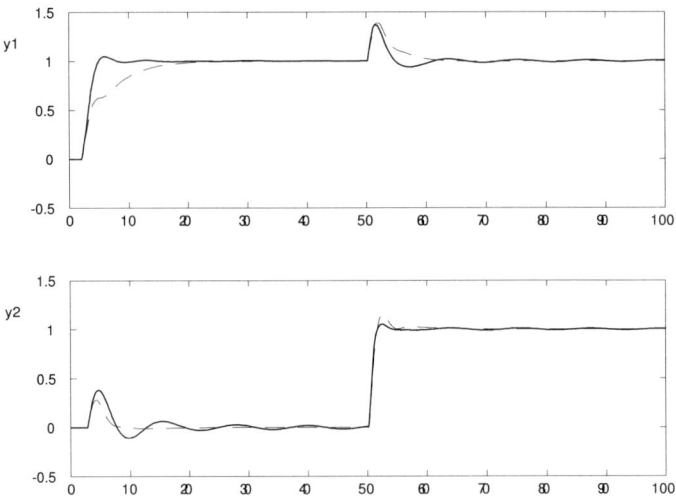

Fig. 12.8. Step response for Example 12.2.2
(solid lines: proposed method; dashed lines: BLT method)

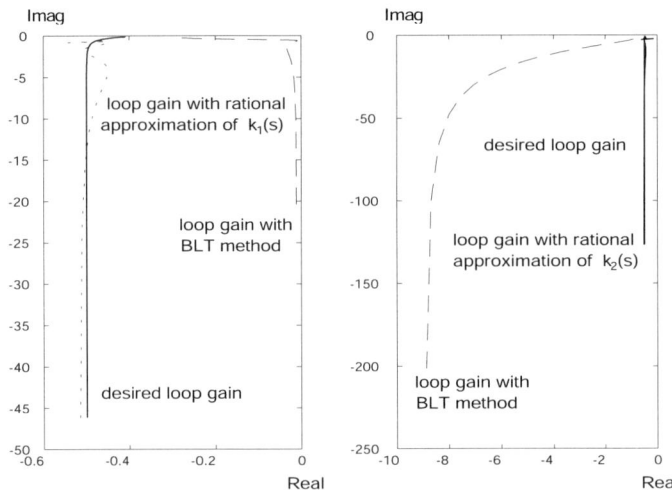

Fig. 12.9. Nyquist plot of equivalent open-loop gains in Example 12.2.2

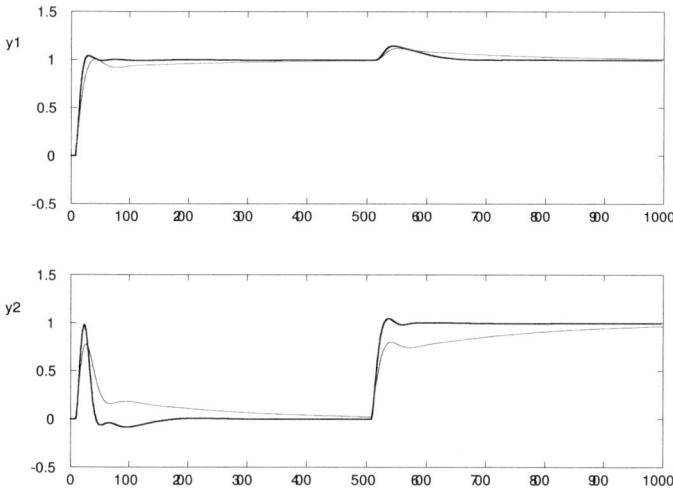

Fig. 12.10. Step response for Example 12.2.3
(solid lines: proposed method; dashed lines: BLT method)

Excellent response is obtained as seen from the figure when compared with the BLT method with

$$K_{BLT}(s) = \text{diag}\left[27.4(1 + \frac{1}{41.4s}), -13.3(1 + \frac{1}{52.9s})\right].$$

The performance index ISE of the proposed and BLT methods is calculated as 39.1 and 56.4, respectively. The approximate open-loop gains using the proposed controller are remarkably close to the desired values as shown in Figure 12.11.

12.4 Controller Reduction and Performance Trade-off

It is observed from the above results that although the individual loop performances have been significantly improved, the closed-loop coupling between the loops may sometimes be still too large. The reason is that our design has focused solely on loop performance and paid no attention to decoupling performance. It is possible to enhance the latter at the expense of the former. This can easily be achieved by increasing the value of ε in (12.16) which gives rise to lower orders of $k_i(s)$. An additional advantage of such a design trade-off

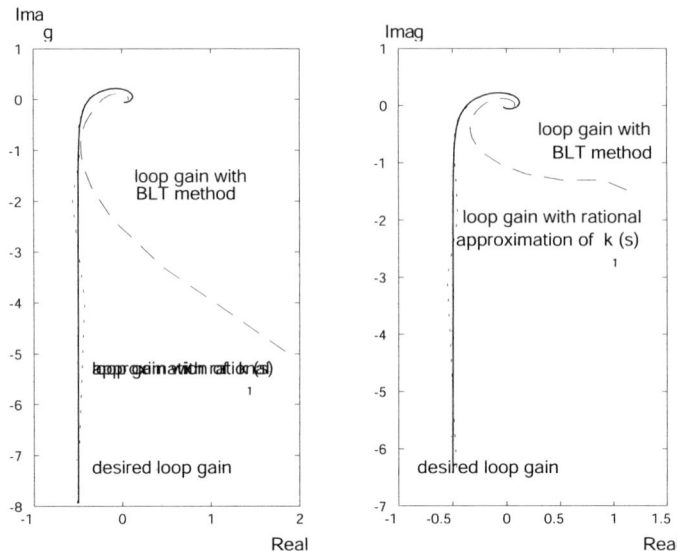

Fig. 12.11. Nyquist plot of equivalent open-loop gains in Example 12.2.3

is that low-order controllers are cheaper and easier to implement. The most widely used and commercially available low-order controllers are the PID type. In our framework of the rational function approximation method described above, multi-loop PID controllers can be obtained simply by specifying the orders of the numerator and denominator as two and one respectively (note that an integrator has already been explicitly implemented in the algorithm). Furthermore, for this particular PID controller structure, we can directly apply the linear least squares method in the frequency domain instead of the recursive least squares approach.

Let the multi-loop controller be of PID type for loop i

$$\widehat{k}_i(s) = k_{Pi} + k_{Ii}\frac{1}{s} + k_{Di}s = \begin{bmatrix} 1 & \frac{1}{s} & s \end{bmatrix} \begin{bmatrix} k_{Pi} \\ k_{Ii} \\ k_{Di} \end{bmatrix}. \qquad (12.17)$$

Its parameters are determined to match $k_i(s)$ as closely as possible in the frequency domain. As shown in Section 2 of Chapter 9, the standard linear least squares method can then be applied to give the required PID parameters. Based on extensive simulations, the objective transfer functions q_{di} in (12.5) and (12.6) in this case are recommended to have the following settings:

$$\left.\begin{array}{l}\omega_i = \omega_{ci} \\ \xi_i = 0.707 \\ L_i = \tau_i\end{array}\right\} \quad i = 1, 2. \tag{12.18}$$

We now apply the method to the same plants and compare the PID controller performance with that of rational function controllers.

Example 12.2.1 (cont'd). The bandwidths of the objective transfer functions are set to be the critical frequencies of the diagonal elements g_{11} and g_{22}, i.e., $\omega_1 = 1.62$ rad/s and $\omega_2 = 0.58$ rad/s, so that

$$q_{d1} = \frac{\omega_1^2 e^{-sL_1}}{s^2 + 2\xi_1\omega_1 s + \omega_1^2(1 - e^{-sL_1})} = \frac{1.62^2 e^{-s}}{s^2 + 1.414 \times 1.62 s + 1.62^2(1 - e^{-s})},$$

$$q_{d2} = \frac{\omega_2^2 e^{-sL_2}}{s^2 + 2\xi_2\omega_2 s + \omega_2^2(1 - e^{-sL_2})} = \frac{0.58^2 e^{-3s}}{s^2 + 1.414 \times 0.58 s + 0.58^2(1 - e^{-3s})}.$$

We next compute the controller gains k_1 and k_2 in the usual way and approximate them by PID controllers using the linear least squares method introduced above. The approximation results are shown in Figure 12.12. The multi-loop PID controller is obtained as

$$K(s) = \text{diag}\left[0.571 + \frac{0.085}{s} + 0.581s, -0.113 - \frac{0.020}{s} - 0.119s\right].$$

The step response of the feedback system is shown in Figure 12.13. The step response using the BLT method is shown as dashed lines in the figure. The performance index ISE of the proposed and BLT methods is calculated as 8.2 and 18.22, respectively. It is observed that the proposed method (solid lines) gives better loop performance and shorter settling time for the couplings. Although the loop performance is not as good as that using rational function controllers, the decoupling capability is better than the former.

Example 12.2.2 (cont'd). With the proposed method, we obtain k_1 and k_2 as

$$K(s) = \text{diag}\left[-1.929 - \frac{0.416}{s}, \ 2.919 + \frac{0.590}{s} + 0.283s\right],$$

whose Nyquist plots are shown in Figure 12.14. The step response of the feedback system is shown in Figure 12.15. The proposed PID controller shows better loop performance and coupling response than the BLT method in loop 1 and they are comparable in loop 2. The performance index ISE of the proposed and BLT methods is calculated as 2.63 and 3.86, respectively. It is again observed that although the PID controller does not generate loop performance as good as that given by the rational function controller, it can handle interactions better.

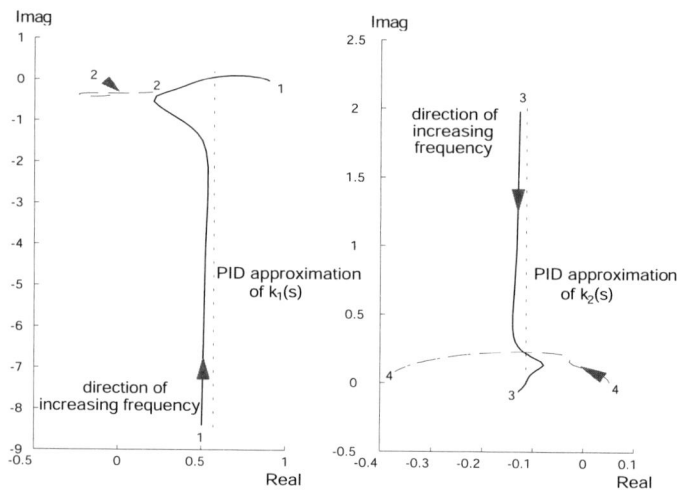

Fig. 12.12. Approximation of controller gains by PID in Example 12.2.1

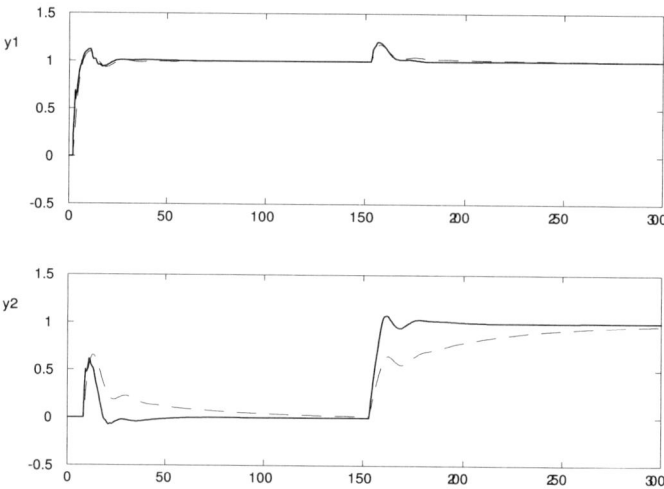

Fig. 12.13. Step response for Example 12.2.1
(solid lines: proposed method; dashed lines: BLT method)

12.4 Controller Reduction and Performance Trade-off

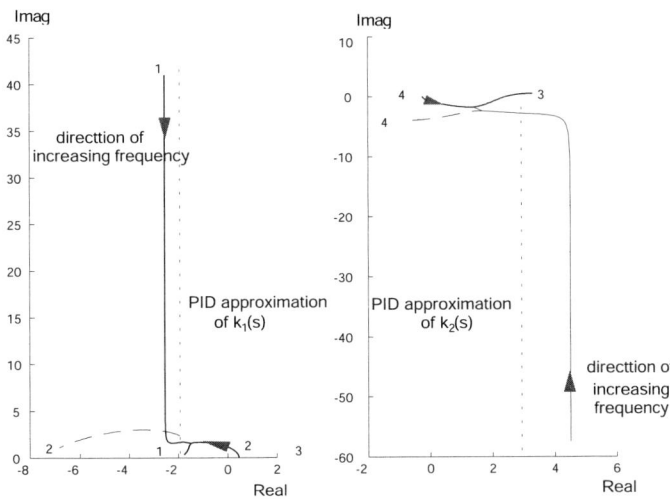

Fig. 12.14. Approximation of controller gains by PID in Example 12.2.2

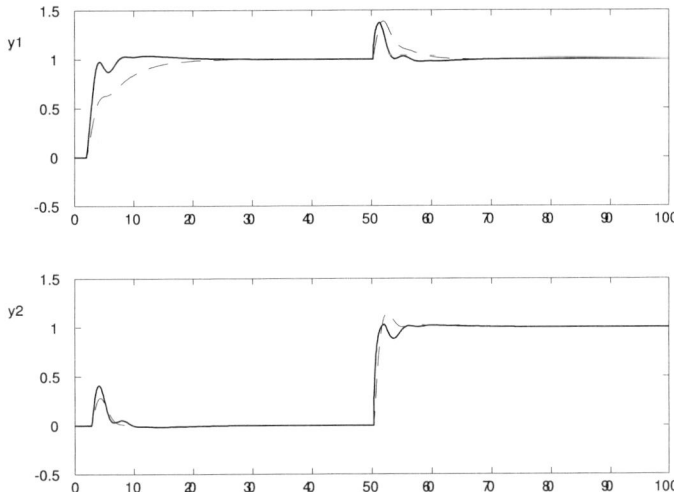

Fig. 12.15. Step response for Example 12.2.2
(solid lines: proposed method; dashed lines: BLT method)

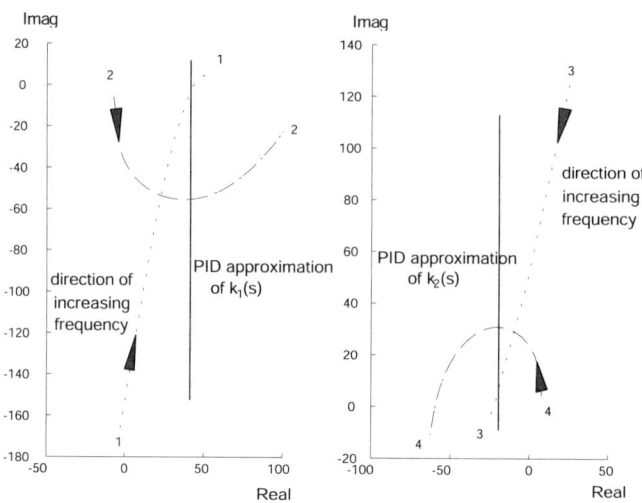

Fig. 12.16. Approximation of controller gains by PID in Example 12.2.3

Example 12.2.3 (cont'd). It follows from our method that

$$K(s) = \text{diag}\left[41.6 + \frac{1.530}{s} + 75.5s, -19.4 - \frac{1.13}{s} - 56.4s\right].$$

Figure 12.16 shows the PID approximation to the exact solutions. Figure 12.17 shows the step response of the feedback. The performance index ISE of the proposed and BLT methods is calculated as 33.2 and 56.4, respectively. More satisfactory response is obtained than that given by the BLT method in both loops.

12.5 Stability Analysis

In the proposed method, we design k_i to stabilize the equivalent process g_i for each i. In this section, we shall investigate whether the stabilization of all g_i by the corresponding k_i leads to an overall stable closed loop, i.e., whether it implies that $K(s)$ stabilizes $G(s)$.

We shall now give a simple counter-example to show that the above implication is in fact incorrect. Suppose that all elements g_{ii} ; $i = 1, 2$, of a 2×2 process are stable but at least one of the elements g_{ij}, $i \neq j$, is unstable. If the controllers $k_1 = k_2 = 0$ are adopted, we shall have $g_i = g_{ii}$ for all i.

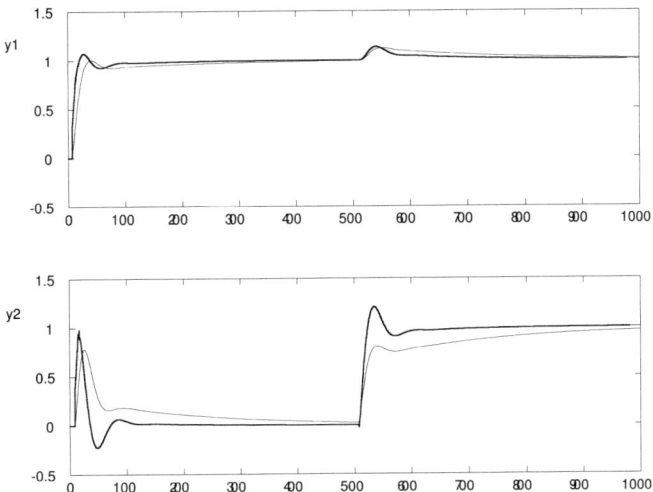

Fig. 12.17. Step response for Example 12.2.3
(solid lines: proposed method; dashed Lines: BLT method)

Then although all k_i stabilize g_i, the whole system is still unstable because the controller holds the system in open loop and the open-loop system is unstable.

Owing to the above observation, it would be of much interest to find out the conditions for closed-loop stability given that k_i stabilizes g_i. Let $\rho(g)$ be the characteristic polynomial of a transfer function $g(s)$ and set $\rho(g) = 1$ if $g(s)$ has no pole at all.

Lemma 12.5.1. *(Rosenbrock, 1974).* For a unity negative feedback system with the plant $G(s)$ and the controller $K(s)$, K stabilizes G if and only if $p_c := \rho(g)\rho(K)det(I + GK)$ has stable roots only.

In our case of $G = \begin{bmatrix} g_{11} & g_{12} \\ g_{21} & g_{22} \end{bmatrix}$, $K = \begin{bmatrix} k_1 & 0 \\ 0 & k_2 \end{bmatrix}$, we have

$$p_c = \rho(G)\rho(k_1)\rho(k_2) \det[I+GK], \tag{12.19}$$

where

$$\det(I+GK) = 1+g_{11}k_1+g_{22}k_2+g_{11}k_1g_{22}k_2-k_1g_{21}g_{12}k_2. \tag{12.20}$$

Now suppose that k_1(respectively k_2) stabilizes g_1 (respectively g_2) where

$$g_1 = g_{11} - \frac{k_2 g_{12} g_{21}}{1 + k_2 g_{22}}, \tag{12.21}$$

$$g_2 = g_{22} - \frac{k_1 g_{21} g_{12}}{1 + k_1 g_{11}}. \tag{12.22}$$

This implies by Lemma 12.5.1 that both

$$p_{c1} = \rho(g_1)\rho(k_1)(1 + g_1 k_1) \tag{12.23}$$

and

$$p_{c2} = \rho(g_2)\rho(k_2)(1 + g_2 k_2) \tag{12.24}$$

are stable. The question is whether or not this individually stabilized system is stable, i.e. whether or not stability of p_{c1} and p_{c2} are equivalent to stability of p_c. Now, bringing (12.21) into (12.23) yields

$$p_{c1} = \rho(g_1)\rho(k_1)\frac{\det[I + GK]}{1 + k_2 g_{22}}. \tag{12.25}$$

Substituting (12.25) into (12.19) yields

$$p_c = p_{c1}\rho(G)\frac{\rho(k_2)(1 + k_2 g_{22})}{\rho(g_1)}. \tag{12.26}$$

Noting that $\rho(k_2)(1 + k_2 g_{22}) = \rho(\frac{k_2}{1+k_2 g_{22}})$, (12.26) can be rewritten as

$$p_c = p_{c1}\rho(G)\frac{\rho(\frac{k_2}{1+k_2 g_{22}})}{\rho(g_1)}. \tag{12.27}$$

Lemma 12.5.2. *Assume that a 2×2 process $G(s)$ is individually stabilized by k_1 and k_2, i.e. p_{c1} and p_{c2} are stable. Then, the resultant feedback system is stable if and only if*

$$\rho(G)\frac{\rho(\frac{k_2}{1+k_2 g_{22}})}{\rho(g_1)} \text{ has stable roots only.} \tag{12.28}$$

By applying this lemma to the previous example where g_{ii} are all stable with $k_1 = k_2 = 0$ while g_{ij} are not, one sees that $\rho(g_1) = \rho(g_{11})$ is stable, $\rho(\frac{k_2}{1+k_2 g_{22}}) = 1$, and $\rho(G)$ is unstable. It follows that (12.28) is violated and the system is unstable.

It is, however, noted that most industrial processes are stable. If G is now restricted to be stable, $\rho(G)$ is stable and it follows from (12.21) that all the unstable factors in $\rho(\frac{k_2}{1+k_2 g_{22}})$ will be cancelled by those of $\rho(g_1)$, leading to the satisfaction of (12.28), provided that

$$\text{None of the unstable poles of } \frac{k_2}{1 + k_2 g_{22}} \text{ are cancelled by } g_{12} g_{21}. \tag{12.29}$$

Theorem 12.5.1. *Assume that a stable 2×2 process is individually stabilized by k_1 and k_2. Then the resultant feedback system is stable if and only if (12.29) holds. In particular, this is the case if (i) both g_{12} and g_{21} are of minimum phase; or (ii) k_2 stabilizes g_{22}.*

It should be pointed out that if we start with k_2 and follow the same development above, we shall arrive at dual results to the above Lemma 12.5.2 and Theorem 12.5.1, in which k_2, g_1 and g_{22} in (12.28) and (12.29) are replaced by k_1, g_2 and g_{11} respectively. Another important observation is that (12.29) generically holds because k_2 is designed for g_2 and the probability of unstable poles of $\frac{k_2}{1+k_2 g_{22}}$ coinciding with unstable zeros of $g_{12}g_{21}$ is almost zero. Therefore, we can conclude that one almost always produces an internally stable system when sequential stabilizing design is used for a stable process. This conclusion is intuitively appealing and practically evidenced. It is the case, indeed, for all three design examples illustrated earlier. But one should be careful if an unstable process is concerned.

A very nice thing with the above stability results is that stability robustness can also be addressed in the same framework. For a stable process with uncertainty, (12.29) is very unlikely to be violated by uncertain $G(s)$ and instead it can be assumed to hold. As a result, for a stable G, the closed loop is robustly stable if and only if k_i stabilizes the uncertain g_i, which can be checked by drawing the Nyquist bands of $g_i(j\omega)k_i(j\omega)$ for a given uncertainty bound on $G(j\omega)$ and counting the number of encirclements of the critical point.

12.6 Extension to the $m \times m$ Case

The above discussions can easily be extended to the general $m \times m$ plants. Equations governing the plant dynamics are

$$\begin{bmatrix} y_1 \\ \vdots \\ y_m \end{bmatrix} = \begin{bmatrix} g_{11} & \cdots & g_{1m} \\ \vdots & & \vdots \\ g_{m1} & \cdots & g_{mm} \end{bmatrix} \begin{bmatrix} u_1 \\ \vdots \\ u_m \end{bmatrix}. \tag{12.30}$$

This is controlled by

$$\begin{bmatrix} u_1 \\ u_2 \\ \vdots \\ u_m \end{bmatrix} = \begin{bmatrix} k_1 & 0 & \cdots & 0 \\ 0 & k_2 & \cdots & 0 \\ \vdots & \vdots & & \vdots \\ 0 & 0 & \cdots & k_m \end{bmatrix} \begin{bmatrix} e_1 \\ e_2 \\ \vdots \\ e_m \end{bmatrix}, \tag{12.31}$$

as shown in Figure 12.18.

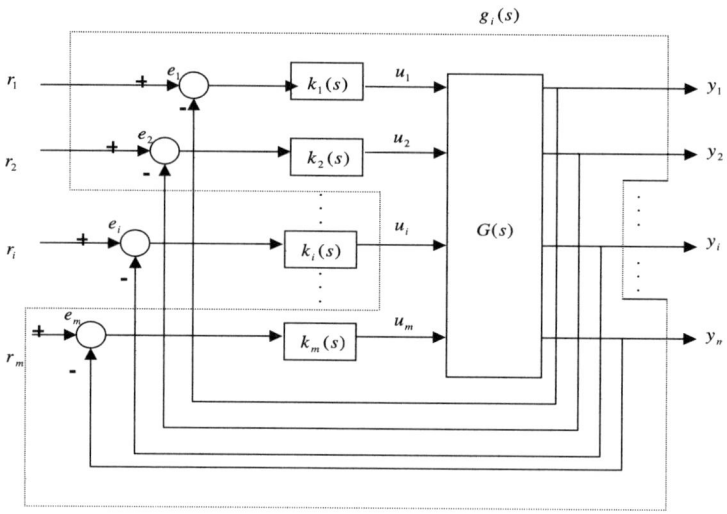

Fig. 12.18. Equivalent $g_i(s)$ for a $m \times m$ plant

Consider loop i separately. One sees

$$y_i = g_{ii}u_i + \begin{bmatrix} g_{i,1} & \cdots & g_{i,i-1} & g_{i,i+1} & \cdots & g_{im} \end{bmatrix} \begin{bmatrix} u_1 \\ \vdots \\ u_{i-1} \\ u_{i+1} \\ \vdots \\ u_m \end{bmatrix}, \qquad (12.32)$$

and

12.6 Extension to the $m \times m$ Case 371

$$\begin{bmatrix} y_1 \\ \vdots \\ y_{i-1} \\ y_{i+1} \\ \vdots \\ y_m \end{bmatrix} = \begin{bmatrix} g_{1i} \\ \vdots \\ g_{i-1,i} \\ g_{i+1,i} \\ \vdots \\ g_{mi} \end{bmatrix} u_i + \begin{bmatrix} g_{11} & \cdots & g_{1,i-1} & g_{1,i+1} & \cdots & g_{1m} \\ \vdots & & \vdots & \vdots & & \vdots \\ g_{i-1,1} & \cdots & g_{i-1,i-1} & g_{i-1,i+1} & \cdots & g_{i-1,m} \\ g_{i+1,1} & \cdots & g_{i+1,i-1} & g_{i+1,i+1} & \cdots & g_{i+1,m} \\ \vdots & & \vdots & \vdots & & \vdots \\ g_{m1} & \cdots & g_{m,i-1} & g_{m,i+1} & \cdots & g_{mm} \end{bmatrix} \begin{bmatrix} u_1 \\ \vdots \\ u_{i-1} \\ u_{i+1} \\ \vdots \\ u_m \end{bmatrix}.$$

(12.33)

Since all loops are closed, we can write

$$\begin{bmatrix} y_1 \\ \vdots \\ y_{i-1} \\ y_{i+1} \\ \vdots \\ y_m \end{bmatrix} = - \begin{bmatrix} \frac{1}{k_1} & & & & & 0 \\ & \ddots & & & & \\ & & \frac{1}{k_{i-1}} & & & \\ & & & \frac{1}{k_{i+1}} & & \\ & & & & \ddots & \\ 0 & & & & & \frac{1}{k_m} \end{bmatrix} \begin{bmatrix} u_1 \\ \vdots \\ u_{i-1} \\ u_{i+1} \\ \vdots \\ u_m \end{bmatrix}. \qquad (12.34)$$

Substitution of (12.34) into (12.33) yields

$$\begin{bmatrix} u_1 \\ \vdots \\ u_{i-1} \\ u_{i+1} \\ \vdots \\ u_m \end{bmatrix} = - \left\{ \begin{bmatrix} g_{11} & \cdots & g_{1,i-1} & g_{1,i+1} & \cdots & g_{1m} \\ \vdots & & \vdots & \vdots & & \vdots \\ g_{i-1,1} & \cdots & g_{i-1,i-1} & g_{i-1,i+1} & \cdots & g_{i-1,m} \\ g_{i+1,1} & \cdots & g_{i+1,i-1} & g_{i+1,i+1} & \cdots & g_{i+1,m} \\ \vdots & & \vdots & \vdots & & \vdots \\ g_{m1} & \cdots & g_{m,i-1} & g_{m,i+1} & \cdots & g_{mm} \end{bmatrix} \right.$$

$$\left. + \begin{bmatrix} \frac{1}{k_1} & & & & & 0 \\ & \ddots & & & & \\ & & \frac{1}{k_{i-1}} & & & \\ & & & \frac{1}{k_{i+1}} & & \\ & & & & \ddots & \\ 0 & & & & & \frac{1}{k_m} \end{bmatrix} \right\}^{-1} \begin{bmatrix} g_{1,i} \\ \vdots \\ g_{i-1,i} \\ g_{i+1,i} \\ \vdots \\ g_{m,i} \end{bmatrix} u_i. \qquad (12.35)$$

Therefore, we have

$$y_i = \left\{ g_{ii} - \begin{bmatrix} g_{i,1} \\ \vdots \\ g_{i,i-1} \\ g_{i,i+1} \\ \vdots \\ g_{i,m} \end{bmatrix}^T \left(\begin{bmatrix} g_{11} & \cdots & g_{1,i-1} & g_{1,i+1} & \cdots & g_{1m} \\ \vdots & & \vdots & \vdots & & \vdots \\ g_{i-1,1} & \cdots & g_{i-1,i-1} & g_{i-1,i+1} & \cdots & g_{i-1,m} \\ g_{i+1,1} & \cdots & g_{i+1,i-1} & g_{i+1,i+1} & \cdots & g_{i+1,m} \\ \vdots & & \vdots & \vdots & & \vdots \\ g_{m1} & \cdots & g_{m,i-1} & g_{m,i+1} & \cdots & g_{mm} \end{bmatrix} \right. \right.$$

$$\left. \left. + \begin{bmatrix} \frac{1}{k_1} & & & & & 0 \\ & \ddots & & & & \\ & & \frac{1}{k_{i-1}} & & & \\ & & & \frac{1}{k_{i+1}} & & \\ & & & & \ddots & \\ 0 & & & & & \frac{1}{k_m} \end{bmatrix} \right)^{-1} \begin{bmatrix} g_{1,i} \\ \vdots \\ g_{i-1,i} \\ g_{i+1,i} \\ \vdots \\ g_{m,i} \end{bmatrix} \right\} u_i, \quad (12.36)$$

or in short

$$y_i = g_i u_i, \quad (12.37)$$

for a suitable g_i. Difficulties arise in designing controllers for each of the equivalent processes g_i for each is a function of $k_1, k_2, \ldots, k_{i-1}, k_{i+1}, \ldots, k_m$, that is

$$g_i = g_i(k_1, k_2, \ldots, k_{i-1}, k_{i+1}, \ldots, k_m). \quad (12.38)$$

However, $k_1, k_2, \ldots, k_{i-1}, k_{i+1}, \ldots, k_m$, can be designed to meet user specifications so that the dependence of g_i on $k_1, k_2, \ldots, k_{i-1}, k_{i+1}, \ldots, k_m$ is released, i.e.,

$$\left. \begin{aligned} g_1(k_2, k_3, \ldots, k_{m-1}, k_m)k_1 &= \tfrac{\omega_1^2 e^{-sL_1}}{s^2 + 2\xi_1 \omega_1 s + \omega_1^2(1 - e^{-sL_1})} = q_{d1}, \\ &\vdots \\ g_i(k_1, \ldots, k_{i-1}, k_{i+1}, \ldots, k_m)k_i &= \tfrac{\omega_i^2 e^{-sL_i}}{s^2 + 2\xi_i \omega_i s + \omega_i^2(1 - e^{-sL_i})} = q_{di}, \\ &\vdots \\ g_m(k_1, k_2, \ldots, k_{m-2}, k_{m-1})k_m &= \tfrac{\omega_m^2 e^{-sL_m}}{s^2 + 2\xi_m \omega_m s + \omega_m^2(1 - e^{-sL_m})} = q_{dm}. \end{aligned} \right\} \quad (12.39)$$

Thus there are m equations and m unknowns, namely $k_1, k_2, \ldots, k_m$, and they can be solved with some manipulations. The separation process starts with expressing k_1 in terms of $k_2, \ldots, k_m$ and q_{d1} using the first equation in (12.39), to obtain

$$k_1 = f_1(k_2, k_3, \ldots, k_{m-1}, k_m, q_{d1}). \tag{12.40}$$

The expression is next substituted into the second equation in (12.39) and k_2 can then be written in terms of $k_3, \ldots, k_m$ and q_{d1}, q_{d2}, i.e.

$$k_2 = f_2(k_3, \ldots, k_{m-1}, k_m, q_{d1}, q_{d2}). \tag{12.41}$$

The same procedure is repeated and we shall subsequently obtain

$$\left.\begin{aligned} k_3 &= f_3(k_4, \ldots, k_{m-1}, k_m, q_{d1}, q_{d2}, q_{d3}), \\ &\vdots \\ k_{m-1} &= f_{m-1}(k_m, q_{d1}, q_{d2}, \ldots, q_{dm-1}), \\ k_m &= f_m(q_{d1}, q_{d2}, \ldots, q_{dm-1}, q_{dm}). \end{aligned}\right\} \tag{12.42}$$

It is noticed that, k_m has been recovered in the last equation of the above set. Independent expressions for $k_2, k_3, \ldots$ and k_{m-1} can be acquired using back substitution as follows. The recovered k_m is substituted into the second last equation in (12.42) for k_{m-1}. The recovered k_m and k_{m-1} are next substituted into the third last equation for k_{m-2}. The same process is repeated until k_2 is finally recovered.

According to the trade-off between loop performance and controller complexity as discussed in Section 12.4, rational function controllers or PID controllers can be designed similarly for $m \times m$ processes. Guidelines for the selection of the values $\{\omega_1, \ldots, \omega_m\}$, $\{L_1, \ldots, L_m\}$ and $\{\xi_1, \ldots, \xi_m\}$ in (12.39) follow exactly those for the 2×2 case. Depending on the type of controllers used, (12.13) or (12.18) is applied accordingly. It should also be noted that the rules for the selection of solutions in Section 12.3 still hold in the $m \times m$ case.

The major increase in workload arising from extension of the proposed method from the 2×2 to the $m \times m$ case lies in computation of the m solutions to the m-degree polynomial equations. The solutions become much more involved but can be solved with the help of well-developed software such as Matlab.

References

Åström, K. J. (1995). Oscillations in systems with relay feedback. In: *Adaptive Control, Filtering, and Signal Processing* (K. J. Åström, G. C. Goodwin and P. R. Kumar, Eds.). Vol. 74. pp. 1–25. Springer. New York.

Åström, K. J. (2000). Limitations on control system performance. *European Journal of Control* **6**(1), 2–20.

Åström, K. J. and B. Writtenmark (1984). *Computer Control Systems : Theory and Design.* Prentice-Hall. Englewood Cliffs, NJ.

Åström, K. J. and T. Hägglund (1984a). Automatic tuning of simple regulators. In: *Proc. 9th IFAC World Congress.* Budapest, Hungary. pp. 267–272.

Åström, K. J. and T. Hägglund (1984b). Automatic tuning of simple regulators with specifications on phase and amplitude margins. *Automatica* **20**(5), 645–651.

Åström, K. J. and T. Hägglund (1988). *Automatic Tuning of PID Controllers.* Instrument Society of America. Research Triangle Park, NC.

Åström, K. J. and T. Hägglund (1995). *PID Controllers: Theory, Design, and Tuning*, 2nd edn. Instrument Society of America. Research Triangle Park, NC.

Åström, K. J., T. Hägglund, C. C. Hang and W. K. Ho (1993). Automatic tuning and adaptation for PID controllers - a survey. *Control Engineering Practice* **1**(4), 699–714.

Atherton, D. P. (1975). *Non-linear Control Engineering.* Van Nostrand Rienhold. London.

Atherton, D. P. (1993). Analysis and design of relay control systems. In: *CAD for Control Systems.* Chap. 15, pp. 367–394. Marcel Dekker. New York.

Barkat, M. (1991). *Signal Detection and Estimation.* Artech House, Inc.. MA, USA.

Barnes, T. J., L. Wang and W. R. Cluett (1993). A frequency domain design method for PID controllers. In: *Proc. American Control Conference.* San Francisco. pp. 890–894.

Bi, Q., Q. G. Wang and C. C. Hang (1997). Relay-based estimation of multiple points on process frequency response. *Automatica* **33**(9), 1753–1757.

Bi, Q., W. J. Cai, Q.-G. Wang and C. C. Hang (1999). Robust identification of first-order plus dead-time model from step response. *Control Engineering Practice* **7**(1), 71–77.

Bi, Q., W.-J. Cai, Q. G. Wang, C.-C. Hang, Eng-Lock Lee, Y. Sun, K.-D. Liu, Y. Zhang and B. Zou (2000). Advanced controller auto-tuning and its application to HVAC systems. *Control Engineering Practice* **8**(6), 633–644.

Bristol, E. H. (1966). On a new measure of interaction for multivariable process control. *IEEE Transactions on Automatic Control* **11**, 133–134.

Broyden, C. G. (1975). *Basic Matrix - An Introduction to Matrix Theory and Practice*. Macmillan. London.

Chen, C.-T. (1984). *Linear System Theory and Design*. Holt, Rinehart and Winston. New York.

Chen, C. T. (1993). *Analog and Digital Control Systems Design: Transfer-function, State-space, and Algebraic Methods*. Saunders College Publishing. Fort Worth.

Chien, I. L. (1988). IMC-PID controller design-an extension. In: *IFAC Proceedings Series*. Vol. 6. pp. 147–152.

Cohen, G. H. and G. A. Coon (1953). Theoretical consideration of retarded control. *Trans. ASME* **75**, 827–834.

Cook, P. A. (1986). *Non-linear Dynamical Systems*. Prentice-Hall. New York.

Cottle, R. W., J.-S. Pang and R. E. Stone (1992). *The Linear Complementarity Problem*. Academic Press. Boston.

Dai, H. and N. K. Sinha (1991). Use of numerical integration methods. In: *Identification of Continuous Systems: Methodology and Computer Implementation* (N. K. Sinha and G. P. Rao, Eds.). pp. 259–289. Kluwer Academic Publishers. Netherlands.

Datta, A., M. T. Ho and S. P. Bhattacharyya (2000). *Structure and Synthesis of PID Controllers*. Springer. London, UK.

Dong, J. and C. B. Brosilow (1997). Design of robust multivariable PID controllers via IMC. In: *Proc. American Control Conference*. Vol. 5. Albuquerque, New Mexico. pp. 3380–3384.

Doukas, N. and W. L. Luyben (1978). Control of sidestream columns separating ternary mixtures. *Instrumentation Technology* **25**(6), 43–48.

Doyle, J. C., J. Wall and G. Stein (1982). Performance and robustness analysis for structured uncertainty. In: *Proc. IEEE Conf. Decision and Control*. Vol. 2. Orlando, FL. pp. 629–636.

Farkas, M. (1994). *Periodic Motions*. Springer-Verlag. New York.

Filippov, A. F. (1988). *Differential Equations with Discontinuous Righthand Sides*. Kluwer Academic Publishers. Dordrecht (Netherlands).

Fisher, D. G. (1991). Process control: An overview and personal perspective. *The Canadian Journal of Chemical Engineering* **69**(1), 5–26.

Forssell, U. and L. Ljung (1999). Closed-loop identification revisited. *Automatica* **35**(7), 1215–1241.

Franklin, G. F., J. D. Powell and A. Emami-Naeini (1994). *Feedback Control of Dynamic Systems*. Addison-Wesley Publishing Company. Wokingham, UK.

Franklin, G. F., J. D. Powell and M. L. Workman (1990). *Digital Control of Dynamic Systems*. Addison-Wesley. Reading, MA.

Freudenberg, J. S. and D. P. Looze (1987). A sensitivity trade-off for plants with time delay. *IEEE Transactions on Automatic Control* **32**, 99–104.

Garcia, C. E. and M. Morari (1982). Internal model control. 1: A unifying review and some new results. *Industrial and Engineering Chemistry: Process Design and Development* **21**, 308–323.

Garcia, C. E. and M. Morari (1985). Internal model control. 2: Design procedure for multivariable systems. *Industrial and Engineering Chemistry: Process Design and Development* **24**, 427–484.

Gawthrop, P. J. and M.T. Nihtila (1985). Identification of time-delays using a polynomial identification method. *Systems and Control Letters* **5**, 267–271.

Georgiou, T. T. and M. C. Smith (2000). Robustness of a relaxation oscillator. *International Journal of Robust and Nonlinear Control* **10**, 1005–1024.

Ghogho, M. and A. Swami (1999). Fast computation of the exact fim for deterministic signals in colored noise. *IEEE Transactions on Signal Processing* **47**(1), 52–61.

Goncalves, J. M., A. Megretski and M. A. Dahleh (1998). Semi-global analysis of relay feedback systems. In: *Proc. IEEE Conf. Decision and Control*. Vol. 2. pp. 1967–1972.

Goncalves, J. M., A. Megretski and M. A. Dahleh (2001). Global stability of relay feedback systems. *IEEE Transactions on Automatic Control* **46**(4), 550–562.

Green, M. and D. J. N. Limebeer (1995). *Linear Robust Control*. Prentice Hall. Englewood Cliffs, NJ.

Guckenheimer, J. and P. Holmes (1983). *Nonlinear Oscillations, Dynamical Systems, and Bifurcations of Vector Fields*. Springer-Verlag. New York.

Hägglund, T. and K. J. Åström (1991). Industrial adaptive controllers based on frequency response techniques. *Automatica* **27**, 599–609.

Hang, C. C., K. J. Åström and W. K. Ho (1993a). Relay auto-tuning in the presence of static load disturbance. *Automatica* **29**, 563–564.

Hang, C. C., T. H. Lee and W. K. Ho (1993b). *Adaptive Control*. Instrument Society of America. NC.

Haykin, S. (1989). *An Introduction to Analog & Digital Communications*. John Wiley & Sons. NY.

Heemels, W. P. M. H., J. M. Schumacher and S. Weiland (1999). The rational complementarity problem. *Linear Algebra and its Applications* **294**(1-3), 93–135.

Holt, B. R. and M. Morari (1985a). Design of resilient processing plants–V: The effect of deadtime on dynamic resilience. *Chemical Engineering Science* **40**, 1229–1237.

Holt, B. R. and M. Morari (1985b). Design of resilient processing plants–VI: The effect of right-half-plane zeros on dynamic resilience. *Chemical Engineering Science* **40**, 59–74.

Hsu, L. (1990). Boundedness of oscillations in relay feedback systems. *International Journal of Control* **52**(5), 1273–1276.

Huang, C. T. and Y. S. Lin (1995). Tuning PID controller for open-loop unstable process with time delay. *Chemical Engineering Communications* **33**, 11–30.

Huang, H. P. and C. L. Chen (1996). Auto-tuning for model-based PID controllers. *AIChE Journal* **42**, 1174–1180.

Huang, H. P. and C. L. Chen (1997). Control-system synthesis for open-loop unstable process with time delay. *IEE Proceedings Part D: Control Theory and Applications* **144**(4), 334–346.

Jerome, N. F. and W. H. Ray (1986). High-performance multivariable control strategies for systems having time delays. *AIChE Journal* **32**, 914–931.

Johansson, K. H., A. Barabanov and K. J. Åström (1997). Limit cycles with chattering in relay feedback systems. In: *Proc. 36th IEEE Conference on Decision and Control*. San Diego, USA. pp. 3220–3225.

Johansson, K. H., A. Rantzer and K. J. Åström (1999). Fast switches in relay feedback systems. *Automatica* **35**(4), 539–552.

Kay, S. M. (1993). *Fundamentals of Statistical Signal Processing: Estimation Theory*. Prentice Hall, Englewood Cliffs. New Jersey.

Kerr, T. H. (1989). Status of CR-like lower bounds for nonlinear filtering. *IEEE Transactions On Aerospace and Electronic Systems* **25**(5), 590–601.

Khamsi, M. A. and W. A. Kirk (2001). *An Introduction to Metric Spaces and Fixed Point Theory*. John Wiley & Sons. New York.

Kong, K. Y. (1995). Feasibility report on a frequency-domain adaptive controller. department of electrical engineering, national university of singapore.

Kuhfitting, P. K. F. (1978). *Introduction to the Laplace Transform*. Plenum Press. New York.

Kuo, B. C. (1991). *Automatic Control Systems*. Prentice-Hall.

Kurz, H. and W. Goedecke (1981). Digital parameter-adaptive control of process with unknown dead time. *Automatica* **17**(1), 245–252.

Kuttler, K. (1998). *Modern Analysis*. CRC Press. Boca Raton, Florida.

Lancaster, P. and M. Tismenetsky (1985). *The Theory of Matrices With Applications*, 2nd edn. Academic Press. New York.

Lawson, C. L. and R. J. Hanson (1974). *Solving Least Squares Problems*. Prentice-Hall.

Leva, A. (1993). PID auto-tuning algorithm based on relay feedback. *IEE Proceedings Part D: Control Theory and Applications* **140**, 328–338.

Li, W., E. Eskinat and W. L. Luyben (1991). An improved auto-tune identification method. *Industrial & Engineering Chemistry Research* **30**, 1530–1541.

Lilja, M. (1989). Controller design by frequency domain approximation. phd thesis, lund institute of technology, lund, sweden.

Lin, C. and Q.-G. Wang (2002). On uniqueness of solutions to relay feedback systems. *Automatica* **38**, 177–180.

Lin, C., Q.-G. Wang and T. H. Lee (2001). General linear relay feedback systems - part II: global stability of limit cycles. *Submitted for publication*.

Lin, C., Q.-G. Wang, T. H. Lee, A. P. Loh and K. H. Kwek (2000). Stability criteria and bounds of limit cycles in relay feedback systems. In: *Proceedings of PSE Asia 2000*. Kyoto, Japan. pp. 291–296.

Linnemann, A. and Q. G. Wang (1993). Block decoupling with stability by unity output feedback - solution and performance limitations. *Automatica* **29**(3), 735–744.

Ljung, L. (1985). On the estimation of transfer functions. *Automatica* **21**(6), 677–696.

Ljung, L. and T. Söderström (1983). *Theory and Practice of Recursive Identification*. The MIT Press. Cambridge, MA.

Loh, A. P. (1994). Necessary conditions for limit cycles in multiloop relay systems. *IEE Proceedings Part D: Control Theory and Applications* **141**(3), 163–168.

Loh, A. P. and U. V. Vasnani (1994). Describing function matrix for multivariable systems and its use in multiloop PI design. *Journal of Process Control* **4**(3), 115–120.

Loh, A. P., C. C. Hang, C. K. Quek and U. V. Vasnani (1993). An approach to multivariable control system design using relay-autotuning. *Industrial & Engineering Chemistry Research* **32**, 1102–1107.

Lootsma, Y. J., A. J. van der Schaft and M. K. Camlibel (1999). Uniqueness of solutions of linear relay systems. *Automatica* **35**(3), 467–478.

Luyben, W. L. (1986). Simple method for tuning SISO controllers in multivariable systems. *Industrial and Engineering Chemistry: Process Design and Development* **25**, 654–660.

Luyben, W. L. (1987). Derivation of transfer functions for highly nonlinear distillation columns. *Industrial & Engineering Chemistry Research* **26**, 2490–2495.

Luyben, W. L. (1990). *Process Modeling, Simulation, and Control for Chemical Engineers*. McGraw-Hill. New York.

Maciejowski, J. M. (1989). *Multivariable Feedback Design*. Addison-Wesley. Reading, MA.

Maffezzoni, C. and P. Rocco (1997). Robust tuning of PID regulators based on step response identification. *European Journal of Control* **3**(2), 125–136.

Majhi, S. and D. P. Atherton (2000). Online tuning of controllers for an unstable FOPDT process. *IEE Proceedings Part D: Control Theory and Applications* **147**(4), 421–427.

Malti, R., S. B. Ekongolo and J. Ragot (1998). Dynamic SISO and MIMO system approximation based on optimal Laguerre models. *IEEE Transactions on Automatic Control* **43**(9), 1318–1323.

Mason, S. J. (1956). Feedback theory - further properties of signal flow graphs. *Proceedings of IRE*. p. 920.

McGillem, C. D. and G. R. Cooper (1984). *Continuous and Discrete Signal and System Analysis*. Holt, Rinehart, and Winston. New York.

Megretski, A. (1996). Global stability of oscillations induced by a relay feedback. In: *Proc. 13th IFAC World Congress*. Vol. E. San Francisco, USA. pp. 49–54.

Melo, D. L. and J. C. Friedly (1992). On-line, closed-loop identification of multivariable systems. *Industrial & Engineering Chemistry Research* **31**(1), 274–281.

Morari, M. and E. Zafiriou (1989). *Robust Process Control*. Prentice Hall. Englewood Cliffs, NJ.

Mori, T. (1990). On the relationship between the spectral radius and stability radius for discrete systems. *IEEE Transactions on Automatic Control* **35**(7), 835.

Morosanov, I. S. (1964). *Extremum-seeking Relay Systems: Approximate Analysis Methods*. Nauka. Moscow.

Morrison, N. (1994). *Introduction to Fourier Analysis*. John Wiley & Sons Inc. New York.

Newton, G. C., L. A. Gould and J. F. Kalser (1957). *Analytical Design of Linear Feedback Controls*. Wiley. New York.

Ninness, B. (1996). Integral constraints on the accuracy of least squares estimation. *Automatica* **32**(3), 391–397.

Noble, B. (1969). *Applied Linear Algebra*. Prentice-Hall. Englewood Cliffs, NJ.

Ogunnaike, B. A. and W. H. Ray (1979). Multivariable controller design for linear systems having multiple time delays. *AIChE Journal* **25**, 1043–1057.

Orgunnaike, B. A. and W. H. Ray (1979). Multivariable controller design for linear systems having mutiple time delays. *AIChE Journal* **25**, 1043–1057.

Palmor, Z. J. (1980). Stability properties of Smith dead-time compensator controllers. *International Journal of Control* **32**(5), 937–949.

Palmor, Z. J. and M. Blau (1994). An auto-tuner for Smith dead time compensator. *International Journal of Control* **60**(1), 117–135.

Palmor, Z. J., Y. Halevi and N. Krasney (1993). Automatic tuning of decentralized PID controllers for TITO processes. *Proc. of 12th IFAC World Congress* **2**, 311–314.

Palmor, Z. J., Y. Halevi and N. Krasney (1995a). Automatic tuning of decentralized PID controllers for TITO processes. *Automatica* **31**, 1001–1010.

Palmor, Z. J., Y. Halevi and N. Krasney (1996). Automatic tuning of decentralized PID controllers for MIMO processes. *Journal of Process Control* **42**, 1174–1180.

Palmor, Z. J., Y. Halevi and T. Efrati (1992). Limit cycles in decentralized relay systems. *International Journal of Control* **56**(4), 755–765.

Palmor, Z. J., Y. Halevi and T. Efrati (1995b). A general and exact method for determining limit cycles in decentralized relay systems. *Automatica* **31**(9), 1333–1339.

Park, J. H., S. W. Sung and I.-B. Lee (1998). An enhanced PID control strategy for unstable processes. *Automatica* **34**(6), 751–756.

Pintelon, R. and J. Schoukens (1990). Real-time integration and differentiation of analog signals by means of digital filtering. *IEEE Transactions On Instrumentation and Measurement* **39**(6), 923–927.

Pintelon, R. and L. V. Biesen (1990). Identification of transfer functions with time delay and its application to cable fault location. *IEEE Transactions On Instrumentation and Measurement* **39**(3), 479–484.

Pintelon, R., P. Guillaume, Y. Rolain, J. Schoukens and H. Van hamme (1994). Parametric identification of transfer functions in the frequency domain – a survey. *IEEE Transactions on Automatic Control* **39**(11), 2245–2260.

Rake, H. (1980). Step response and frequency response methods. *Automatica* **16**(5), 519–526.

Richalet, J. A., A. Rault, J. D. Testud and J. Papon (1978). Model predictive heuristic control: applications to industrial processes. *Automatica* **14**, 413–428.

Rivera, D. E., M. Morari and S. Skogestad (1986). Internal model control. 4. PID controller design. *Industrial and Engineering Chemistry: Process Design and Development* **25**, 252–256.

Robbe, M. and M. Sadkane (2000). Discrete-time lyapunov stability of large matrices. *Journal of Computational and Applied Mathematics* **115**, 479–494.

Rosenbrock, H. H. (1974). *Computer-aided Control System Design*. Academic Press. NY.

Rotstein, G. E. and D. R. Lewin (1991). Simple PI and PID tuning for open-loop unstable systems. *Industrial & Engineering Chemistry Research* **30**(8), 1864–1869.

Rotstein, G. E. and D. R. Lewin (1992). Control of an unstable batch chemical reactor. *Computers & Chemical Engineering* **16**(1), 27–49.

Sagara, S. and Z.Y. Zhao (1990). Numerical integration approach to on-line identification of continuous-time systems. *Automatica* **26**(1), 63–74.

Schoukens, J. and R. Pintelon (1991). *Identification of Linear Systems: A Practical Guideline to Accurate Modelling*. Pergamon Press. Oxford.

Shafiei, Z. and A. T. Shenton (1994). Tuning of PID-type controllers for stable and unstable systems with time delay. *Automatica* **30**(10), 1609–1615.

Skogestad, S. and I. Postlethwaite (1996). *Multivariable Feedback Control: Analysis and Design*. Wiley. New York.

Skogestad, S. and M. Morari (1989). Robust performance of decentralized control systems by independent designs. *Automatica* **25**(1), 119–125.

Slotine, J.-J. E. and W. Li (1991). *Applied Nonlinear Control*. Prentice Hall. NJ.

Smart, D. R. (1974). *Fixed Point Theorems*. Cambridge University Press. London.

Smith, O. J. M. (1957). Closer control of loops with dead time. *Control Engineering Practice* **53**, 217–219.

Söderström, T. and P. G. Stoica (1983). *Instrumental Variable Methods for System Identification*. Springer Verlag. Berlin.

Söderström, T. and P. Stoica (1989). *System Identification*. Prentice Hall. New York.

Souza, C. E. D., G. C. Goodwin, D. Q. Mayne and M. Palaniswami (1992). An adaptive control algorithm for linear systems having unknown time delay. *Automatica* **24**(3), 327–341.

Strejc, V. (1980). Least squares parameter estimation. *Automatica* **16**(5), 535–550.

Strejc, V. (1981). Trends in identification. *Automatica* **17**(1), 7–21.

Suganda, P., P. R. Krishnaswamy and G. P. Rangaiah (1998). On-line process identification from closed-loop tests under PI control. *Chemical Engineering Research & Design, Transactions of the Institute of Chemical Engineers, Part A* **76**(A4), 451–457.

Sung, S. W. and I. Lee (1996). Limitations and countermeasures of PID controllers. *Industrial & Engineering Chemistry Research* **35**, 2596–2610.

Sung, S. W., J. O. Lee, I. B. Lee, J. Lee and S. Yi (1996). Automatic tuning of PID controller using second-order plus time delay model. *Journal of Chemical Engineering of Japan* **29**(6), 990–999.

Taiwo, O. (1999). Cheap computation of optimal reduced-order models for systems with time delay. *Journal of Process Control* **9**(4), 365–371.

Tan, K. K., Q.-G. Wang and C. C. Hang (1999). *Advances in PID Control*. Springer. London.

Tsypkin, Ya. Z. (1984). *Relay Control Systems*. Cambridge University Press. New Jork.

Tugnait, J. K. and C. Tontiruttananon (1998). Identification of linear systems via spectral analysis given time-domain data: consistency, reduced-order approximation, and performance analysis. *IEEE Transactions on Automatic Control* **43**(10), 1354–1373.

Tyreus, B. D. (1982). Paper presented at the Lehigh University distillation control short course, Bethlehem, PA.

Unbehauen, H. and G. P. Rao (1987). *Identification of Continuous Systems*. Elsevier Science. Netherlands.

Van der Schaft, A. J. and J. M. Schumacher (1998). Complementarity modeling of hybrid systems. *IEEE Transactions on Automatic Control* **43**(4), 483–490.

Varigonda, S. and T. T. Georgious (2001). Dynamics of relay relaxation oscillators. *IEEE Transactions on Automatic Control* **46**(1), 65–77.

Vasnani, V. U. (1994). Towards relay feedback auto-tuning of multi-loop systems. ph.D thesis. national university of singapore.

Wahlberg, B. and L. Ljung (1992). Hard frequency-domain model error bounds from least-squares like identification techniques. *IEEE Transactions on Automatic Control* **37**(7), 900–912.

Walton, K. and J. E. Marshall (1987). Direct method for time delay system stability analysis. *IEE Proceedings Part D: Control Theory and Applications* **134**, 101–107.

Wang, G., V. Sreeram and W. Q. Liu (1998a). A new frequency-weighted optimal Hankel norm model reduction and an error bound. In: *Proc. IEEE Conf. Decision and Control*. Vol. 2. Tampa. FL. pp. 2185–2188.

Wang, Q. G. and Y. Zhang (2000). Frequency response identification from cascade relay feedback. *International Journal of Control* **73**(18), 1647–1656.

Wang, Q. G. and Y. Zhang (2001a). A novel FFT-based robust multivariable process identification method. *Industrial & Engineering Chemistry Research* **40**(11), 2485–2494.

Wang, Q. G. and Y. Zhang (2001b). Robust identification of continuous time delay systems from step responses. *Automatica* **37**, 377–390.

Wang, Q. G., B. Zou, T. H. Lee and Q. Bi (1997a). Auto-tuning of multivariable PID controllers from decentralized relay feedback. *Automatica* **33**(3), 319–330.

Wang, Q. G., C. C. Hang and B. Zou (2000a). Multivariable process identification and control from decentralized relay feedback. *International Journal of Modelling and Simulation* **20**(4), 341–348.

Wang, Q. G., C. C. Hang and Q. Bi (1997b). Process frequency response estimation from relay feedback. *Control Engineering Practice* **5**(9), 1293–1302.

Wang, Q. G., C. C. Hang and Q. Bi (1999a). Implementation and testing of an advanced relay auto-tuner. *Journal of Process Control* **9**(4), 291–300.

Wang, Q. G., C. C. Hang and Q. Bi (1999b). A technique for frequency response identification from relay feedback. *IEEE Transactions on Control Systems Technology* **7**(1), 122–128.

Wang, Q. G., C. C. Hang and X. P. Yang (2001a). Single-loop controller design via IMC principles. *Automatica* **37**, 2041–2048.

Wang, Q. G., H. W. Fung and Y. Zhang (2000b). Independent design of multi-loop controllers taking into account multivariable interactions. *Journal of Chemical Engineering of Japan* **33**(3), 427–439.

Wang, Q. G., H.-W. Fung and Yu Zhang (1999c). Robust estimation of process frequency response from relay feedback. *ISA Transactions* **38**(1), 3–9.

Wang, Q.-G., K. K. Tan and T. H. Lee (1996). An enhanced automatic tuning procedure for pi/pid controllers for process control. *AIChE Journal* **42**(9), 2555–2562.

Wang, Q. G., Q. Bi and B. Zhou (1997c). Use of FFT in relay feedback systems. *Electronics Letters* **33**(12), 1099–1100.

Wang, Q. G., Q. Bi and X. P. Yang (2001b). High performance conversions between continuous and discrete systems. *Signal Processing* **81**, 1865–1877.

Wang, Q. G., Q. Bi and Y. Zhang (2001c). Partial internal model control. *IEEE Industrial Electronics* **48**(5), 976–982.

Wang, Q. G., T. H. Lee and H. W. Fung (1999d). PID tuning for improved performance. *IEEE Transactions on Control Systems Technology* **7**(4), 457–465.

Wang, Q. G., T. H. Lee and J. B. He (1997d). Towards minimal-order stabilizers for all-pole plants. *Systems and Control Letters* **31**, 49–57.

Wang, Q. G., T. H. Lee and J. B. He (1999e). Internal stability of interconnected systems. *IEEE Transactions on Automatic Control* **44**(3), 593–597.

Wang, Q. G., T. H. Lee and K. K. Tan (1998b). *Finite Spectrum Assignment for Time-delay Systems*. In the series of Lecture Notes in Control and Information Science, No. 239. Springer Verlag. London.

Wang, Q. G., Y. Zhang and M. S. Chiu (2002). Decoupling internal model control for multivariable systems with multiple time delays. *Chemical Engineering Science* **57**, 115–124.

Wang, Q. G., Y. Zhang and X. Guo (2001d). Robust closed-loop process identification with application to auto-tuning. *Journal of Process Control* **11**, 519–530.

Wood, R. K. and M. W. Berry (1973). Terminal composition control of a binary distillation column. *Chemical Engineering Science* **28**, 1707–1717.

Yamanaka, K. and E. Shimemura (1987). Effects of mismatched smith controller on stability in systems with time delay. *Automatica* **23**(6), 787–791.

Yan, W.-Y. and J. Lam (1999). An approximate approach to H_2 optimal model reduction. *IEEE Transactions on Automatic Control* **44**(7), 1341–1358.

Young, P. C. (1970). An instrumental variable method for real-time identification of a noisy process. *Automatica* **6**(2), 271–287.

Yu, C. C. (1999). *Autotuning of PID Controllers*. Springer. London.

Zakian, V. (1979). New formulation for the method of inequalities. *Proceedings of the Institution of Electrical Engineers* **126**, 579–584.

Zhu, H. A., C. L. Teo, A. N. Poo and G. S. Hong (1995). An enhanced internal model structure. *Control Theory and Advanced Technology* **10**, 1115–1127.

Zhuang, M. and D. P. Atherton (1993). Automatic tuning of optimum PID controllers. *IEE Proceedings Part D: Control Theory and Applications* **140**(3), 216–224.

Ziegler, J. G. and N. B. Nichols (1942). Optimum settings for automatic controllers. *Transactions ASME* **64**, 759–768.

Index

biased relay, 90
cascade relay, 115
data screening, 195
decentralized control, 347
decentralized relay feedback, 126, 160
decentralized test, 218
decomposition method, 140
decoupling, 275
describing function method, 91
disturbance, 111
Fast Fourier transform, 136
filter, 240, 246, 296
FOPDT, 95
Fourier series method, 94
Fourier transform, 136, 146
frequency response, 211
high-order controller, 253, 269, 303
hysteresis, 90, 93
IMC, 239, 240, 274, 296
independent single relay feedback, 126
independent test, 218
input constraint, 241, 298
limit cycle, 96, 97, 99
MIMO, 126
model order, 169, 194
multivariable systems, 273
noises, 93, 111
noise-to-signal ratio, 106, 112, 143, 196

nonlinearity, 124
offset errors, 195
parameter estimation, 99, 103
parasitic relay, 108
PID, 239, 242, 244, 266, 299, 357
PIMC, 319, 323
process frequency response, 221
recursive least squares, 171
regulation, 326
relay feedback system, 90, 136
relay function, 90, 219
right half-plane zeros, 240
sequential relay feedback, 126, 219
sequential test, 218
single-loop control system, 242
stability, 255, 276, 311, 324, 331, 366
standard relay, 90, 92
static gain, 95
stationary, 137
step function, 219
step response, 174, 210, 225
symmetric, 90
tracking, 326
transfer function, 210
transfer function modeling, 169
transient, 137
unstable processes, 230, 269
weighting method, 146